# FOOD BIOTECHNOLOGY

# FOOD BIOTECHNOLOGY

## TECHNIQUES AND APPLICATIONS

# Gauri S. Mittal, Ph.D., P.Eng.

SCHOOL OF ENGINEERING
UNIVERSITY OF GUELPH
GUELPH, ONTARIO, CANADA

TECHNOMIC
PUBLISHING CO., INC.
LANCASTER · BASEL

**Food Biotechnology**
a **TECHNOMIC**® publication

*Published in the Western Hemisphere by*
Technomic Publishing Company, Inc.
851 New Holland Avenue
Box 3535
Lancaster, Pennsylvania 17604 U.S.A.

*Distributed in the Rest of the World by*
Technomic Publishing AG

Printed in the United States of America
10  9  8  7  6  5  4  3  2  1

Main entry under title:
  Food Biotechnology – Techniques and Applications

A Technomic Publishing Company book
Bibliography: p. 337
Includes Index p. 367

Library of Congress Card No. 92-60559
ISBN No. 0-87762-888-2

# CONTENTS

*Preface* . . . . . . . . . . . . . . . . . . . . . . . . . . . . . . . . . *ix*

## 1. Food Biotechnology      1

1.1 Introduction . . . . . . . . . . . . . . . . . . . . . . . .1
1.2 Importance . . . . . . . . . . . . . . . . . . . . . . 2
1.3 Advances and Trends . . . . . . . . . . . . . . . . 5
1.4 Techniques and Applications . . . . . . . . . . . . 6

## 2. Genetic Engineering Techniques      7

2.1 Introduction . . . . . . . . . . . . . . . . . . . . .7
2.2 Restriction Enzymes . . . . . . . . . . . . . . . . 11
2.3 DNA Cloning Vectors . . . . . . . . . . . . . . . 18
2.4 Isolation of High Molecular Weight DNA . . . . . . . 31
2.5 Determination of the Concentration of DNA
    or RNA . . . . . . . . . . . . . . . . . . . . . . . . 34
2.6 Separation of DNA by Gel Electrophoresis . . . . . . 35
2.7 Molecular Cloning Techniques . . . . . . . . . . . 39
2.8 Southern Blotting and Hybridization . . . . . . . . . 53
2.9 RNA Analysis . . . . . . . . . . . . . . . . . . . . 58
2.10 DNA Sequencing . . . . . . . . . . . . . . . . . 62
2.11 Analysis of Proteins by SDS-Polyacrylamide Gel
     Electrophoresis (SDS-PAGE) . . . . . . . . . . . . . 78
2.12 Cloning in Yeast . . . . . . . . . . . . . . . . . . 83
2.13 Ti Plasmid-Assisted Cloning in Plants . . . . . . . . 84
2.14 Gene Transfer into Mammalian Cells . . . . . . . . 85
2.15 Automation of DNA Synthesis . . . . . . . . . . . 86

## 3. Plant Tissue/Cell Culture Techniques      87

3.1 Introduction . . . . . . . . . . . . . . . . . . . . 87
3.2 Tissue Culture Techniques . . . . . . . . . . . . . 89
3.3 Cell Cultures . . . . . . . . . . . . . . . . . . . . 94
3.4 Protoplast Fusion . . . . . . . . . . . . . . . . . . 98
3.5 Plant Cell Synthesis . . . . . . . . . . . . . . . . .101
3.6 Culturing Plant Cells for Flavor . . . . . . . . . . .101

## 4. Microbial Synthesis and Production    105

4.1 Introduction . . . . . . . . . . . . . . . . . . . . . . . . . . .105
4.2 Flavor . . . . . . . . . . . . . . . . . . . . . . . . . . . . . .105
4.3 Color . . . . . . . . . . . . . . . . . . . . . . . . . . . . . . .107
4.4 Vitamins . . . . . . . . . . . . . . . . . . . . . . . . . . . . .107
4.5 Single-Cell Proteins (SCP) . . . . . . . . . . . . . . . .108
4.6 Antimicrobial Substances . . . . . . . . . . . . . . . . .110
4.7 Polysaccharides and Biopolymers . . . . . . . . . . .111
4.8 Fat and Oil . . . . . . . . . . . . . . . . . . . . . . . . . . .113

## 5. Mutagenesis and Protein Engineering Techniques    119

5.1 Introduction . . . . . . . . . . . . . . . . . . . . . . . . . . .119
5.2 Generation of Deletion Mutants by Bal-31 . . . . . .119
5.3 Oligonucleotide-Mediated Mutagenesis . . . . . . . .121
5.4 Site-Directed Mutagenesis without Phenotypic
    Selection . . . . . . . . . . . . . . . . . . . . . . . . . . . .125
5.5 Random Mutagenesis . . . . . . . . . . . . . . . . . . . .126
5.6 Mutagenesis Using Degenerate Oligonucleotides    128
5.7 Protein Engineering . . . . . . . . . . . . . . . . . . . . .128

## 6. Enzyme Engineering and Immobilization Techniques for Enzymes and Cells    133

6.1 Introduction . . . . . . . . . . . . . . . . . . . . . . . . . . .133
6.2 Enzymes in Food Processing . . . . . . . . . . . . . . .134
6.3 Use and Selection of Enzymes . . . . . . . . . . . . . .138
6.4 Enzyme Immobilization . . . . . . . . . . . . . . . . . . .140
6.5 Microbial and Animal Cell Immobilization . . . . . .148
6.6 Plant Cell Immobilization . . . . . . . . . . . . . . . . . .159
6.7 Reactors for Immobilized Enzymes and Cells . . . .163
6.8 Changes in Cell after Immobilization . . . . . . . . . .168
6.9 Membrane Binding . . . . . . . . . . . . . . . . . . . . . .169

## 7. Biosensor Techniques    171

7.1 Introduction . . . . . . . . . . . . . . . . . . . . . . . . . . .171
7.2 Techniques . . . . . . . . . . . . . . . . . . . . . . . . . . .172
7.3 Microbial Count Sensor . . . . . . . . . . . . . . . . . . .177
7.4 Enzyme Electrode Probes . . . . . . . . . . . . . . . . .178
7.5 Microbe and Organelle Probes . . . . . . . . . . . . . .183
7.6 DNA Probe . . . . . . . . . . . . . . . . . . . . . . . . . . .183
7.7 Miscellaneous Sensors . . . . . . . . . . . . . . . . . . .185

## 8. Down-Stream Processing Techniques    189

8.1 Introduction . . . . . . . . . . . . . . . . . . . . . . . . . . .189

8.2 Cell Separation and Disruption or Rupture . . . . . . .191
8.3 Mechanical Methods . . . . . . . . . . . . . . . . . .194
8.4 Membrane Separation Methods . . . . . . . . . . . .200
8.5 Electric Methods . . . . . . . . . . . . . . . . . . . .211
8.6 Extraction Methods . . . . . . . . . . . . . . . . . .215
8.7 Thermal Methods . . . . . . . . . . . . . . . . . . .241
8.8 Technique Selection . . . . . . . . . . . . . . . . . .243

**9. Fermentation Techniques**         **245**

9.1 Introduction . . . . . . . . . . . . . . . . . . . . . .245
9.2 Factors Affecting the Fermentation . . . . . . . . .245
9.3 Microorganisms . . . . . . . . . . . . . . . . . . . .251
9.4 Equipment . . . . . . . . . . . . . . . . . . . . . . .252
9.5 Agitation and Mixing . . . . . . . . . . . . . . . . .261
9.6 Oxygen Transfer . . . . . . . . . . . . . . . . . . . .263
9.7 Media Sterilization . . . . . . . . . . . . . . . . . .268
9.8 Asepsis in Fermenter Design . . . . . . . . . . . . .269
9.9 Instrumentation and Control . . . . . . . . . . . . .270
9.10 Other Considerations . . . . . . . . . . . . . . . . .278
9.11 Design . . . . . . . . . . . . . . . . . . . . . . . . .279

**10. Scale-Up Techniques**         **283**

10.1 Introduction . . . . . . . . . . . . . . . . . . . . . .283
10.2 Scale-Up . . . . . . . . . . . . . . . . . . . . . . . .284
10.3 Rules of Thumb . . . . . . . . . . . . . . . . . . . .291
10.4 Reactor Scale-Up . . . . . . . . . . . . . . . . . . .292

**11. Applications**         **295**

11.1 Introduction . . . . . . . . . . . . . . . . . . . . . .295
11.2 Dairy Processing . . . . . . . . . . . . . . . . . . .295
11.3 Meat Processing . . . . . . . . . . . . . . . . . . . .305
11.4 Beverage Processing . . . . . . . . . . . . . . . . .306
11.5 Vegetables and Fruits Processing . . . . . . . . . .310
11.6 Cereals Processing . . . . . . . . . . . . . . . . . .312
11.7 Oil and Fat Processing . . . . . . . . . . . . . . . .314
11.8 Immobilized Cell Applications . . . . . . . . . . . .315
11.9 Plant Genetic Engineering Applications . . . . . .317
11.10 Miscellaneous Applications . . . . . . . . . . . . .317

*List of Symbols* . . . . . . . . . . . . . . . . . . . . . . . *323*

*Glossary* . . . . . . . . . . . . . . . . . . . . . . . . . . . *327*

*References* . . . . . . . . . . . . . . . . . . . . . . . . . . *337*

*Index* . . . . . . . . . . . . . . . . . . . . . . . . . . . . *367*

# PREFACE

*The* importance of biotechnology techniques for research, development and education needs very little discussion. Although biotechnology techniques have been used in food processing since ancient times, intensive research and study in this area have only been initiated during recent years. These new techniques are being used in many food processes, and will be used on a wider scale in the future to improve food quality, safety, nutritional value and palatability, and to develop new food products.

Many books have been published in recent years on food biotechnology, based on papers presented in various regional, national and international conferences. In addition to these, many books are available on biotechnology techniques, but each book only covers a particular field, e.g., tissue culture, genetic engineering, and protein engineering. The present book fills the gap and illustrates recent biotechnology techniques with applications in food processing.

This book introduces food scientists and engineers to biotechnology techniques in food processing and production. The book is divided into eleven chapters, with a glossary of important terms used in this book. Chapter 1 introduces the subject of food biotechnology, its importance and present trends. The genetic engineering principles, including recombinant DNA techniques, are discussed in Chapter 2. This chapter introduces the reader to the basics of restriction enzymes, DNA cloning, RNA analysis, DNA sequencing, and cloning in yeast. This chapter concludes with techniques on gene transfer into mammalian cells. In the remainder of the book, I do not merely discuss techniques, but I attempt to demonstrate their applications. Chapter 3 surveys plant tissue or cell culture techniques including embryogenesis, organogenesis, and protoplast fusion.

Chapter 4 presents microbial synthesis and production techniques, and thus deals with the single-cell protein, single-cell oil, antimicrobial substances, and cloning. Mutagenesis and protein engineering techniques are covered in Chapter 5. Various types of mutagenesis — random, *in vitro*, site-directed, and oligonucleotide-mediated are explained, and their techniques presented. The chapter concludes with a discussion of the techniques of protein engineering. Needless to say, this field has a great potential for future expansion.

Chapter 6 provides an in-depth view of the immobilization techniques for enzymes and cells. This concentrates upon immobilization, but use and selection of enzymes for food processing, and reactors are also discussed. Techniques of biosensor development are described in Chapter 7. These include potentiometric, amperometric, calorimetric, optical, conductimetric, and piezoelectric biosensors. DNA, microbe, and enzyme probes are also explained.

Down-stream processing techniques are treated in Chapter 8. Included in this discussion are mechanical, membrane separation, electrical, thermal, and extraction methods. Chapter 9 presents fermentation techniques including equipment, agitation, oxygen transfer, asepsis, instrumentation and control, and fermenter design.

Scale-up techniques are discussed in Chapter 10, and applications are explained in Chapter 11. The list of glossary presents a description of important terms used in the book. A list of references completes the book.

In conclusion, one might regard this book as a blend of many of the most up-to-date techniques. Each of the techniques may have some limited area where it could stand alone, but for the most part, several techniques are used in conjunction, each one reinforcing and supporting the others.

Acknowledgement is due to my colleagues, Dr. V. J. Davidson, and Dr. P. Saxena (Dept. of Horticulture) for reading the first drafts of a few chapters and for their suggestions. I also acknowledge the encouragement and support in preparation of this book from my colleagues, family, and friends. I sincerely appreciate the work of Mr. Ming Zhang for preparing drawings using Autosketch™, and my daughter, Miss Charu Mittal, for typing using Wordperfect™. I am deeply indebted to my brother, Dr. S. K. Mittal, Dept. of Biology, McMaster University, Hamilton, Ontario, for writing most of Chapters 2 and 5, as well as for supplying references and reading the manuscript.

Assistance provided by the staff of Technomic Publishing Company including J. Eckenrode, T. Deraco, K. Finlayson, E. Kladky, M. Margotta, and L. Motter is highly appreciated. I acknowledge all the organizations and individuals for their permission to use their published work. My thanks to everyone.

Gauri S. Mittal

CHAPTER 1

# Food Biotechnology

## 1.1 **INTRODUCTION**

*Biotechnology* is defined as the integrated use of biochemistry, microbiology, and biochemical, genetic, and process engineering to manufacture products from microorganisms and cell cultures (Figure 1.1). This technology utilizes biologically derived molecules, cells, or organisms to complete a process.

Biotechnology utilizes bacteria, yeasts, fungi, algae, plant cells, or cultured mammalian cells as constituents of industrial processes.

The role of biotechnology in food processing is very broad. According to the Canadian Committee on Food Biotechnology (Anon, 1988c),

> Food biotechnology involves process manipulations of enzyme in the free and immobilized forms, and in the cells for (1) producing compounds to enhance food quality, (2) to lengthen the shelf life of fresh fruits and vegetables, and (3) analyzing for food constituents and toxicants. Food biotechnology includes microbial fermentation processes (intracellular and extracellular enzyme reactions) for flavorful fermented foods, for preservation of foods, and for biomass production; plant cell culture processes (intracellular enzyme reactions); post-harvest modulation of metabolic processes for prolonging shelf life of fresh fruits and vegetables; purified enzyme processes for improving quality attributes of foods; and bioprocessing such as membrane processing.

Biotechnology in a certain sense is one of the oldest industries. Activities such as baking, brewing, and wine making are known to date back several millennia. The ancient Sumerians and Babylonians were drinking beer by 6000 B.C., the Egyptians were baking leavened bread by 4000 B.C., while wine was known in the Near East by the time of the Book of Genesis (Smith, 1985). The production of fermented milk products (cheese and yogurt) and various oriental foods (soy sauce, etc.) can equally claim distinct ancestry. Many centuries elapsed before a reasonable understanding of these processes surfaced.

Although biotechnology has been around for centuries, the secret code of DNA was not cracked until 1953 (Joglekar et al., 1983). It took 20 years for two Stanford University scientists to recombine pieces of DNA to produce a hybrid—giving birth to the biotechnology industry as we know it today.

**1**

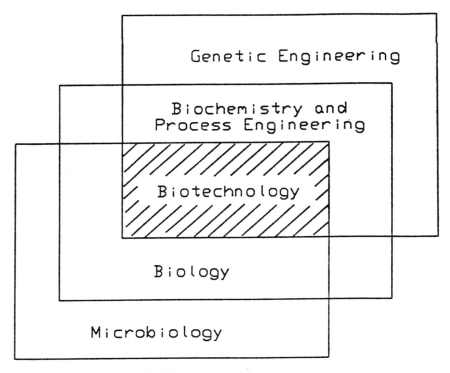

**FIGURE 1.1.** Scope of biotechnology.

Biotechnology is characterized by five major technological areas (Slotin, 1984):

(1) Recombinant DNA (rDNA) – the insertion of foreign DNA into a new system, where it is cloned and expressed

(2) Enzyme and enzyme technology – the catalytic capabilities of enzymes used either in solution or in immobilized state

(3) Plant cell culture – the *in vitro* culturing of plant cells for required substances and biotransformation

(4) Fused cell techniques – the fusing of two different cells producing hybrids exhibiting characteristics of each of the parents

(5) Process engineering and fermentation technology

## 1.2 **IMPORTANCE**

Biotechnology is an area of expansion and opportunity involving many sectors of industry, including agriculture, food and feedstuffs. It is moving toward commercial reality in the food industry. Many common foods and beverage products are based on natural fermentation (bread, wine, pickles, yogurt, cheese, etc.) or are based on the use of enzymes (beer, tenderized meat, and cheese) (Neidleman, 1986). Table 1.1 lists important fermented food products. Beer is the largest product by value in all the biotechnology industry.

Biotechnology in food processing is broadly used in two ways: (1) to design microorganisms that transform inedible biomass into food for human consumption or into feed for animals; and (2) to use biological systems to aid in food processing, either by acting directly on the food itself or by providing materials that can be added to food (Slotin, 1984). Biotechnology can customize raw materials and thus provide food scientists with a powerful tool. For every 1 % increase in tomato solids, the tomato processing industry would save about $80 M per year (Newell and Gordon, 1986). This value was obtained by estimating the savings a tomato processor would realize from reduced raw product volumes, raw material transportation costs and processing energy costs.

The developments in biotechnology are likely to affect the food industry in the following ways (King and Cheetham, 1987):

(1) The modification of food components will provide new and/or improved functional properties. For example, the bio-modification of milk fat in butter-making could result in unsaturation of the fat and subsequent higher polyunsaturated fatty acid levels and improved spreadability of refrigerated butter (Anon, 1985a).

(2) New methods of assaying food constituents, such as immobilized enzyme sensors, will be developed.

TABLE 1.1. *Important food biotechnology products.*

| Product | Base Material | Organism |
|---------|--------------|----------|
| Baked goods | Cereal flour | *Saccharomyces cerevisiae* |
| Beer | Malt | *Saccharomyces cervisiae* |
| Brandy/Whisky | Wine/cereals | *Saccharomyces cerevisiae* |
| Cheese (hard) | Milk | *Streptococcus* and *Lactobacillus* |
| Cheese (mold) | Milk | *Penicillium camembertii* or *roquefortii* |
| Coffee | Coffee beans | Lactic acid bacteria |
| Kefir | Milk | *Candida kefir* and *Lactobacillus kefir* |
| Sauerkraut | White cabbage | *Lactobacillus plantarum* |
| Sausage | Beef, mutton, pork | *Lactobacillus* and *Staphylococcus* |
| Schneidebohnen | French beans | Lactic acid bacteria |
| Soy sauce | Soybeans | *Aspergillus oryzae* |
| Vinegar | Malt | *Acetobacter aceti* |
| Wine | Grape juice | *Saccharomyces cerevisiae* |
| Yogurt | Milk | *Streptococcus* and *Lactobacillus* |

(3) New processes for the production of foods and components, e.g., the use of plant cell cultures for the production of flavors, and the detoxification of certain food products, will be perfected. Specific examples are removal of erucic acid from rapeseed oil, removal of caffeine from coffee, or removal of bitterness from some food products (Kosaric, 1984).

These will improve the cost, quality and acceptability of food products. Table 1.2 lists some biotechnology products used in the food industry.

Enzymes will be engineered to better withstand processes such as heat. For example, rDNA technology applied to microbes producing glucose isomerase could lower the pH optimum to minimize browning reactions (Morris, 1986). The yield and purity of an enzyme can be increased through genetic engineering, and properties such as pH, temperature optima, stability, and resistance to chemicals can be altered (Taylor, 1985). Genetically altered yeast strains, increased productivity, and reduced energy costs have been realized by some major brewers.

The scientific development of complex biotechnology disciplines may be the only avenue left open in today's world of shrinking natural resource supply and growing market demand (Anon, 1985a).

TABLE 1.2. *Some biotechnology products for food processing.*

| Product | Use | Organism |
|---|---|---|
| Acetic acid | Pickles | *Acetobacter aceti* |
| Alginate | Icre cream, meat, pudding | *Acetobacter vinelandii* |
| Ascorbic acid | Food enrichment | *Gluconobacter oxydans* |
| Beta-carotene | Food enrichment | *Blakeslea trispora* |
| Citric acid | Beverages, dairy product | *Aspergillus niger* |
| Curdlan | Puddings | *Alcaligenes faecalis* |
| Enzymes | ''Various uses'' | ''Many organisms'' |
| Fumaric acid | Dairy and meat products | *Rhizopus* sp. |
| Glutamic acid | Flavor enhancer | *Corynebacterium glutamicum* |
| Inosinic acid | Packaged dishes | *Corynebacterium glutamicum* |
| Lactic acid | Desserts, juice | *Lactobacillus delbrueckii* |
| Lysine | Food enrichment | *Corynebacterium glutamicum* |
| Malic acid | Jam, jelly, beverages | *Aspergillus* sp. |
| Methionine | Protein enrichment | *Corynebacterium glutamicum* |
| Nisin | Canned food and meat | *Streptococcus lactis* |
| Riboflavin | Food enrichment | *Ashbya gossipii* |
| Single cell protein | Food supplement | *Kluyveromyces fragilis* and *Fusarium graminearum* |
| Tryptophan | Antioxidant | *Corynebacterium glutamicum* |
| Xanthan | Beverages, cheese, emulsions | *Xanthomonas campestris* |

## 1.3 **ADVANCES AND TRENDS**

Developments in molecular biology (e.g., recombinant DNA technology) and process engineering (e.g., continuous reaction using immobilized enzymes) are the basis for the growth of new markets through the introduction of new and higher-quality products, and better process economics for existing products (Joglekar et al., 1983). The biotechnology will also exert a strong impact on the nature of the raw materials for food industry.

The potential for the production of new and unique low-calorie foods is tremendous. Development of low-calorie fats and oils was initiated by inducing the production of shorter-chain fatty acids in commonly used vegetable oils. Efforts are being made to develop better non-nutritive sweetener (Newell and Gordon, 1986). Taste-active proteins act as sweeteners and/or flavor modifiers such as aspartame, thaumatin, monellin, and steviosiole. Thaumatin, a sweetener known under the trade name Talin®, is the sweetest compound in the world (2500 times as sweet as sucrose). Biotechnology is being used to develop specialized yeasts and enhanced enzymes that will aid in the production of low-calorie and high-quality beer.

Xenozymes (engineered enzymes) can be created from natural enzymes via chemical modification, random mutation, or site-specific genetic and protein engineering (Morris, 1986). Chemical modifications are of three types: (1) reactions with ions or small molecules, (2) reactions with water-soluble polymers, and (3) reactions with water-insoluble polymers and matrices, as with immobilized enzymes. Commercial applications of biologically altered and improved fermentation enzymes for both the cheese manufacturing and the brewing industries have been tested, with improved flavors and faster fermentation processes. Efforts are being made to change the structure of the main egg white protein (ovalbumin) and to determine how these structural changes affect the protein's functional properties in baked goods (Anon, 1985a).

Plant tissue culture and recombinant DNA biotechnology can provide both variations and improvements in plants. Plant tissue culture techniques can be used to develop plants with stress tolerance, disease resistance, and nutritional self-sufficiency (Anon, 1985a). Utilization of recombinant DNA technology can provide disease resistance, increased yield, herbicide resistance, efficient fertilizer assimilation, or enhanced oil and protein quality and quantity in crops.

Biotechnology will be the most dynamic area of industrial innovation in future, analogous to the emergence of computers in the 1970s. World markets for biotechnology products are projected to be on the order of hundreds of billions of dollars by the end of the century (Smith, 1985; Joglekar et al., 1983).

## 1.4 **TECHNIQUES AND APPLICATIONS**

Major advances have been made in bioreactor designs, in process monitoring techniques, and in computer control of fermentation processes with biosensors. There is an increasing need to design efficient recovery processes, in particular for high value products. Enzymes have long been a part of many biotechnological processes and their catalytic properties are being further utilized with the development of suitable immobilization techniques.

Recombinant DNA techniques involve breaking living cells, extracting DNA, purifying it, and subsequently selectively fragmenting it through highly specific enzymes; the sorting, analysis, selection, and purification of a fragment containing a required gene; chemical bonding to the DNA of a carrier molecule and the introduction of the hybrid DNA into a selected cell for reproduction and cellular synthesis (Smith, 1985).

Computers are used to store and analyze biological sequence data; organic chemistry is used for the synthesis of proteins and polynucleotides; small oligonucleotides can be used as probes to trap larger molecules through hybridization techniques; and peptides are employed to produce monoclonal antibodies for the isolation of proteins (deRosnay, 1985).

Therefore, the following techniques are discussed in this book:

- genetic engineering techniques
- plant cell/tissue culture
- microbial synthesis and production
- protein engineering techniques
- mutagenesis and selection techniques
- immobilization techniques for enzymes and cells
- biosensor techniques
- separation and downstream processing techniques
- fermentation techniques
- reactor design and selection techniques
- scale-up techniques

Applications of these techniques in meat, dairy, cereal, fruits and vegetables, beverages, and oil and fat processing are also briefly discussed.

# CHAPTER 2

# Genetic Engineering Techniques

## 2.1 INTRODUCTION

*Genetic* engineering is the methodology that employs recombinant deoxyribonucleic acid (DNA) techniques. This allows the incorporation of the foreign DNA into a host organism where the DNA does not naturally occur, but where it is capable of continued propagation. This technology can alter the hereditary material, the DNA, of a cell so that cells can produce more or different chemicals or perform completely new functions (Anon, 1985b).

At the molecular level, the life can be reduced to the nucleotide sequences in DNA. The basic building blocks of the nucleic acid are the bases, the sugars, and the phosphates (Figure 2.1).

As shown in Figure 2.2, there are five kinds of bases: two purines and three pyrimidines. The two purines that are present in both DNA and RNA (ribonucleic acid) are adenine (A) and guanine (G). The three pyrimidines are cytosine (C), thymine (T), and uracil (U). The thymine is found only in DNA, and uracil appears in RNA. Like purines, the cytosine is present in both DNA and RNA. The sugars are ribose and deoxyribose; RNA contains ribose while DNA contains deoxyribose. Nucleosides are composed of a base and a sugar. The structure of the guanosine (guanine + ribose) is shown in Figure 2.2. The other nucleosides are adenosine (adenine + ribose), thymidine (thymine + deoxyribose), cytidine (cytosine + ribose) and uridine (uracil + ribose). The nucleotide is formed when a phosphate group is attached to a nucleoside. These bases are linked together to form polynucleotides. The phosphate of one nucleotide is linked to the deoxyribose or ribose of another, and so forth. The number of nucleotides in the DNA of different organisms varies greatly from about 5500 in a virus to $10^{10}$ in human cells.

The two nucleotide chains of DNA are bound to each other by hydrogen bonding between specific base pairs to form a double helix structure (Figure 2.3). However, the majority of RNAs are single-stranded. Various classes of RNA are: mRNA (messenger RNA), tRNA (transfer RNA), and rRNA (ribosomal RNA). These RNAs have specific functions.

The genetic information is present on the DNA in the form of genes. The genes are transcribed in copies of mRNA by the enzyme RNA polymerase.

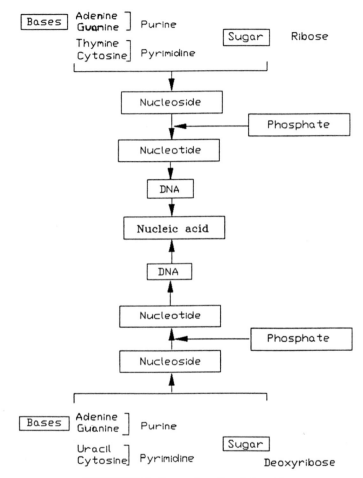

**FIGURE 2.1.** Formation of nucleic acid.

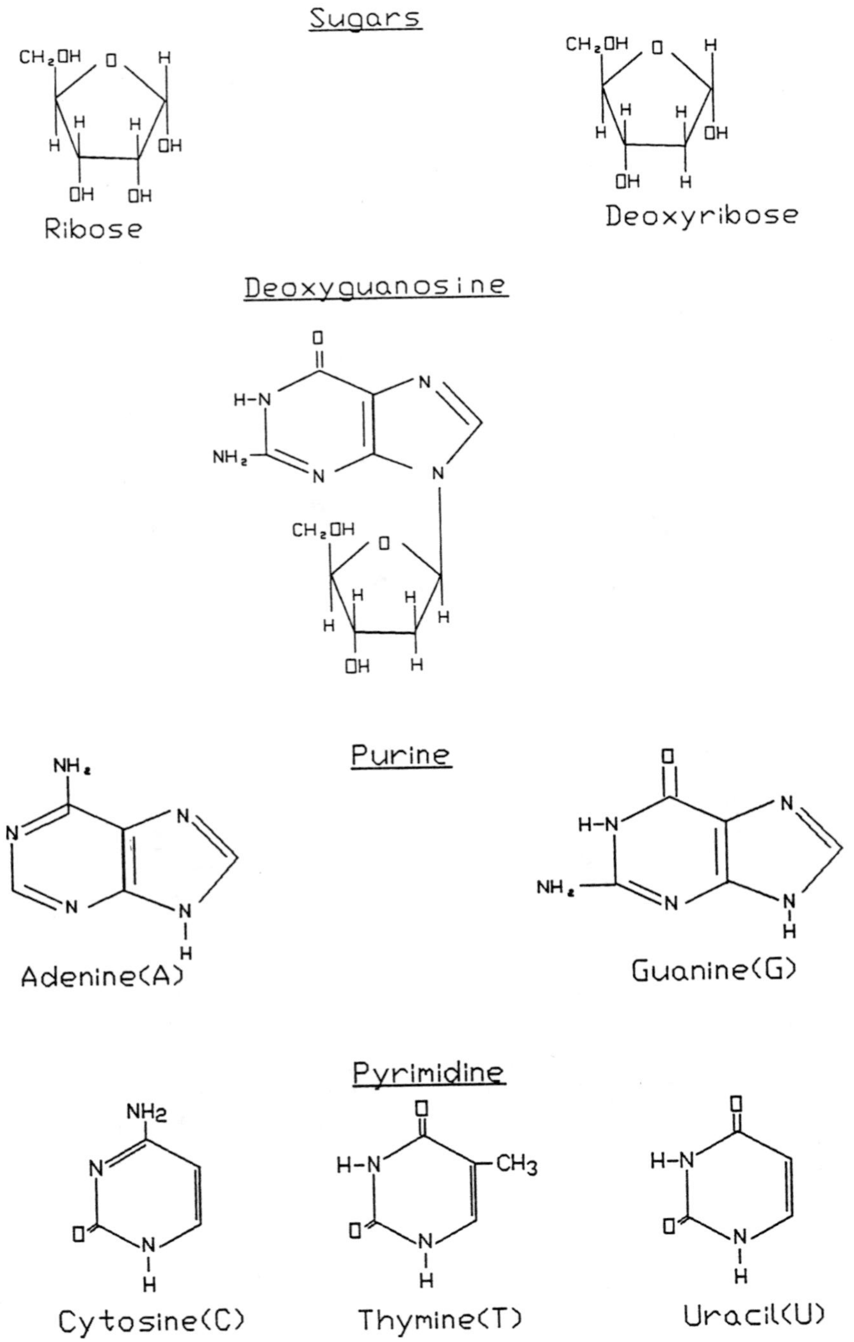

## Sugars

**Ribose**

**Deoxyribose**

## Deoxyguanosine

## Purine

**Adenine(A)**

**Guanine(G)**

## Pyrimidine

**Cytosine(C)**

**Thymine(T)**

**Uracil(U)**

**FIGURE 2.2.** Basic building blocks for DNA and RNA, and linkage of a base with a sugar.

9

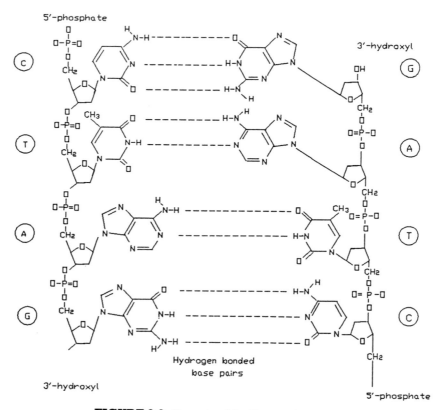

**FIGURE 2.3.** Structure of double-stranded DNA.

On the ribosomes, mRNAs are translated into polypeptides; the process is known as translation. Many polypeptides undergo secondary modification, such as glycosylation, phosphorylation, and sulphation, to form functional proteins (Figure 2.4).

Life programs are written in genetic codes of A, T, G, and C, which are known as bases. Three bases form a "codon." The combination of four bases results in a total of 64 codons. A few codons are "stop" codons, such as TAA, TAG, and TGA, and the other codons represent the 20 amino acids that form different proteins, e.g., the "ATG" codon represents the amino acid methionine. The amino acids that constitute proteins are alanine, arginine, asparagine, aspartic acid, cysteine, glutamic acid, glutamine, glycine, histidine, isoleucine, leucine, lysine, methionine, phenylalanine, proline, serine, threonine, tryptophan, tyrosine, and valine.

On ribosomes, the mRNA dictates the synthesis of a particular polypeptide. Amino acids for the growing chain are carried to the ribosomes by tRNA. A specific codon on mRNA is identified by an anti-codon on tRNA, e.g., the tRNA carrying methionine has UAC as anti-codon.

Basically, there are three important tools in genetic engineering: (1) enzymes to alter DNA, (2) DNA cloning vectors, and (3) hosts for the large-scale production of the recombinant vectors. The gram negative bacterium *Escherichia coli* has been the host used most widely in recombinant DNA research.

## 2.2 **RESTRICTION ENZYMES**

### 2.2.1 Introduction

The restriction enzymes and other DNA/RNA modifying enzymes are important tools in manipulating DNA. The present advancement in genetic engineering might not have been possible without the advent of such enzymes.

Restriction endonucleases are DNAases that recognize specific oligonucleotide sequences in a duplex DNA molecule, and make double-stranded cleavages within or adjacent to the recognition sequences.

Restriction endonucleases are classified into three classes: I, II, and III. Classes I and III restriction endonucleases are not commonly used in molecular cloning. The binding and cleavage of the duplex DNA by class II restriction endonucleases are highly sequence specific.

A large number of class II restriction enzymes have been isolated from different prokaryotes, and many of them are extensively used in molecular cloning (Roberts, 1988). The recognition sites of the majority of class II restriction enzymes are four, five, or six nucleotides in length, and show two-fold symmetry (Table 2.1).

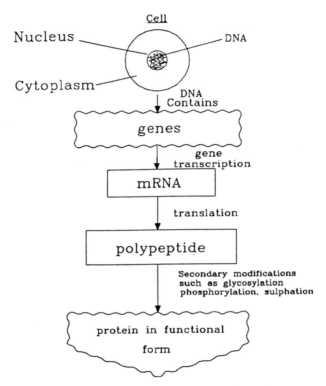

**FIGURE 2.4.** From DNA to functional proteins.

TABLE 2.1. *Some of the commonly used restriction enzymes (modified after Kessler et al., 1985; Roberts, 1988).*

| Enzyme | Source | Recognition Sequence (5′ – – – 3′) (3′ – – – 5′) |
|---|---|---|
| AatII | *Acetobacter aceti* | G ACGT$^\downarrow$C <br> C$_1$TGCA G |
| AccII | *Acinetobacter calcoaceticus* | CG$^\downarrow$CG <br> GC$_1$GC |
| AluI | *Arthrobacter luteus* | AG$^\downarrow$CT <br> TC$_1$GA |
| ApaI | *Acetobacter pasteurianus* | G GGCC$^\downarrow$C <br> C$_1$CCGG G |
| Asp718 | *Achromobacter* sp. 718 | G$^\downarrow$GTAC C <br> C CATG$_1$G |
| AvaI | *Anabaena variabilis* | C$^\downarrow$PyCGPu G <br> G PuGCPy$_1$C |
| AvaII | *Anabaena variabilis* | G$^\downarrow$G(A or T)C C <br> C C(A or T)G$_1$G |
| BamHI | *Bacillus amyloliquefaciens* H | G$^\downarrow$GATC C <br> C CTAG$_1$G |
| BanII | *Bacillus aneurinolyticus* | G$^\downarrow$PuGCPy C <br> C PyCGPu$_1$G |
| BclI | *Bacillus caldolyticus* | T$^\downarrow$GATC A <br> A CTAG$_1$T |
| BglII | *Bacillus globigii* | A$^\downarrow$GATC T <br> T CTAG$_1$A |
| Blu | *Brevibacterium luteum* | C$^\downarrow$TCGA G <br> G AGCT$_1$C |
| CfoI (HhaI) | *Clostridium formicoaceticum* | G CG$^\downarrow$C <br> C$_1$GC G |
| ClaI | *Caryophanon latum* | AT$^\downarrow$CG AT <br> TA GC$_1$TA |
| DdeI | *Desulfovibrio desulfuricans* | C$^\downarrow$TNA G <br> G ANT$_1$C |
| DraI (AhaIII) | *Deinococcus radiophilus* | TTT$^\downarrow$AAA <br> AAA$_1$TTT |
| EcoRI | *Escherichia coli* BS5 | G$^\downarrow$AATT C <br> C TTAA$_1$G |
| EcoRV | *Escherichia coli* J62pLG74 | GAT$^\downarrow$ATC <br> CTA$_1$TAG |
| HaeII | *Haemophilus aegyptius* | Pu GCGC$^\downarrow$Py <br> Py$_1$CGCG Pu |
| HaeIII | *Haemophilus aegyptius* | GG$^\downarrow$CC <br> CC$_1$GG |

TABLE 2.1. *(continued)*.

| Enzyme | Source | Recognition Sequence (5' – – – 3') (3' – – – 5') |
|--------|--------|--------------------------------------------------|
| HINDII (HincII) | *Haemophilus influenzae* Rd com-10 | GTPy$^\downarrow$PuAC CAPu$_\uparrow$PyTG |
| HindIII | *Haemophilus influenzae* | A$^\downarrow$AGCT T T TCGA$_\uparrow$A |
| HinfI | *Haemophilus influenzae* | G$^\downarrow$ANT C CTNA$_\uparrow$G |
| HpaI | *Haemophilus parainfluenzae* | GTT$^\downarrow$AAC CAA$_\uparrow$TTG |
| HpaII (MspI) | *Haemophilus parainfluenzae* | C$^\downarrow$CG G G GC$_\uparrow$C |
| KpnI | *Klebsiella pneumoniae* OK8 | G GTAC$^\downarrow$C C$_\uparrow$CATG G |
| MaeI | *Methanococcus aeolicus* PL-15/H | C$^\downarrow$TA G G AT$_\uparrow$C |
| MaeII | *Methanococcus aeolicus* PL-15/H | A$^\downarrow$CG T T GC$_\uparrow$A |
| MluI | *Micrococcus luteus* | A$^\downarrow$CGCG T T GCGC$_\uparrow$A |
| MspI | *Moraxella* sp. | C$^\downarrow$CG G G GC$_\uparrow$C |
| NaeI | *Nocardia aerocolonigenes* | GCC$^\downarrow$GGC CGG$_\uparrow$CCG |
| NcoI | *Nocardia corallina* | C$^\downarrow$CATG G G GTAC$_\uparrow$C |
| NdeI | *Neisseria denitrificans* | CA$^\downarrow$TA TG GT AT$_\uparrow$AC |
| NdeII (MboI) | *Neisseria denitrificans* | $^\downarrow$GATC CTAG$_\uparrow$ |
| NheI | *Neisseria mucosa* | G$^\downarrow$CTAG C C GATC$_\uparrow$G |
| NotI | *Nocardia otitidis-caviarum* | GC$^\downarrow$GGCC GC CG CCGG$_\uparrow$CG |
| NruI | *Nocardia rubra* | TCG$^\downarrow$CGA AGC$_\uparrow$GCT |
| NsiI | *Neisseria sicca* | A TGCA$^\downarrow$T T$_\uparrow$ACGT A |
| PstI | *Providencia stuartii* | C TGCA$^\downarrow$G G$_\uparrow$ACGT C |

**14**

TABLE 2.1. *(continued)*.

| Enzyme | Source | Recognition Sequence (5' − − − 3') (3' − − − 5') |
|--------|--------|--------------------------------------------------|
| PvuI | *Proteus vulgaris* | CG AT$^{\downarrow}$CG<br>GC$_{\uparrow}$TA GC |
| PvuII | *Proteus vulgaris* | CAG$^{\downarrow}$CTG<br>GTC$_{\uparrow}$GAC |
| RsaI | *Rhodopseudomonas sphaeroides* | GT$^{\downarrow}$AC<br>CA$_{\uparrow}$TG |
| SacI<br>(SstI) | *Streptomyces achromogenes* | G AGCT$^{\downarrow}$C<br>C$_{\uparrow}$TCGA G |
| SalI | *Streptomyces albus* G | G$^{\downarrow}$TCGA C<br>C AGCT$_{\uparrow}$G |
| Sau3A | *Staphylococcus aureus* 3A | $^{\downarrow}$GATC<br>CTAG$_{\uparrow}$ |
| ScaI | *Streptomyces caespitosus* | AGT$^{\downarrow}$ACT<br>TCA$_{\uparrow}$TGA |
| SmaI | *Serratia marcescens* | CCC$^{\downarrow}$GGG<br>GGG$_{\uparrow}$CCC |
| SnaBI | *Sphaerotilus natans* | TAC$^{\downarrow}$GTA<br>ATG$_{\uparrow}$CAT |
| SpeI | *Sphaerotilus natans* | A$^{\downarrow}$CTAG T<br>T GATC$_{\uparrow}$A |
| SphI | *Streptomyces phaeochromogenes* | G CATG$^{\downarrow}$C<br>C$_{\uparrow}$GTAC G |
| SspI | *Sphaerotilus* | AAT$^{\downarrow}$ATT<br>TTA$_{\uparrow}$TAA |
| SstI | *Streptomyces stanfordii* | G AGCT$^{\downarrow}$C<br>C$_{\uparrow}$TCGA G |
| StuI | *Streptomyces tubercidicus* | AGG$^{\downarrow}$CCT<br>TCC$_{\uparrow}$GGA |
| TaqI | *Thermus aquaticus* YT1 | T$^{\downarrow}$CG A<br>A GC$_{\uparrow}$T |
| XbaI | *Xanthomonas badrii* | T$^{\downarrow}$CTAG A<br>A GATC$_{\uparrow}$T |
| XhoI | *Xanthomonas holcicola* | C$^{\downarrow}$TCGA G<br>G AGCT$_{\uparrow}$C |

($^{\downarrow}$)Breaks generated by a restriction enzyme in the recognition ($_{\uparrow}$) sequence, A = adenine, G = guanine, T = thymine, C = cytosine, N = any nucleotide, Pu = purine, Py = pyrimidine.

The location of cleavage sites with the recognition sequence differs from enzyme to enzyme. The enzymes that cleave both strands at the axis of symmetry result in DNA fragments with blunt ends (Figure 2.5).

Other restriction enzymes cleave each strand at similar sites on opposite sides of the axis of symmetry resulting in DNA fragments with protruding ends (Figure 2.6).

In most of the cases, different restriction enzymes would bind different sequences. However, there are a number of enzymes isolated from different sources that cleave within the same recognition sequence; such enzymes are known as isoschizomers. For instance, SacI and SstI cleave within the sequence

5' G AGCT¹C 3'
3' C₁TCGA G 5'

## 2.2.2 Cleaving DNA with Restriction Enzymes

For the digestion of DNA with a restriction enzyme, digestion conditions are very important. It is always better to follow manufacturers' instructions. In some cases, even for the same restriction enzyme, different manufacturers recommend very different digestion conditions. The manufacturers optimize the reaction conditions for their products. Now most of the manufacturers usually supply concentrated restriction enzyme buffer with each batch of the restriction enzyme. It is always better to use the appropriate restriction enzyme buffer supplied by the manufacturer and follow manufacturer's instructions for digesting DNA with a particular enzyme.

Potassium glutamate buffer (KGB) can be used to digest most of the commonly used restriction enzymes with reasonable success (Hanish and McClelland, 1988).

**2× KGB**
200 mM potassium glutamate
20 mM magnesium acetate
50 mM Tris-acetate (pH 7.5)
1 mM β-mercaptoethanol
100 μg/mL bovine serum albumin (fraction V)

In the situation where DNA needs to be digested with two or more restriction enzymes, the reactions can be carried out simultaneously if enzymes (two or more) work well in one buffer. However, if the enzymes need different digestion conditions, first digest with an enzyme that works well in the low ionic strength buffer, then add sodium chloride to bring the salt concentration to the optimum for digestion with the second enzyme. Add the second enzyme and continue incubation.

```
5'-ACC⌐GTTAAC⌐TCG-3'        double stranded
3'-TGG⌐CAATTG⌐AGC-5'        DNA
```

Digestion with restriction enzyme
Hpal under appropriate conditions

```
5'-ACCGTT-3'      5'-AACTCG-3'
3'-TGGCAA-3'      3'-TTGAGC-5'
```

**FIGURE 2.5.** Generation of DNA fragments with blunt ends by restriction enzyme HpaI.

For typical reaction conditions, in a microfuge tube, mix:

DNA 0.2 to 2 $\mu$g
$10\times$ restriction 2 $\mu$L
Enzyme buffer (from manufacturer)
*or*
$2\times$ KGB 10 $\mu$L
Sterile distilled water to make 19 $\mu$L
Restriction enzyme 2 to 4 units

One unit of the enzyme is defined as the amount of the enzyme required to digest 1 $\mu$g of DNA (usually a bacteriophage) to completion in one hour in the recommended buffer and under the recommended incubation conditions.

(1) Incubate at the appropriate temperature for the required time.
(2) Bring the concentration of EDTA to 10 mM by adding 0.5 M EDTA (pH 8.0) to stop the reaction.

```
5'-GTTAA⌐GGATCC⌐GTAA-3'     double stranded
3'-CAATT⌐CCTAGG⌐CATT-5'     DNA
```

Digestion with restriction enzyme
BamHI under appropriate
conditions

```
5'-GTTAAG-3'          5'-GATCCGTAA-3'
3'-CAATTCCTAG-5'          3'-GCATT-5'
```

**FIGURE 2.6.** Generation of DNA fragments with 5' protruding ends by restriction enzyme BamHI.

To purify DNA, extract once with equal volume of phenol:chloroform (1:1), and precipitate with ethanol after adding 1/10 volume of 3 M sodium acetate (pH 4.8).

If the DNA needs to be analyzed in a gel, add 2 $\mu$L of loading mix, and load the sample into the slot of the agarose gel submerged in the gel electrophoresis buffer.

**Loading mix**
0.125 M EDTA (pH 8.0)
10% SDS
0.25% bromophenol blue
50% glycerol

The enzymes, other than restriction endonucleases, which are widely used in molecular cloning are given in Table 2.2.

## 2.3 **DNA CLONING VECTORS**

### 2.3.1 Plasmids

Bacterial plasmids are extra-chromosomal double-stranded circular DNA molecules found in a number of bacterial species. They are self-replicating and are stably inherited independently of the host DNA. However, they depend on host proteins for their replication. Some of the host properties are specified by plasmids' genes, such as resistance to antibiotics, production of antibiotics, resistance to heavy metal, degradation of complex organic compounds, production of toxin, and production of bacteriocin. Some of the plasmids have a functional set of genes that specifies bacterial sexuality.

A plasmid carries an origin of DNA replication "replicon" that controls the plasmid copy number in the host. In most cases, the replicon is derived from the plasmid pMBI or ColEI (Hershfield et al., 1974, Bolivar et al., 1977a); only a small portion of a naturally occurring plasmid is needed for its replication *in vivo*. This property allows the reconstruction of the naturally occurring plasmids by *in vitro* recombinant DNA techniques to develop a number of cloning vectors for effective cloning of DNA from various sources.

Plasmids are maintained in bacteria in low copy number (stringent control of replication) or high copy number (relaxed control of replication). The copy number of a plasmid carrying the pMBI or ColEI replicon can be amplified many-fold by the growth of the bacterial cells carrying plasmids in the presence of antibiotics such as chloramphenicol, which inhibits protein synthesis and replication of the bacterial chromosome.

TABLE 2.2. *Commonly used enzymes in molecular cloning.*

| Enzymes | Activity | Use |
|---|---|---|
| E. coli DNA polymerase I | 1. Adds dNTPs in dsDNA template having recessed 3'OH<br>2. 5' → 3' exonuclease activity<br>3. 3' → 5' exonuclease activity | Labeling DNA with radioactive dNTPs by nick translation |
| Large fragment of E. coli DNA polymerase (Klenow fragment) | 1. Adds dNTPs in dsDNA template having recessed 3'OH<br>2. 3' → 5' exonuclease activity | 1. Filling of recessed 3' OH<br>2. Labeling 3' termini of DNA fragments with radioactive NTPs<br>3. Sequencing DNA by dideoxynucleotide technique<br>4. Synthesis of 2nd strand in cDNA cloning |
| Calf intestinal alkaline phosphatase | Catalyzes removal of 5' phosphate group from either DNA or RNA | Removal of 5' phosphate from either DNA or RNA |
| Bacteriophage T4 DNA ligase | Ligates 5' phosphate to 3' OH in DNA | 1. Joining of either cohesive or blunt ends generated by restriction enzymes<br>2. Repairing of nicks in DNA |
| Bacteriophage T4 DNA polymerase | 1. Adds dNTPs in ds template having recessed 3' OH (i.e., 5' → 3' polymerase activity)<br>2. 3' → 5' exonuclease activity | 1. Labeling of DNA by "replacement synthesis"<br>2. 3' end labeling of DNA in the presence of labeled dNTPs |
| Reverse transcriptase | Catalyzes synthesis of second strand from RNA or ssDNA | 1. Used in cDNA cloning<br>2. Synthesis of ss radiolabeled probes |
| Bacteriophage T4 polynucleotide kinase | 1. Catalyzes transfer of phosphate from NTP to 5' OH of DNA or RNA<br>2. Catalyzes transfer of phosphate from 5' end of DNA or RNA<br>3. Removes 3' phosphoryl groups | 5' labeling of DNA or RNA by exchange reaction. Phosphate from 5' end of DNA or RNA are transferred to NDP and replaced with labeled phosphate from $\gamma$-labeled NTP. |
| Terminal deoxynucleotidyl transferase | 1. Adds dNTPs to 3' OH ends of DNA | 1. Labeling of 3' end of DNA<br>2. Adding complementary homopolymer tails to vector and cDNA |

**19**

TABLE 2.2. *(continued)*.

| Enzymes | Activity | Use |
|---|---|---|
| ExoIII | Degrades ssDNA and RNA at 3′ and 5′ ends | 1. Helps in mapping the location of exons and introns when used with S1 or mung bean nuclease<br>2. Generation of nested sets of deletions for directed sequencing by dideoxy-nucleotide sequencing technique |
| S1 nuclease | Degrades ssDNA or RNA from nicks or gaps | 1. Helps in mapping the location of exons and introns<br>2. Helps in locating transcribed regions of DNA<br>3. Cleaves ssDNA regions of dscDNA |
| Mung bean nuclease | Degrades ssDNA or RNA from gaps | Same as S1 nuclease |
| Nuclease Bal-31 | Degrades 5′ and 3′ strands of DNA in a progressive manner | 1. Restriction mapping<br>2. Making deletions in a DNA clone |
| Bacteriophage SP6, T3 or T7 polymerases | Transcribe from SP6, T3 or T6 promoters | Making ssRNA probe |
| DeoxyribonucleaseI (DNAaseI) | Hydrolyzes single-stranded or double-stranded DNA:<br>1. In the presence of $Mg^{++}$ there is random cleavage of each strand of DNA.<br>2. In the presence of $Mn^{++}$ there is cleavage of both strands of DNA at approximately the same place. | 1. Generation of random nicks into double-stranded DNA<br>2. Generation of random clones for sequencing<br>3. Protein:DNA complexes analysis (DNAase footprinting) |
| Bacteriophage λ exonuclease | Degrades DNA at 5′ ends (degrades dsDNA 100-fold more efficiently than ssDNA) | Alter the 5′-phosphate termini of dsDNA |

dNTP = deoxynucleotide-triphosphate, NTP = nucleotide-triphosphate, NDP = nucleotidediphosphate, ss = single-stranded, ds = double-stranded, OH = hydroxy group, cDNA = complementary DNA.

**20**

## 2.3.1.1 Maps of Plasmid Vectors

*pBR322:* For many years pBR322 (Bolivar et al., 1977b; Sutcliffe, 1979) has been extensively used as a cloning vector. The introduction of improved vectors has superseded pBR322.

**Characteristics of pBR322 (Figure 2.7)**

Size – 4.36 kb
Replicon – pMB1
Copy number – 15 to 20
Resistance genes – ampicillin (amp$^r$), tetracycline (tet$^r$)
Cloning sites – BamHI, ClaI, HindIII, PstI, PvuI, SalI

This plasmid has the genes for tetracycline and ampicillin resistance. Insertion of foreign DNA into one of the antibiotic resistance genes results in its inactivation and thus provides an excellent selection for the identification of bacterial colonies carrying recombinant plasmids.

*pUC18/pUC19:* These vectors carry (1) PvuII/EcoRI fragment of pBR322 containing the origin of replication and ampicillin resistance gene, and (2) a HaeII fragment containing the α-peptide of LacZ gene and a cloning site of one of the M13mp vectors (Vieira and Messing, 1982; Yanisch-Perron et al., 1985). pUC18 and pUC19 vary only in the orientation of the polyclonal site (Figure 2.8).

The LacZ gene codes the *N*-terminal portion of β-galactosidase, which can be induced by isopropyl-thio β-D-galactoside (IPTG). The *N*-terminal position of β-galactosidase is capable of intra-allelic α-complementation with a defective β-galactosidase encoded by the bacteria. Bacteria containing pUC18 or pUC19, when exposed to IPTG, synthesize both fragments of the β-galactosidase, and thus result in blue colonies when plated on medium containing the substrate 5-bromo-4-chloro-3-indolyl-β-galactoside (X-gal). Bacteria carrying recombinant plasmids give rise to white colonies, because insertion of the foreign DNA into the polyclonal site of the plasmid inhibits the production of the *N*-terminal portion of β-galactosidase, and thus α-complementation is abolished.

## 2.3.2 Bacteriophage λ

There are two types of bacteriophage λ vectors:

- "insertion" vectors having a single site for the insertion of the foreign DNA fragment
- "replacement" vectors having a pair of sites flanking a portion of non-essential λDNA, which can be replaced by the foreign DNA fragment

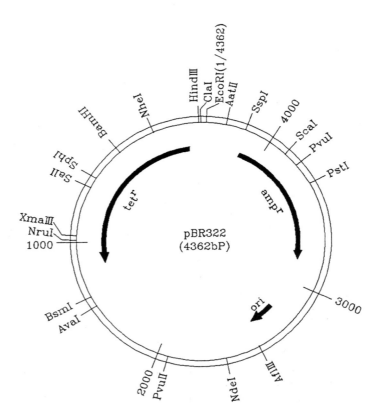

**FIGURE 2.7.** Plasmid pBR322.

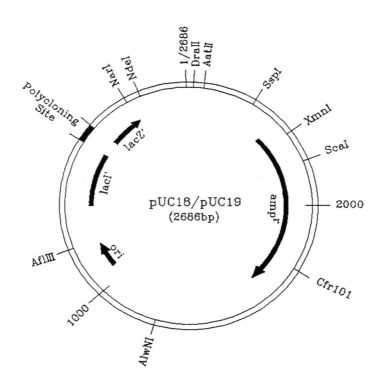

## Polycloning Sites

### pUC18

455

5'-GAATTCGAGCTCGGTACCCGGGGATCCTCTAGAGTCGACCTGCAGGCATGCAAGCTT-3'

399

EcoRI   SstI   KpnI   XmaI   BamHI   XbaI   SalI   PstI   SphI   HindIII

BanII   SmaI   HincII

AccI

### pUC19

455

5'-AAGCTTGCATGCCTGCAGGTCGACTCTAGAGGATCCCCGGGTACCGAGCTCGAATTC-3'

399

HindIII   SphI   PstI   SalI   XbaI   BamHI   XmaI   KpnI   SstI   EcoRI

HincII   SmaI   BanII

AccI

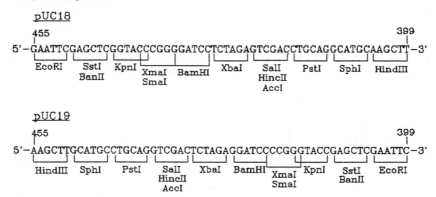

**FIGURE 2.8.** Plasmid pUC18/pUC19.

For the selection of a suitable bacteriophage λ vector for cloning DNA fragments, there are a number of points to be considered, such as:

- the size of the foreign DNA fragment
- restriction site(s)

However, there is no single bacteriophage λ vector suitable for cloning DNA fragments from a variety of sources.

Approximately 60% of the λ genome is essential for lytic infection, which consists of the left arm (approximately 20 kb) including the head and tail genes A−J, and the right arm (approximately 10 kb) from pR to CosR site. The ability of the bacteriophage λ DNA to be packaged into viable particles decreases dramatically when the λ DNA is more than 105% or less than 80% of the wild-type bacteriophage λ genome.

Bacteriophage λ vectors, which carry the bacterial β-galactosidase gene, form blue plaques on lac-negative bacterial lawn in the presence of 5-bromo-4-chloro-3-indolyl-β-D-galactoside (X-gal). The recombinant bacteriophage λ forms colorless plaques on lac-negative bacterial lawn in the presence of X-gal, because cloning of the foreign DNA fragment into such vectors results in the replacement of a large segment of the β-galactosidase gene.

To remove the nonrecombinant bacteriophages from the population, in many bacteriophage vectors genetic selections for recombinant bacteriophages can be used, e.g., bacteriophage vector λgt10 (Figure 2.9). The λgt10 is CI+ and thus undergoes lysogeny (integration of the bacteriophage DNA into the chromosome of its bacterial host, and thus the lytic cycle is inhibited) with high frequency in *E. coli* strains having hfl (high frequency of lysogenization) mutation. The vector λgt10 carries a single EcoRI site present in the CI gene for cloning the foreign DNA. The insertion of the foreign DNA at the EcoRI site results in the inactivation of the CI gene and, therefore, the growth of the recombinant bacteriophage is not inhibited in hfl cells (Huynh et al., 1985). Only recombinant λgt10 will form plaques when grown on hfl cells. Such vectors are particularly useful for constructing genomic libraries.

## 2.3.2.1  λgt11

A number of bacteriophage λ vectors have been constructed that allow replication as well as expression of foreign DNA sequences in bacterial cells, e.g., λgt11 (Young and Davis, 1983). This vector has a portion of the *E. coli* β-galactosidase gene, which also includes the upstream control elements essential for expression. The single EcoRI site present within the carboxy-terminal coding region of the β-galactosidase gene is being used for the insertion of the foreign DNA into λgt11.

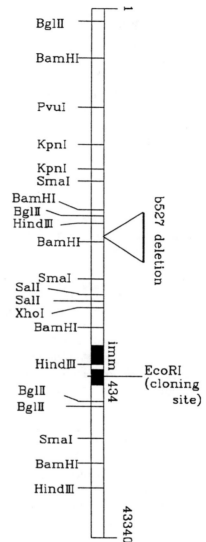

b527 deletion: deletion of the part of <u>att</u> site to prevent lysogenization

imm434: substitution from bacteriophage φ434

cloning site: EcoRI
size of insert: upto 6.0kb

**FIGURE 2.9.** Bacteriophage λgt10 vector.

The recombinant λgt11 will express the foreign gene as a fusion protein consisting of the *N*-terminal portion of the β-galactosidase. Therefore, cDNA libraries constructed in the λgt11 can be screened immunologically for the expression of specific antigens in the bacterial cells. These vectors have isolated a number of genes coding for a variety of proteins for which only specific antisera were available. The vector characteristics are:

- cloning site — EcoRI
- type of cloning — insertion
- size of right arm — 24.2 kb
- size of left arm — 19.5 kb
- size of insert — up to 7.2 kb
- recombinant plaque — white
- deletion from the original bacteriophage — shndIII2-3, nin5
- preferable bacterial host — y1090hsdR
- amber mutations — Sam100

## 2.3.3 Cosmids

Cosmid vectors, like plasmids, usually carry a ColEI origin of replication and a drug resistance gene. They also carry a copy of cos sequences, which is required for packaging recombinant cosmid DNA into bacteriophage λ particles. Cosmids, when introduced into a bacterial host, replicate as plasmids.

Cosmids are used to clone comparatively large fragments of foreign DNA (35 to 45 kb long). They are particularly useful (1) in cloning an entire eukaryotic gene as a single recombinant cosmid, e.g., murine dihydrofolate reductase (Nunberg et al., 1980), chicken proα2 collagen (Ohkubo et al., 1980; Vogeli et al., 1980), and many other eukaryotic genes that have a large number of interons, and (2) in cloning a segment of eukaryotic DNA containing a family of genes. If the sizes of such regions of the genome are too large to be cloned into a cosmid vector, then usually a series of overlapping recombinants are isolated.

The cosmid vector pJB8 (Ish-Horowicz and Burke, 1981) is illustrated in Figure 2.10, and its properties are given below:

- size — 5.4 kb
- origin of replication — ColEI
- number of *cos* sites — one
- antibiotic resistance genes — ampicillin (amp$^r$)
- cloning sites — HindIII, ClaI, EcoRI, BamHI
- cloning capacity — 33 to 46.5 kb

The following procedure is usually followed to construct a genomic library into a cosmid vector (Figure 2.11):

(1) Digest a cosmid vector with a suitable restriction enzyme. The linearized vector may be treated with alkaline phosphatase to remove 5′ phosphate groups.

(2) Partially digest eukaryotic DNA with restriction enzyme (usually MboI or Sau3A1) to produce termini compatible with the linearized cosmid vector.

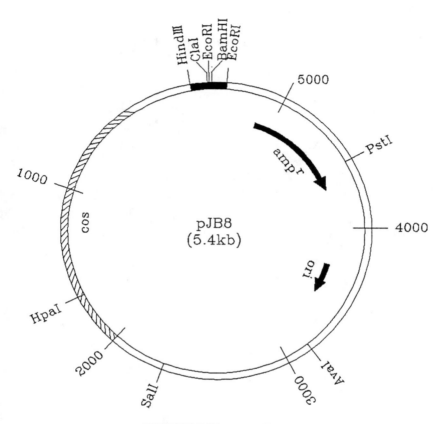

**FIGURE 2.10.** Cosmid pJB8.

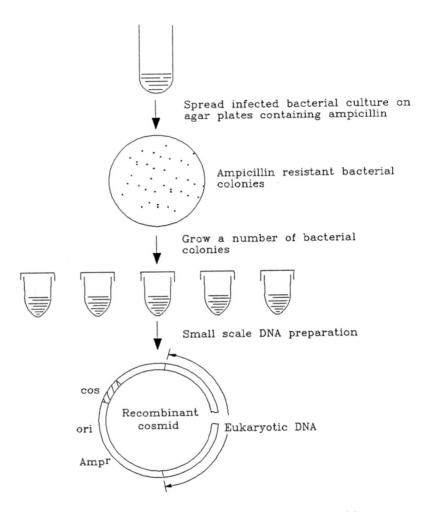

Spread infected bacterial culture on agar plates containing ampicillin

Ampicillin resistant bacterial colonies

Grow a number of bacterial colonies

Small scale DNA preparation

cos

ori

Amp<sup>r</sup>

Recombinant cosmid

Eukaryotic DNA

Digestion with appropriate restriction enzyme(s) to analyse recombinant cosmid

**FIGURE 2.11.** Cloning in cosmid vector.

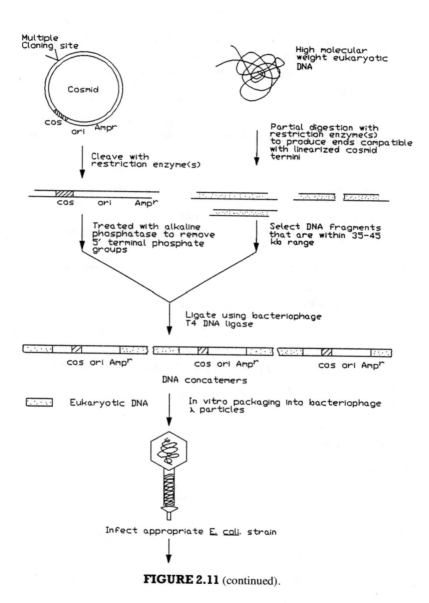

**FIGURE 2.11** (continued).

(3) Select the DNA fragment size range appropriate for the cosmid vector (i.e., 35 to 45 kb) by sucrose density gradient sedimentation.

(4) Ligate the partially digested eukaryotic DNA with the linearized cosmid vector by using T4 DNA ligase. The ligation step should result in the formation of the high molecular weight concatemers' DNA structures. In these structures, the foreign DNA will be flanked by cosmid vectors in which two *cos* sites are arranged in the same orientation.

(5) Package the ligated DNA into bacteriophage λ particles. In *in vitro* packaging reaction concatemers are cleaved at *cos* sites by the *ter* function of the bacteriophage λ gene A product. The DNA between two *cos* sites is packaged into bacteriophage λ particles.

(6) Mix a small aliquot from step 5 with a recA⁻ strain of *E. coli* under appropriate conditions. Incubate the infected culture at 37°C to allow the adsorption of the bacteriophage particles.

(7) Incubate at 37°C to allow the expression of the antibiotic resistance gene.

(8) Plate the bacterial culture on agar plates containing appropriate antibiotic (usually ampicillin).

(9) Incubate the plates overnight at 37°C for the development of bacterial colonies.

(10) Pick a number of isolated bacterial colonies, and grow them overnight for small-scale DNA preparation by alkaline lysis method.

(11) Digest the recombinant DNA with appropriate restriction enzyme, and electrophorese the resulting DNA fragments onto an agarose gel to analyze their sizes.

### 2.3.4 Filamentous Bacteriophages

The filamentous bacteriophages, e.g., M13 (Hofschneider, 1963) contain single-strand closed circular DNA molecules. A large quantity of single-stranded DNA molecules are produced when a filamentous bacteriophage infects *E. coli*.

These single-stranded DNA molecules serve as the templates for:

- DNA sequencing by Sanger's dideoxynucleotide chain termination technique
- site-directed mutagenesis using synthetic oligonucleotides as primer for the synthesis of second strand
- single-stranded DNA probes for the hybridization

The infecting single-stranded bacteriophage is converted into a double-stranded circular form known as replicative form (RF) DNA molecules, which serve as a template for further rounds of transcription within the host cells. When approximately 100 to 200 copies of the RF DNA accumulate within the infected cells, bacteriophage-induced proteins prevent the newly synthesized strands from converting into RF DNA, and thus single-stranded DNA molecules are produced.

Single-stranded DNA molecules are difficult to manipulate, therefore double-stranded foreign DNA fragments are cloned into the RF DNA of bacteriophage.

Like the plasmid DNA, the double-stranded closed circular RF DNA can be purified from bacteriophage-infected cells. The RF DNA containing double-stranded foreign DNA can easily be introduced into competent bacterial cells using transformation procedures such as those used for plasmids. When RF DNA containing double-stranded foreign DNA is introduced into bacterial cells, the DNA replication cycles eventually produce single-stranded bacteriophage containing only one of the two strands, which is excreted out of the bacterial cells into the supernatant.

Bacteriophage M13 as a vector is illustrated in Figure 2.12 (Messing and Vieira, 1982; Messing, 1983; Norrander et al., 1983; Yanish-Perron et al., 1985). A series of M13mp vectors have been derived by inserting a short fragment of the *E. coli lac* regulatory region into the wild-type bacteriophage M13. This segment carries the regulatory sequences and the region coding for the first 146 amino acids of the $\alpha$-peptide of $\beta$-galactosidase. Synthetic oligonucleotides containing a number of unique restriction enzyme sites have been inserted within the *LacZ* sequences. Multiple cloning sites within the *LacZ* sequences still retain the $\alpha$-peptide reading-frame of the $\beta$-galactosidase. When such M13 vectors infect the detective *E. coli* in which the F' plasmid codes for an enzymatically inactive $\beta$-galactosidase lacking amino acids 11-41, the $\alpha$-peptide of $\beta$-galactosidase produced by a M13 vector carries out $\alpha$-complementation to form an enzymatically active protein. When M13 vectors are grown on a bacterial lawn in the presence of IPTG and X-gal, X-gal is hydrolyzed by the $\beta$-galactosidase to bromochloroindole, which gives a blue color to the M13 plaques. Insertion of foreign DNA into one of the restriction sites in the polyclonal site generally interrupts the $\alpha$-peptide coding region, which destroys $\alpha$-complementation. Therefore, the recombinant bacteriophage will produce colorless plaques. This makes an excellent system for the detection of recombinant bacteriophage M13.

## 2.4 **ISOLATION OF HIGH MOLECULAR WEIGHT DNA**

Mammalian or plant DNA can easily be isolated by digesting cells with proteinase-K in the presence of EDTA and SDS, followed by extractions with phenol:chloroform. The DNA isolated using such methods is good enough for constructing genomic DNA libraries in bacteriophage $\lambda$ and for Southern blot analysis (see Section 2.8). The size of such DNA is approximately 100 to 150 kb in length.

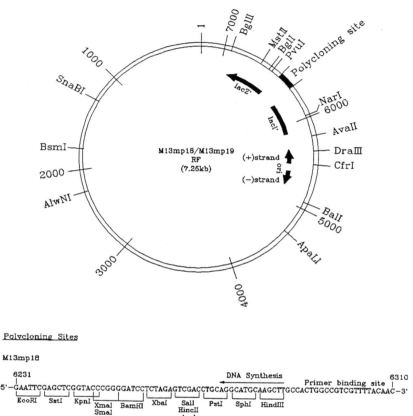

## 2.4.1 Isolation of High Molecular Weight DNA from Plant Cells

The method given by Robert et al. (1989) is summarized below:

(1) Freeze the plant leaves or seedlings at $-80°C$. Crush about 1 g of the tissue to a fine powder in a precooled motor, and pestle in the presence of liquid nitrogen.

(2) Mix the frozen powder with 5 mL of DNA extraction buffer and 0.1 mg/mL proteinase-K. Incubate at 37°C for 1 h with occasional mixing.

**DNA extraction buffer**

0.1 M Tris-HCl (pH 8.0)
0.1 M NaCl
50 mM EDTA (pH 8.0)
2% SDS
*Proteinase-K (stock solution):* 20 mg/mL in water.

(3) Extract the mixture twice with an equal volume of phenol:chloroform: isoamyl alcohol (25:24:1), and separate the phases by centrifugation at 3500 rpm for 20 min.

(4) Mix the aqueous phase with two volumes of cold ethanol. Spool out the precipitated DNA on the blunt pasteur pipet, wash with 70% ethanol, air dry, and resuspend in TE (10 mM Tris-HCl; pH 8.0 and 1 mM EDTA; pH 8.0).

(5) Add RNAase A to a final concentration of 10 $\mu g/mL$ and incubate at 37°C for 1 h to get rid of RNA.

(6) Extract with an equal volume of phenol:chloroform:isoamyl alcohol solution (25:24:1), and separate the phases by centrifugation at 3500 rpm for 20 min.

(7) Add 1/10 volume of 3 M sodium acetate to the aqueous phase, and precipitate the DNA by adding two volumes of cold ethanol.

(8) Pellet the DNA by centrifugation, wash with 70% ethanol, air dry, resuspend in TE and store at 4°C.

(9) The purity and the size of the DNA can be judged by the agarose gel electrophoresis. Use bacteriophage $\lambda$ DNA as a marker.

## 2.4.2 Isolation of High Molecular Weight DNA from Mammalian Cells

The method given by Mittal (1989) is described (with permission):

(1) Harvest the cells by scraping off the confluent monolayers, and pellet the cells by centrifugation at 2000 rpm for 5 min.

(2) Wash the cell pellet with TE (20 mM Tris; pH 8.0 and 100 mM EDTA) and resuspend the cells in TE ($5 \times 10^7$ cells/mL).

Note: For animal tissue samples, put small pieces of tissue samples into liquid nitrogen in a stainless steel cup of a blender, and blend at the top speed to grind the tissue to a powder form. Alternatively, tissue samples are crushed to a fine powder in the presence of liquid nitrogen with the help of a mortar and pestle. After evaporating the liquid nitrogen, add the finely powdered tissues to approximately 10 volumes of TE.

(3) Add SDS to a final concentration of 0.5%, and pronase to a final concentration of 1 mg/mL. Incubate the mixture at 37°C for 6 h.

*Pronase (stock solution):* 20 mg/mL in TE (10 mM Tris HCl, pH 8.0; and 1 mM EDTA, pH 8.0). Incubate at 56°C for 30 min and then store at −20°C in small aliquots.

(4) Add NaCl to a final concentration of 0.1 M and extract twice with equal volume of TE-saturated phenol:chloroform:isoamyl alcohol (25:24:1).

(5) Precipitate the DNA with two volumes of ethanol, and spool out on the tip of a sealed pasteur pipet.

(6) Wash the DNA with 70% ethanol, air dry and resuspend in TE (10 mM Tris-HCl, pH 8.0; and 1.0 mM EDTA, pH 8.0).

(7) Add RNAase A to a final concentration of 10 $\mu$g/mL, and incubate at 37°C for 1 h.

(8) Extract with an equal volume of phenol:chloroform:isoamyl alcohol (25:24:1).

(9) Add 1/10 volumes of 3 M sodium acetate, and precipitate the DNA with 2 volumes of cold ethanol.

(10) Pellet the DNA by centrifugation, wash with 70% ethanol, air dry, resuspend in TE and store at 4°C.

## 2.5 DETERMINATION OF THE CONCENTRATION OF DNA OR RNA

The concentration of DNA or RNA extracted from different sources can be determined spectrophotometrically. The optical density (O.D.) of each sample should be taken at 260 nm and 280 nm. An O.D. at 260 nm is approximately equal to 50 $\mu$g/mL for double-stranded DNA, 40 $\mu$g/mL of RNA or single-stranded DNA and 20 $\mu$g/mL for single-stranded oligonucleotides. The pure DNA and RNA samples should have the O.D. ratio at 260 nm and 280 nm (O.D. 260/O.D. 280) equal to 1.8 and 2.0, respectively. If the DNA or RNA preparations are contaminated with phenol or protein, the ratio (O.D. 260/O.D. 280) will be significantly less than 1.8 and 2.0, respectively (Maniatis et al., 1982).

The DNA concentrations can also be estimated by comparing the test sample with different concentrations of the standard DNA (e.g., bacteriophage λ DNA) on ethidium bromide stained agarose gels over UV transilluminator after electrophoresis (Table 2.3).

## 2.6 **SEPARATION OF DNA BY GEL ELECTROPHORESIS**

This simple and rapid technique is very helpful in analysis and purification of DNA fragments. Based on the sizes of DNA fragments, agarose or polyacrylamide gels are used to fractionate DNA by electrophoresis. Different percentages of polyacrylamide gels are used to effectively separate small fragments of DNA ranging from 5 to 500 bp in length or single-stranded nucleotides of similar length. However, the conventional agarose gels, having different concentrations of agarose, are used to separate DNA fragments ranging from >200 bp to approximately 50 to 60 kb in length (Table 2.4).

Through the introduction of pulsed-field gel electrophoresis technique (Schwartz and Cantor, 1984), in which the direction of electric current is changed periodically, it is possible to separate larger pieces of DNA (up to 10,000 kb in length).

Polyacrylamide gels are usually polymerized between two glass plates, and are run in a vertical position in an electric field; whereas agarose gels are

TABLE 2.3. *Bacteriophage λDNA markers for agarose gel electrophoresis (DNA fragment sizes are in kb).*

| λHindIII | λ EcoRI | λ EcoRI and HindIII |
|---|---|---|
| 23.1 | 21.2 | 21.2 |
| 9.4 | 7.4 | 5.15 |
| 6.6 | 5.8 | 5.0 |
| 4.4 | 5.65 | 4.3 |
| 2.3 | 4.9 | 3.5 |
| 2.0 | 3.5 | 2.0 |
| 0.56 | | 1.9 |
| 0.125 | | 1.6 |
| | | 1.4 |
| | | 0.95 |
| | | 0.83 |
| | | 0.56 |
| | | 0.125 |

TABLE 2.4. *Separation of DNA fragments on different concentrations of agarose gels.*

| % Agarose in Gel | Range of DNA Fragments in kb |
|:---:|:---:|
| 0.3 | >5.0−60 |
| 0.6 | 2.0−20 |
| 1.0 | 0.5−8 |
| 1.5 | 0.2−4 |
| 2.0 | 0.1−2 |

usually allowed to solidify on plastic trays, and are run in a horizontal position in an electric field. The DNA is negatively charged at neutral pH, therefore it migrates towards the anode in an electric field applied across the agarose or polyacrylamide gel.

There are a number of factors that affect the rate of migration of DNA fragments through the gel. Some of these factors are as follows:

- Concentration of agarose in gel: If the concentration of agarose is increased, the rate of migration of DNA fragments will be slower.
- Length of the DNA fragment: The rate of migration is inversely proportional to the logarithm of the number of base pairs.
- Type of the DNA: The linear, circular, and supercoiled circular forms of the same DNA migrate differently.
- Others: Examples of other significant factors are the composition of the electrophoresis buffer and the presence of intercalating dye (e.g., ethidium bromide).

## 2.6.1 Agarose Gel Preparation

There are different sizes and types of agarose gel electrophoresis apparatus commercially available. The choice depends on the type of analysis.

(1) For the preparation of an agarose gel, seal the edges of a plastic tray with clean tape and place on a horizontal surface.

(2) Mix agarose with 1× TAE or 0.5 TBE buffer (the amount of agarose depends on the percentage of the agarose gel needed) in a glass flask. Melt the agarose by heating the flask in a boiling water bath or in a microwave.

**20× TAE buffer**

96.8 g Tris
13.6 g sodium acetate
2.8 g EDTA

Bring volume to 900 mL, adjust pH to 7.9 with glacial acetic acid, and bring volume to 1 L with distilled water.

**10× TBE buffer**

54.0 g Tris
24.0 g boric acid
4.75 g EDTA
Bring final volume to 500 mL, and pH to 7.9.

(3) Cool the agarose solutions to $45-50°C$

Note: Ethidium bromide, a fluorescent dye that intercalates between stacked base pairs of DNA fragments, is used to visualize DNA bands on a UV illuminator after electrophoretic separation of DNA fragments onto an agarose gel. Care must be taken when handling ethidium bromide or solution containing ethidium bromide because it is a powerful mutagen.

Ethidium bromide may be added at this stage from the stock solution (10 mg of ethidium bromide/mL in water) to a final concentration of 0.5 $\mu$g/mL, or agarose gel can be stained in a solution containing ethidium bromide (0.5 $\mu$g/mL in water) for 30 min after electrophoresis.

(4) Place the comb on the gel tray 1.0 to 1.5 mm above the plate and pour the warm agarose solution into the mold. Allow 30 to 40 min to solidify the gel. The gel should be 4 to 6 mm thick.

(5) Remove the tape and the comb, and place the gel tray in the electrophoresis tank.

(6) Add electrophoresis buffer (same as used for making the gel) just enough to cover the gel (1 to 2 mm above the gel).

(7) Prepare the DNA samples for loading by mixing with 1/10 volume of the loading mix.

**Loading mix**

0.125 M EDTA (pH 8.0)
10% SDS
0.25% bromophenol blue
50% glycerol

(8) Carefully load the samples into the well slots using disposable tips and a micropipet. In one slot, run HindIII digested bacteriophage λ DNA as a marker.

The amount of DNA that can be loaded into each well depends on the width of the well and the type of DNA. In a 5 mm wide slot, if the plasmid DNA forms a single band, then 100 to 200 ng DNA in a slot should be enough. The digestion of mammalian DNA with a restriction

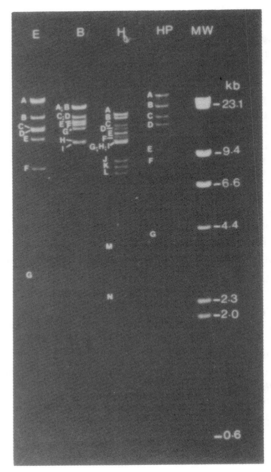

**FIGURE 2.13.** Restriction enzyme profiles of bovine herpesvirus-1 (strain "6660") DNA. The restriction enzyme profiles are of 2 μg samples of the purified BHV-1 DNA digested with EcoRI (lane E), BamHI (lane B), HindIII (lane H), and HpaI (lane HP), resolved on a 0.7% agarose gel and stained with ethidium bromide. The fragments are numbered alphabetically in decreasing order with increase in their relative mobility (i.e., with decrease in molecular weights). Sizes of fragments generated after cleavage of bacteriophage λ DNA by restriction enzyme HindIII are shown in kb (lane MW) as molecular weight markers.

enzyme generates a large number of DNA fragments of various sizes. 20 to 25 $\mu$g of such DNA samples may be loaded into each slot.

(9) Connect the electrophoresis tank with the power supply and run the gel at a constant voltage of 1 to 5 V/cm. Switch off the power supply after a certain period depending on the type of gel and source of DNA samples.

(10) Remove the tray along with the gel from the electrophoresis tank. If ethidium bromide was added in the gel, visualize the gel by ultraviolet light on a UV illuminator. Otherwise, stain the gel first with ethidium bromide as described previously.

(11) Photograph the gel with a Polaroid camera using a black and white film (Figure 2.13).

## 2.7 **MOLECULAR CLONING TECHNIQUES**

Some of the commonly used techniques in molecular cloning are described below.

### 2.7.1 Removal of Terminal Phosphate Groups from Linear Plasmid Vector DNA

To minimize the recirculation of the vector DNA during cloning, the removal of the 5'-phosphate groups from both termini of the linear vector DNA is usually recommended. During ligation, the DNA ligase catalyzes the formation of phosphodiester bonds between 5'-phosphate group and 3'-hydroxyl group of adjacent nucleotides. The removal of 5'-phosphate groups is achieved by treating linearized vector DNA with the alkaline phosphatase. In the absence of 5'-phosphate groups neither strand of the linear DNA can form a phosphodiester bond in the presence of the DNA ligase. However, the foreign DNA fragment having 5'-phosphates can be ligated to the dephosphorylated plasmid DNA (Figure 2.14). Such ligation will result in an open circular DNA molecule containing two nicks. The nicked circular DNA molecule transforms bacteria efficiently.

The procedure is as follows (modified after Sambrook et al., 1989):

(1) Digest 10 $\mu$g plasmid DNA with the restriction endonuclease of choice. Run a small amount of the DNA on 1% agarose gel to check its digestion. Extract the rest of the DNA with phenol/chloroform and precipitate with 2 volumes of cold ethanol after adding 1/10 volume of 3 M sodium acetate (pH 4.8).

(2) Dissolve the DNA in 50 $\mu$L of water.

(3) Add 5 mL of 10× phosphatase buffer.

5 g sodium chloride
Distilled water to make 1 L
Sterilize by autoclaving at 105 kPa (15 psi) for 15 min.

(2) Dilute 0.3 mL of overnight culture in 30 mL of SOB and allow to grow to an absorbency of 0.4 at 600 nm on a shaker at 37°C (usually for 2−2.5 h).

**SOB**

Tryptone 2%
Yeast extract 0.5%
NaCl 10 mM
KCl 2.5 mM

Sterilize by autoclaving at 105 kPa (15 psi) pressure for 15 min. Just before use add:

$MgCl_2$ − 10 mM
$MgSO_4$ − 10 mM

(3) Keep bacterial suspension on ice for 15 min.

(4) Pellet the bacteria at 2000 rpm for 10 min at 4°C, and resuspend the pellet gently in 6.7 mL of TFB.

**TFB**

KCl 100 mM
$MnCl_2 \cdot 2H_2O$ 45 mM
$CaCl_2 \cdot 2H_2O$ 10 mM
Hexaminecobaltic chloride 3 mM
($HACoCl_2$)
K-MES 10 mM

Sterilize by filtration

0.5 M K-MES: 0.5 M 2(*N*-morpholino)ethone sulphonic acid
Adjust pH to 6.3 with concentrated KOH.

(5) Keep the bacterial suspension on ice for 10 to 15 min and centrifuge at 2000 rpm and 4°C for 10 min.

(6) Resuspend the pellet gently in 1.6 mL of TFB.

(7) Add 56 $\mu$L of dimethylsulfoxide (DMSO) and incubate on ice for 5 min.

(8) Add 56 $\mu$L of DTT solution and incubate for 10 min on ice. DTT solution: 2.25 M dithiothreitol in 40 mM potassium acetate (pH 6.0).

(9) Repeat DMSO treatment.

## 2.7.3.1 Plasmid Transformation

(1) After the second DMSO treatment, add 50 $\mu$L of the bacterial suspension to the chilled microfuge tubes containing DNA (20 to 40 ng).

(2) Incubate on ice for 30 min.

(3) Keep the mixture at 42°C for 2 min and then at room temperature for 5 min.

(4) Add 200 $\mu$L of SOB supplemented with 20 mM glucose and incubate at 37°C for 30 min.

(5) Spread 125 $\mu$L of the bacterial mixture on 2TY agar plates containing appropriate antibiotic. (25 $\mu$L of IPTG and 25 $\mu$L of X-gal should also be added on 2TY agar plates if the plasmid contains *lacZ* gene for the production of $\beta$-galactosidase).

*2 TY agar plates:* 1.5% agar in 2 TY.

*IPTG:* 25 mg/mL of isopropyl-$\beta$-D-thio-galactoside (IPTG) in sterile distilled water.

*X-gal:* 25 mg/mL of 5-bromo-4-chloro-3-indolyl-$\beta$-D-galactoside (X-gal) in dimethyl formamide.

(6) Incubate the plates at 37°C overnight for the development of bacterial colonies containing the recombinant vector. Pick up bacterial colonies (white colonies if IPTG and X-gal are used), and test for the presence of desired insert by colony hybridization or small-scale plasmid DNA preparation.

## 2.7.3.2 Bacteriophage Transformation

(1) After the second DMSO treatment (preparation of bacteria for transformation), add 200 $\mu$L of the bacterial suspension to the chilled microfuge tubes containing DNA (10 to 20 ng) and incubate on ice for 30 min.

(2) Hold the mixture at 42°C for 2 min and then keep at room temperature for 5 min.

(3) Keep 4 mL aliquots of top agar (0.7% agar in 2 TY) at 48°C. (For M13 transformation, add 25 $\mu$L of 1PTG and 25 $\mu$L of X-gal to the top agar.)

(4) Add bacteria/DNA mixture to a top agar aliquot.

(5) Pour onto pre-warmed 2 TY agar plate.

(6) After the solidification of the top agar, incubate the plate at 37°C overnight in a sealed box.

(7) Pick up the appropriate plaques and test for the presence of the desired insert.

## 2.7.4 Rapid Isolation of Plasmid DNA by Alkaline Extraction Method

Alkaline extraction method for the isolation of covalently closed circular (CCC) DNA (i.e., plasmid) from bacterial cells is simple and reliable for screening recombinant plasmids. The plasmid DNA obtained by this method is sufficiently pure to be digested with restriction endonucleases. Fifty to 100 or more samples could be extracted in a few hours. A modified method is used for the isolation of large quantities of highly purified plasmid DNA (Birnboim and Doly, 1979).

For the isolation of plasmid DNA, the separation of host-cell chromosomal DNA and other macromolecular components from plasmid DNA is needed. In the alkaline extraction procedure, exposure of a cell extract to an alkaline pH > 11.5 results in the denaturation of linear (chromosomal) DNA, but not of CCC DNA. Neutralization of the extract with high concentration of salt leads to the precipitation of chromosomal DNA and most of the cellular RNA and protein; however, the CCC DNA remains in the soluble fraction. The supernatant containing CCC DNA could be precipitated with ethanol.

## 2.7.4.1 Small-Scale Plasmid DNA Preparations

(1) Pick up a large number of single bacterial colonies from original plates with sterile cocktail sticks and add into tubes containing 2 mL of 2 TY supplemented with appropriate antibiotic.

**2 TY**

16 g tryptone
10 g yeast extract
5 g sodium chloride
Distilled water to make 1 L
Sterilize by autoclaving at 105 kPa (15 psi) pressure for 15 min.

(2) Shake tubes at 37°C overnight on an orbital shaker.
(3) Centrifuge 1.5 mL bacterial cultures in microfuge tubes at 13,000 rpm for 30 s.
(4) Suspend bacterial pellets in 100 mL bacterial lysis buffer and leave on ice for 30 min.

**Bacterial lysis buffer**

50 mM sucrose

20 mM Tris-HCl (pH 7.6)
10 mM EDTA
5 $\mu$g/mL lysozyme added just before use

(5) Add 200 $\mu$L of alkaline SDS and continue incubation on ice further for 5 min.

**Alkaline SDS**

0.2 M NaOH
1% SDS

(6) To the bacterial lysate, add 150 mL of 3 M sodium acetate (pH 4.8), vortex and keep on ice for 1 h with occasional mixing.

(7) Centrifuge at 13,000 rpm for 10 min in a microfuge.

(8) Collect 400 $\mu$L of supernatant and precipitate with 2.5 volumes of cold ethanol.

(9) Leave at $-20°$C for 20 min and centrifuge for 5 min to pellet the plasmid DNA.

(10) Dissolve the pellet in 100 $\mu$L of 0.1 M sodium acetate (pH 6.0) and reprecipitate with 2 volumes of cold ethanol.

(11) Centrifuge for 5 min and dissolve the DNA pellet in 50 $\mu$L of TE.

(12) Digest 5 $\mu$L of the plasmid DNA with appropriate restriction endonuclease and electrophorese on agarose gel to check the presence of the right insert (Figure 2.15).

## 2.7.4.2 Large-Scale Plasmid Preparation

(1) Inoculate 400 mL of 2TY containing appropriate antibiotic with a bacterial colony harboring plasmid.

(2) Grow bacterial culture at 37°C overnight with continuous shaking on a rotary shaker.

(3) Centrifuge at 10,000 rpm for 10 min to pellet bacteria.

(4) Resuspend the bacterial pellet in 12 mL of ice-cold 25% sucrose made in 50 mM Tris-HCl (pH 8.0).

(5) Add 1 mL of freshly prepared lysozyme (10 $\mu$g/mL), mix the bacterial suspension gently, and keep on ice for 5 min.

(6) Add 1 mL of 0.5 M EDTA (pH 8.0) and 5 mL of Triton solution.

**Triton solution**

1% Triton X-100
62.5 mM EDTA (pH 8.0)
50.0 mM Tris-HCl (pH 8.0)

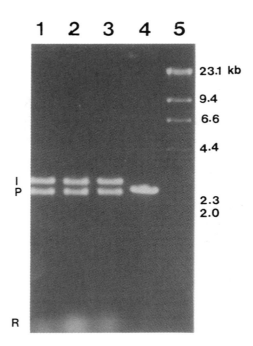

**FIGURE 2.15.** Identification of recombinant plasmid containing the desired insert by separation of restriction enzyme-digested recombinant plasmid DNA on an agarose gel. An aliquot of 5 μL from each recombinant plasmid DNA sample, obtained by small-scale plasmid DNA preparation from three different bacterial colonies, is digested with restriction enzyme BamHI (lanes 1, 2, and 3), resolved on a 1.0% agarose gel and stained with ethidium bromide. The lane 4 represents the purified pUC18 plasmid DNA digested with restriction enzyme BamHI. Sizes of fragments generated after cleavage of bacteriophage λ DNA by restriction enzyme HindIII are shown in kb (lane 5) as molecular weight markers. I = insert DNA, P = pUC18 plasmid DNA, and R = RNA. Reproduced with permission from Mittal, 1989.

(7) Mix well and keep at $-70°C$ for 1 h. Freeze and thaw the mixture twice.

(8) Centrifuge at 21,000 rpm for 45 min at 4°C and collect the supernatant.

(9) Isolate supercoiled plasmid DNA by caesium chloride (CsCl) density gradient centrifugation. For this purpose add CsCl (1.1 g/mL) to the supernatant and mix gently to dissolve all of the salt.

(10) Add ethidium bromide (0.3 mL/10 mL from 10 mg/mL stock solution).

(11) Transfer the CsCl solution to a tube suitable for centrifugation in a vertical type rotor. Fill the tube with the light paraffin oil.

(12) Seal the tube and centrifuge at 50,000 rpm for 16 h at 20°C.

(13) After centrifugation, remove the tube carefully and fix it vertically on a stand. In ordinary light two bands should be visible. The lower thick band consists of closed circular plasmid DNA, and the upper band consists of linear plasmid DNA or nicked circular plasmid DNA.

(14) Put 1/3 of syringe needle (#18) in a paraffin layer after puncturing the tube's wall. Collect the lower band with the help of a syringe and a needle (Figure 2.16).

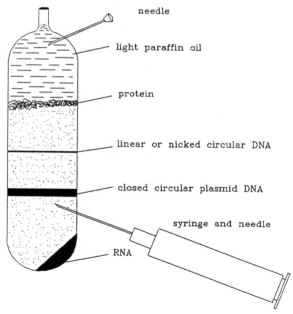

**FIGURE 2.16.** Collection of CsCl density gradient, purified, closed circular plasmid DNA.

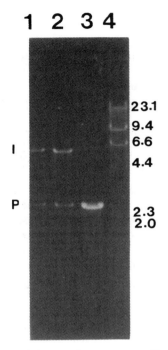

**FIGURE 2.17.** Separation of the restriction enzyme-digested recombinant plasmid DNA on an agarose gel. 0.1 μg (lane 1) and 0.2 μg (lane 2) of the recombinant plasmid DNA, obtained by large-scale plasmid preparation, digested with restriction enzyme BamHI, resolved on a 1.0% agarose gel and stained with ethidium bromide. The lane 3 represents the purified pUC18 plasmid DNA digested with restriction enzyme BamHI. Sizes of fragments generated after cleavage of bacteriophage λ DNA by restriction enzyme HindIII are shown in kb (lane 4) as molecular weight markers. I = insert DNA, and P = pUC18. Reproduced with permission from Mittal, 1989.

(15) Extract the DNA with water-saturated butanol to remove ethidium bromide.

(16) Dialyze the aqueous phase extensively against TE and check the concentration of the DNA spectrophotometrically.

(17) Digest a small aliquot of the plasmid DNA with appropriate restriction enzyme and electrophorese on agarose gel to check the presence of the right insert (Figure 2.17).

## 2.7.5 Recovery of DNA Fragments from Agarose Gel

The extraction and purification of DNA fragment from agarose gels could be performed by following the technique described by Dretzen et al. (1981). This procedure enables purification of DNA fragments up to 20 kb in size with a yield of 50 to 80%.

DNA fragment recovered by following this technique is good enough for the:

- restriction enzyme digestion
- cloning in suitable vector
- nick translation
- end labeling
- DNA sequencing

The procedure for the recovery of a DNA fragment from the agarose gel is as follows:

(1) Digest the DNA with an appropriate restriction enzyme.

(2) Separate the DNA fragments on an agarose gel by electrophoresis.

(3) Strain the gel with ethidium bromide and visualize DNA fragments over the UV transilluminator.

(4) Make a small slit just below the desired band and insert a strip of DEAE cellulose paper (DE81, Whatman) into the slit.

Soak DE81 strip in 2.5 M NaCl overnight and wash with the electrophoresis buffer just before use. Take care not to leave any air bubble in between the gel and DE81 paper.

(5) Electrophorese the gel again to allow the transfer of DNA onto the DE81 paper strip.

(6) Carefully remove the DE81 paper strip from the gel and incubate with TEN buffer at 37°C for 2 h.

**TEN buffer**

10 mM Tris-HCl (pH 8.0)
1 mM EDTA
1.5 M NaCl

(7) Pass the mixture through a plug of siliconized glass wool in a 1 mL syringe to get rid of shredded DE81 paper.

(8) Extract the TEN buffer containing DNA with water-saturated butanol.

(9) Extract once with TE-saturated phenol and chloroform.

(10) Precipitate the DNA with 2.5 volumes of cold ethanol. Dry the DNA pellet and dissolve in TE.

(11) Electrophorese a sample of the eluted DNA onto a "mini" gel to check the recovery.

### 2.7.6  Filling Recessed 3' Ends of Double-Stranded DNA

For filling recessed 3' ends, generated by the digestion of DNA with restriction endonucleases, a large fragment of *E.coli* DNA polymeraseI (Klenow fragment) is needed.

(1) Digest DNA with appropriate restriction enzyme.
(2) Extract once with phenol:chloroform (1:1).
(3) Precipitate the DNA with 2 volumes of ethanol after adding 1/10 volume of 3 M sodium acetate (pH 4.8).
(4) Wash the DNA pellet with 70% ethanol and dissolve in a minimum volume of TE buffer or water. The DNA is ready for filling recessed 3' ends.

The sequence of 5' overhang (i.e., protruding 5' ends) determines which deoxynucleotidetriphosphates (dNTPs) should be added to the reaction mixture. All four dNTPs may be added even if all of them are not needed. For end-labeling reaction, one of the dNTPs required to fill the recessed 3' ends is labeled with $[\alpha - ^{32}p]$.

For filling recessed 3' ends, mix:

- 1 $\mu$g of DNA having recessed 3' termini
- 1 $\mu$L (each) of 2 mM stock solution of unlabeled dNTPs (all or according to the requirement)
- 20 $\mu$Ci of $[\alpha - ^{32}p]$ dNTP (if end-labeling is needed)
- 2 $\mu$L of 10$\times$ nick translation buffer
- 1 unit of Klenow fragment
- $H_2O$ sufficient to make 20 $\mu$L

(1) Incubate at room temperature for 30 to 40 min.
(2) Add 1 $\mu$L of 0.5 M EDTA to stop the reaction.
(3) Extract once with phenol:chloroform (1:1) and precipitate the DNA with ethanol after adding 1/10 volume of 3 M sodium acetate (pH 4.8). The DNA can be used for blunt ends cloning.

If the DNA is end-labeled, pass the reaction mixture after step 2 through a small column of Sephadex G-50 to get rid of unincorporated dNTP.

### 2.7.7  In vitro Labeling of DNA by Nick Translation

For the identification of a particular gene or a DNA fragment from a genomic library or from a cDNA library, a DNA probe of the same gene or from a related source is needed.

The DNA probes are labeled *in vitro* with radioactive deoxynucleotidetriphosphates (dNTPs) by nick translation for hybridization experiments.

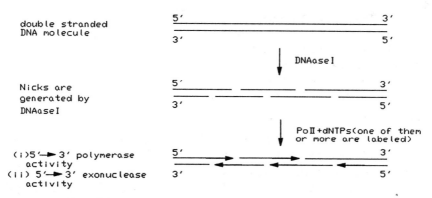

**FIGURE 2.18.** *In vitro* labeling of DNA by nick translation.

Two enzymes are used in the nick translation reaction: (1) pancreatic deoxyribonucleaseI (DNAaseI), and (2) *E. coli* DNA polymeraseI (PolI). DNAaseI generates nicks in double-stranded DNA molecules, whereas PolI adds nucleotide residues to the 3'-hydroxyl terminus of a nick (created by DNAaseI) (Figure 2.18). The nucleotides are eliminated from the 5'-phosphoryl terminus due to 5'→3' exonuclease activity of PolI.

By this technique double-stranded DNA molecule can be labeled to specific activities $> 10^8$ Cpm/$\mu$g of the DNA with $[\alpha - {}^{32}p]$ dNTPs (Rigby et al., 1977; Maniatis et al., 1982).

In the double-stranded DNA molecules nicks are generated at random by DNAaseI, which is why populations of radioactive fragments generated partially overlap each other. The size of DNA fragments labeled by nick translation depends on the concentration of DNAaseI. It is usually adequate to use one dNTP labeled with $[\alpha - {}^{32}p]$ and three unlabeled dNTP's in the nick translation reaction.

## 2.7.7.1 Nick Translation Reaction

**10× nick translation buffer**

0.5 M Tris-HCl (pH 7.4)
0.1 M MgSO$_4$
1 mM dithiothreitol
500 $\mu$g/mL bovine serum albumin (fraction V)
Store at $-20°$C

*Dilution of DNAaseI:* Ten-thousand-fold dilution of a stock solution of DNAaseI (1 mg/mL) in ice-cold 1× nick translation buffer containing 50% glycerol.

In a 0.5 mL microfuge tube add:

- 0.5 to 1 $\mu$g of DNA (to be nick-translated)
- 5 $\mu$L of 10× nick translation buffer
- 1 $\mu$L (each) of a 2 mM solution of three unlabeled dNTPs
- 20 $\mu$Ci of [$\alpha - {}^{32}$p] dNTP
- Sterile distilled water to make 45 $\mu$L
- 1.0 $\mu$L of diluted DNAaseI

(1) Mix by vortexing and incubate on ice for 1 min.

(2) Add 5 units of PolI, mix well and incubate at 16°C for 1 h.

   Stop the reaction by adding 2 $\mu$L of 0.5 M EDTA.

(3) Separate nick-translated DNA from unincorporated dNTPs by centrifugation through a small column of Sephadex G-50 as described below.

*In vitro* radiolabeling of DNA to high specific activity can also be achieved by the use of random oligonucleotides as primers (Feinberg and Vogelstein, 1983).

## 2.7.8 Separation of Labeled DNA from Unincorporated dNTPs

The technique for the separation of labeled DNA from unincorporated dNTPs described by Maniatis et al. (1982) is summarized below.

(1) Plug the bottom of a 1 mL disposable syringe with siliconized glass wool.

(2) Prepare a 0.9 mL bed volume column of Sephadex G-50 equilibrated in TE.

   **TE**

   10 mM Tris-HCl (pH 8.0)

   1 mM EDTA (pH 8.0)

(3) Place a decapped microfuge tube into a centrifuge tube. After inserting the syringe into a centrifuge tube, centrifuge at 2000 rpm for 3 min.

(4) Add 100 mL of TE and centrifuge at 2000 rpm for 3 min.

(5) Replace the decapped microfuge tube with a fresh one.

(6) Add 50 $\mu$L of TE to nick translation reaction mixture and apply the sample to the column.

(7) Centrifuge at 2000 rpm for 3 min.

(8) Collect the effluent from the decapped microfuge tube.

(9) Denature the labeled DNA collected from the decapped microfuge tube in a boiling water bath for 3 min prior to its use in hybridization experiments.

## 2.8 **SOUTHERN BLOTTING AND HYBRIDIZATION**

### 2.8.1 Southern Blotting

This technique is widely used to identify the location of particular sequences within genomic DNA. Restriction-enzymes-digested genomic DNA is run on an agarose gel by electrophoresis. The agarose gel is soaked in an alkaline solution to facilitate *in situ* denaturation of DNA. The denatured DNA fragments from the gel are transferred to a nitrocellulose filter or nylon membrane by Southern transfer techniques (Southern, 1975). The filter is hybridized to radiolabeled DNA or RNA probes, and the positions of the DNA bands, complementary to the probe, can easily be identified by autoradiography. The technique is as follows.

(1) Usually 10 $\mu$g of the genomic DNA is digested with one or more restriction enzymes, under appropriate conditions, to identify single copy gene with a probe of several hundred nucleotides in length. If oligonucleotides are used as probes, 30−40 $\mu$g of genomic DNA may be loaded into each 5 mm-wide slot. If the sequences of interest are present in high copy number within genomic DNA, the amount of DNA per slot may be decreased proportionately.

(2) Separate the restriction fragments of genomic DNA by electrophoresis through 0.7% agarose gel. Photograph the gel after staining with ethidium bromide solution on a UV transilluminator.

(3) Remove the unwanted portion of the gel with a razor blade.

(4) The transfer of the DNA fragments > 15 kb in length onto a membrane is comparatively poor. This problem can be overcome by soaking the gel in 0.25 N HCl for 10 min, which results in nicking of the DNA due to depurination.

(5) Soak the gel in the denaturation solution for 30 min with constant shaking, and then rinse with deionized water. It will separate the two strands of DNA, which is desired for hybridization with a radioactive probe.

**Denaturation solution**

0.5 N NaOH
1.5 M NaCl

(6) To neutralize the alkali, soak the gel in the neutralization solution for 1 h with constant shaking.

**Neutralization solution**

0.5 M Tris-HCl (pH 7.4)
3.0 M NaCl

(7) The transfer of DNA fragment from the gel to a membrane can be achieved by the following three methods.

- Capillary transfer: This method of DNA transfer was originally developed by Southern (1975) and hence the technique is popularly known as Southern blotting. The DNA fragments from the gel are transferred to the membrane (kept over the gel in close contact) with the flow of the liquid by capillary action.
- Vacuum transfer: The DNA fragments from a gel are transferred to a membrane under vacuum. This transfer technique is more efficient than capillary transfer. The transfer is achieved within 1 to 2 h.
- Electrophoretic transfer: The transfer of the DNA fragments from a gel to a membrane is achieved by electrophoresis. However, the capillary transfer technique is widely used, the transfer by this method is described here.

For the purpose of capillary transfer, place 3 MM Whatman chromatographic paper soaked in $10 \times$ SSC over a glass plate. Place the glass plate on a platform in the center of a trough filled with $10 \times$ SSC. Keep the paper in contact with $10 \times$ SSC on both sides.

$$20 \times \text{ SSC} = 3 \text{ M NaCl, } 0.3 \text{ M sodium citrate}$$

(8) Place the gel on the top of the glass plate covered with 3 MM Whatman paper.

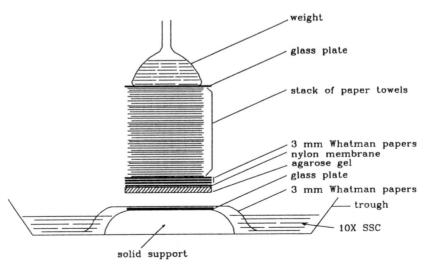

**FIGURE 2.19.** Capillary transfer of DNA fragments from agarose gel to nylon membrane.

(9) Soak the nylon membrane (a number of nylon membranes from different manufacturers are commercially available) of the same size as the gel in deionized water and then float the membrane in 6× SSC for 5 min.

(10) Place the nylon membrane on the top of the gel. Remove air bubbles in between the gel and the membrane, if there are any. Cover the Whatman paper around the gel with Parafilm.

(11) Soak three pieces of 3 MM Whatman paper of the same size as the gel in the 2× SSC, and place them on top of the nylon membrane.

(12) Put a stack of paper towels slightly smaller than the gel, 8 to 10 cm high on the 3 MM Whatman paper.

(13) Place a glass plate on the top of paper towels, and put a 0.5 to 1 kg weight on the top of the glass plate to allow the flow of the liquid from the trough through the gel and nylon membrane by capillary action (Figure 2.19).

(14) The wet paper towels should be replaced to maintain the capillary transfer.

(15) Continue the transfer of DNA to the membrane for 16 to 24 h.

(16) Remove the membrane after dismantling the assembly and rinse in the 2× SSC.

(17) Remove the membrane from the 2× SSC, drain away the excess solution, and dry the membrane at room temperature.

(18) Expose the membrane to UV light for 3 min to cross-link the transferred DNA onto the membrane. Store the membranes in sealed bags at room temperature.

## 2.8.2 Hybridization

The following technique is modified from Sambrook et al. (1989):

(1) Soak the nylon membrane containing cross-linked DNA fragments in the 6× SSC for 2 min.

(2) Put the wet filter into a heat-sealable bag. Add 0.1 mL of prehybridization mix for each $cm^2$ of nylon membrane. Care should be taken to squeeze out as many air bubbles as possible from the bag. With heat sealer, seal the open end of the bag and incubate the bag at 65°C for 2 to 3 h in a shaking water bath.

**Prehybridization mix**

6× SSC
5× Denhardt's solution
0.5% SDS
200 $\mu$g/mL denatured salmon sperm DNA

**100 × Denhardt's solution**

2% bovine serum albumin
2% ficoll
2% polyvinylpyrrolidone

Add 10 mg/mL in water of salmon sperm DNA (stock solution). Boil for 10 min in a water bath. Cool at room temperature and store at −20°C.

(3) Prepare *in vitro* [$^{32}$p]-labeled DNA probe as described previously (section 2.7.7). The double-stranded radiolabeled probe should be denatured by heating for 3 to 5 min in a boiling water bath just before use.

(4) Remove the bag containing the nylon membrane from the water bath. Replace the prehybridization mix with hybridization mix by cutting off one corner of the bag. Add the denatured probe to the hybridization mix.

**Hybridization mix**

6× SSC
5× Denhardt's solution
0.5% SDS
100 µg/mL denatured salmon sperm DNA

Squeeze air bubbles from the bag and reseal. The resealed bag should be kept inside another clean bag to avoid [$^{32}$p] contamination of the water bath. The hybridization is continued at 65°C for 16 h.

(5) To remove the unhybridized probe from the nylon membrane, take out the membrane from the bag after pouring the hybridization mix in a sink suitable for radioactive disposal.

(6) Wash the membrane with the 2 × SSC, and 0.5% SDS solution at room temperature.

(7) Wash three times with the 0.1 × SSC, and 0.1% SDS solution for 30 min at 65°C in a shaking water bath.

(8) Dry the membrane on a piece of 3 MM Whatman paper.

(9) Expose the membrane to X-ray film with an intensifying screen (placed behind the film) at −70°C for 24 to 36 h.

(10) Develop the X-ray film to obtain an autoradiographic image of the membrane (Figure 2.20).

## 2.8.2.1 Important Notes

• To increase the rate of reassociation of nucleic acids during hybridization, dextran sulphate (Wahl et al., 1979) or polyethylene glycol (Amasino, 1986) can be added in the hybridization mix.

- Formamide can be used in the prehybridization and hybridization mixes to reduce the temperature of incubation.
- Bacterial colonies containing recombinant plasmids can be lysed on nitrocellulose filters or nylon membranes and hybridization is performed as described earlier.

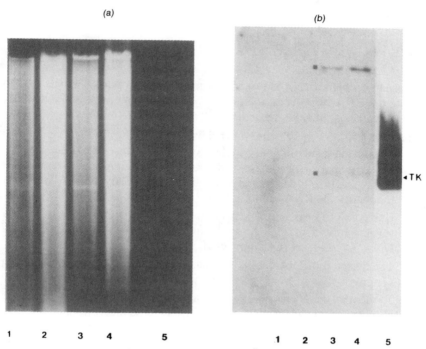

**FIGURE 2.20.** Identification of the herpes simplex virus-1 (HSV-1) thymidine kinase (*tk*) gene-specific sequences in DNA fragments generated by the restriction enzyme BamHI digestion of the transformed-cell DNA. a) BamHI restriction enzyme profiles of DNA from L(TK⁻) cells and L(TK⁻) cells transformed with the HSV-1 *tk* gene. 5 and 10 μg DNA of L(TK⁻) cells (lanes 1 and 2, respectively) and HSV-1 *tk* gene-transformed L(TK⁻) cells (lanes 3 and 4, respectively) digested with the restriction enzyme BamHI. 50 ng of the clone containing HSV-1 *tk* gene was also digested with BamHI (lane 5). The restriction enzyme-digested DNA samples were fractionated on a 0.7% agarose gel and stained with ethidium bromide. The DNA fragments from the gel were transferred onto a Hybond-*N* membrane and analyzed by Southern-blot hybridization. b) Southern-blot hybridization of Figure 2.20(a), probed with the purified, cloned 3.5 kb BamHI "P" DNA fragment of HSV-1 (containing the *tk* gene). The hybridization was carried out at 65°C and the filter washed at 65°C in a solution containing 0.1 × SSC and 0.1% SDS. Lane 5 was cut out for shorter autoradiographic exposure of the same blot. The DNA fragments that hybridized to this probe are shown by rectangles beside the relevant bands. The hybridization to the HSV-1 DNA fragment containing *tk* gene is shown in lane 5. Reproduced with permission from Mittal, 1989.

- Similarly, genomic or cDNA libraries in bacteriophage are screened to identify recombinant bacteriophages carrying sequences of interest by hybridization with $^{32}$p-labeled probes. The imprints of bacteriophage plaques onto nitrocellulose filter or nylon membrane are obtained by layering the membrane onto the plaques. From plaques, DNA and bacteriophage particles are transferred to the membrane by capillary action.

## 2.9 **RNA ANALYSIS**

Instead of DNA, RNA is required for a number of purposes, such as: to isolate the RNA of interest, to map the 5' and 3' ends of a specific mRNA, cDNA synthesis, *in vitro* translation, etc.

(1) By RNA blotting (Northern hybridization), RNAs are separated onto a denaturing agarose gel by electrophoresis, transferred to a nitro-cellulose paper (Thomas, 1980) or nylon membrane (Bresser and Gillespie, 1983), and then hybridized with radiolabeled RNA or DNA to identify the RNA of interest.

(2) Using poly(dT) as synthetic primer, the cDNA synthesis is achieved by reverse transcriptase, RNAaseH and the Klenow fragment of *E. coli* DNA polymeraseI (Okayama et al., 1988).

(3) The selected mRNA can be translated in *in vitro* protein synthesizing systems.

(4) To identify the 5' end of a mRNA, a small DNA fragment or synthetic oligonucleotide is hybridized to the mRNA, and this primer is extended by reverse transcriptase to the extreme 5' end of the mRNA. The size of the product in polyacrylamide gel electrophoresis indicates the number of nucleotides from the position of the primer binding site to the 5' end of the mRNA.

(5) The precise locations of the 5' and 3' ends of a selected mRNA and the positions of splice junctions can easily be analyzed by S1 nuclease mapping (Berk and Sharp, 1977). The mRNAs of interest are hybridized to usually labeled DNA probes derived from different fragments of the genomic DNA under conditions that favor the formation of RNA:DNA hybrids (Casey and Davidson, 1977). These are then digested with S1 nuclease to remove single-stranded nucleic acids. The S1 nuclease digested products are analyzed by gel electrophoresis and DNA sequencing.

The ribosomal RNA (rRNA) (28S, 15S and 5S) forms more than 80% of the total RNA present in a typical mammalian cell. The low molecular weight

RNA species (small nuclear RNAs and transfer RNAs) constitute 15 to 20% of the total cellular RNA. The mRNA is about 1 to 5% of the total cellular RNA. Most of the eukaryotic mRNAs have poly(A) tails at their 3' ends, which help in their purification on oligo(dT)-cellulose column by affinity chromatography.

## 2.9.1 Extraction of Total Mammalian Cell RNA

The isolation of mRNA from cultured cells or tissue by using guanidine thiocyanate was described by Chilgwin et al. (1978).

(1) Mix $5 \times 10^8$ cells or homogenize 2 to 4 g of tissue material with 100 mL of the 5.5 M GT solution.

**5.5 M GT solution**
5.5 M guanidine thiocyanate
25 mM sodium citrate
0.5% sodium lauryl sarcosine
0.2 M 2-mercaptoethanol
Adjust pH to 7.0

(2) Shear the DNA by passing the cell lysate through an 18 gauge needle 5 to 6 times.

(3) Centrifuge at 2500 rpm for 10 min at 4°C to remove the cell debris.

(4) Overlay the supernatant onto a 17 mL CT solution in sterile SW28 centrifuge tubes and spin at 25,000 rpm for 24 h at 15°C.

**CT solution**
Cesium trifluoroacetate (density 1.5 g/mL)
0.1 M EDTA

(5) Carefully remove the supernatant and DNA band, then invert the centrifuge tube on tissue paper to drain for 5 to 10 min.

(6) Dissolve the RNA pellet in 0.4 mL of 4 M GT solution on ice and centrifuge to remove the insoluble debris.

(7) Precipitate the RNA with 300 $\mu$L of ethanol after adding 10 $\mu$L of 1 M acetic acid at $-20$°C for 3 to 4 h. Pellet the RNA by centrifuging in a microfuge for 10 min at 4°C.

(8) Dissolve the RNA pellet in 1 mL of TE.

**TE**
10 mM Tris-HCl (pH 7.5)
1 mM EDTA (pH 8.0)

(9) Reprecipitate the RNA by adding 100 μL of 2 M NaCl and 3 mL of ethanol at −20°C for 3 to 4 h. The RNA can be stored as a wet precipitate.

## 2.9.2 cDNA Cloning

The complementary DNA (cDNA) cloning is one of the most important steps in isolating and analyzing eukaryotic genes. In cDNA synthesis, poly-(A)⁺mRNA are converted to double-stranded DNA by using a combination of enzymes.

Steps involved in making cDNA library (Figures 2.21 and 2.22):

(1) RNA is isolated from eukaryotic cells (as above).

(2) Poly(A)⁺ mRNA is separated from the total cellular RNA with the help of oligo(dT)-cellulose column chromatography.

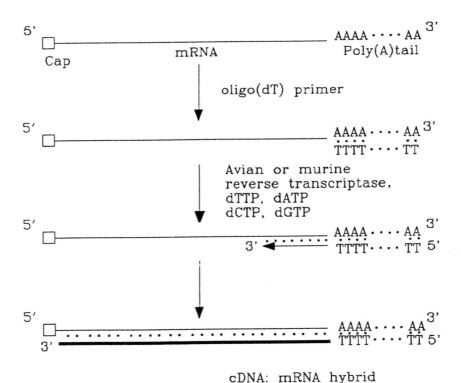

cDNA: mRNA hybrid

**FIGURE 2.21.** Synthesis of first strand of cDNA.

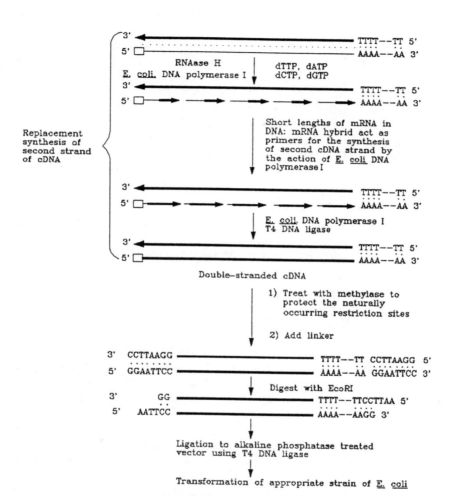

**FIGURE 2.22.** cDNA cloning.

(3) A synthetic oligo(dT) primer (usually 12 to 18 nucleotides in length) is used to bind to the poly(A) tail present at the 3' end of eukaryotic mRNA. RNA-dependent DNA polymerase (reverse transcriptase) is used to initiate the synthesis of the first strand of cDNA from the oligo(dT) primer.

(4) There are a number of methods available for the synthesis of the second strand of cDNA. One of the methods to synthesize the second strand of cDNA is by replacement synthesis (Okayama and Berg, 1982; Gubler and Hoffman, 1983) in which a DNA:mRNA hybrid is used as a template for the introduction of nicks and gaps in the mRNA strand of the DNA:mRNA hybrid with RNAaseH. A number of RNA primers produced by RNAaseH are utilized by *E. coli* DNA polymerase for the synthesis of the second strand of cDNA.

(5) Double-stranded cDNA termini are modified to enhance their cloning efficiency.

(6) Double-stranded cDNA is cloned in an appropriate vector.

(7) A cDNA library is screened to isolate the clone of interest. This is usually achieved by hybridization or by using specific antibodies if the cDNA library is made in an expression vector.

The technique of cDNA cloning developed by Okayama and Berg (1982) is described in Figure 2.23.

## 2.10 DNA SEQUENCING

Protein sequences are predicted from the nucleotide sequences of genes or cDNAs. The site-directed mutagenesis to modify a protein or to understand the role of a particular residue in a functional domain of a protein, is carried out once the nucleotide sequence of the gene coding for that particular protein is known.

There are two sequencing techniques that are currently in use: (1) the chemical degradation method of Maxam and Gilbert (1977), and (2) the enzymatic dideoxynucleotide chain termination method of Sanger et al. (1977) (Figure 2.24). However, the basic principle of these two techniques is different, and both of these methods generate different species of radiolabeled oligonucleotides that start from a fixed point but terminate randomly at a fixed residue. These randomly generated oligonucleotides are resolved by electrophoresis for different nucleotides loaded into adjacent lanes of a sequencing gel. The nucleotide sequence can be read directly from the image of the gel on an X-ray film from the bottom of the gel towards the top.

By using bacteriophage M13 as a cloning vector, the dideoxy chain termination technique of DNA sequencing is considered to be the easy and quick method of sequencing large DNA fragments. DNA fragments are cloned into

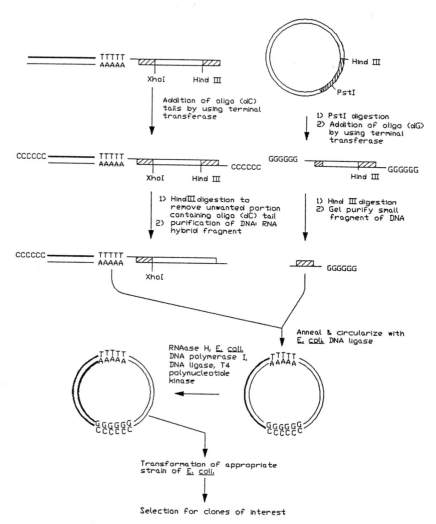

**FIGURE 2.23.** Okayama-Berg method of cDNA cloning.

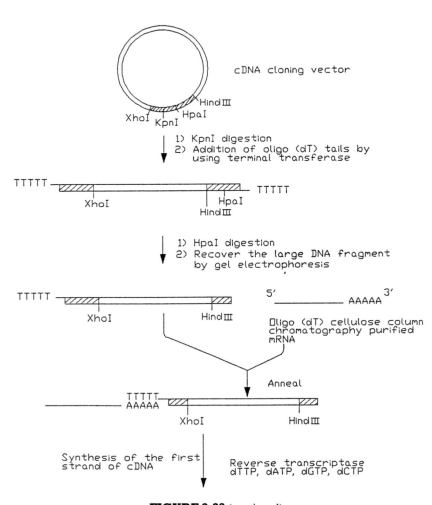

**FIGURE 2.23** (continued).

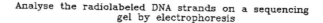

Analyse the radiolabeled DNA strands on a sequencing gel by electrophoresis

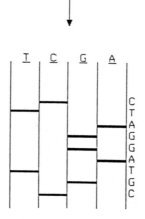

dTTP = deoxythymidinetriphosphate, dGTP = deoxyguanosinetriphosphate,
dCTP = deoxycytidinetriphosphate,  dATP = deoxyadenosinetriphosphate,
ddTTP = dideoxyTTP, ddGTP = dideoxyGTP, ddCTP = dideoxyCTP,
ddATP = dideoxyATP, T = thymine, C = cytosine, G = guanine, A = adenine,
T = dideoxyT, C = dideoxyC, G = dideoxyG, A = dideoxyA,
A* = radiolabeledA, NTPs = nucleotidetriphosphates,
ddNTP = dideoxynucleotidetriphosphate

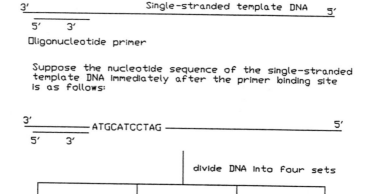

FIGURE 2.24. Sequencing by Sanger's dideoxynucleotide chain termination technique.

**65**

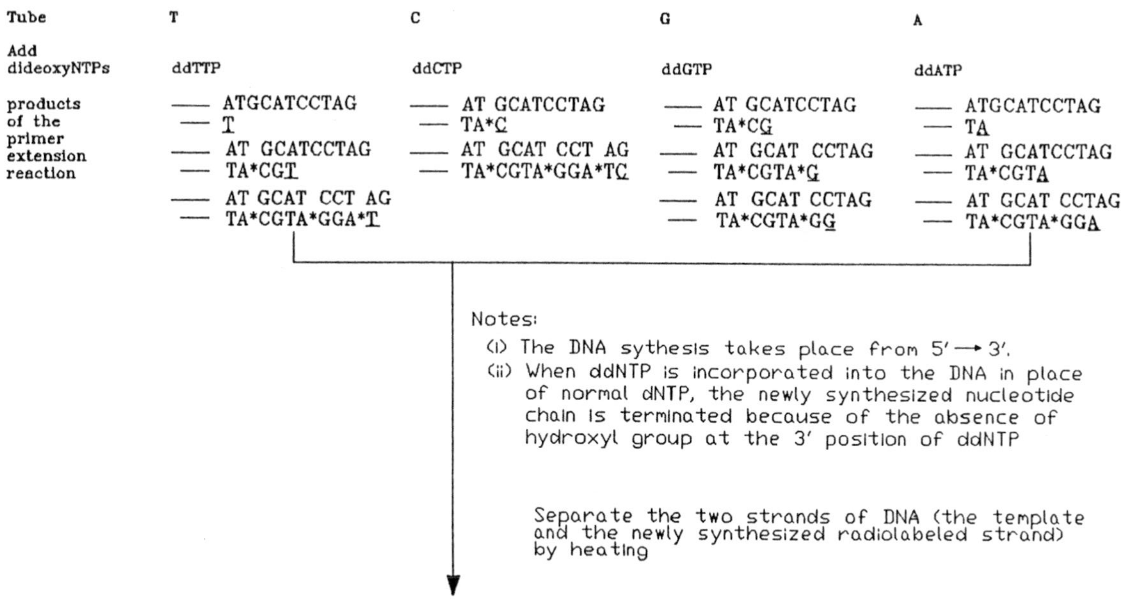

| Tube | T | C | G | A |
|---|---|---|---|---|
| Add dideoxyNTPs | ddTTP | ddCTP | ddGTP | ddATP |
| products of the primer extension reaction | — ATGCATCCTAG<br>— T<br>— AT GCATCCTAG<br>— TA*CGT<br>— AT GCAT CCT AG<br>— TA*CGTA*GGA*T | — AT GCATCCTAG<br>— TA*C<br>— AT GCAT CCT AG<br>— TA*CGTA*GGA*TC | — AT GCATCCTAG<br>— TA*CG<br>— AT GCAT CCTAG<br>— TA*CGTA*G<br>— AT GCAT CCTAG<br>— TA*CGTA*GG | — ATGCATCCTAG<br>— TA<br>— AT GCATCCTTAG<br>— TA*CGTA<br>— AT GCAT CCTAG<br>— TA*CGTA*GGA |

Notes:
(i) The DNA sythesis takes place from 5' ──→ 3'.
(ii) When ddNTP is incorporated into the DNA in place of normal dNTP, the newly synthesized nucleotide chain is terminated because of the absence of hydroxyl group at the 3' position of ddNTP

Separate the two strands of DNA (the template and the newly synthesized radiolabeled strand) by heating

**FIGURE 2.24** (continued).

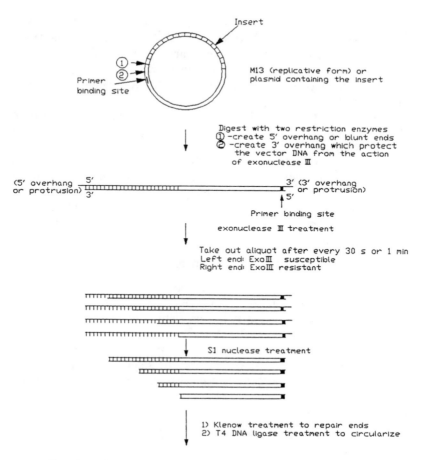

**FIGURE 2.25.** Generation of deletion mutants with exonuclease III.

M13, and single-stranded DNA templates are prepared and used in a primer extension sequencing reaction. From a single sequencing reaction, about 500 bases can be read. To determine the sequence of a DNA fragment more than 500 nucleotides in length, different strategies can be used: (1) Synthetic oligonucleotide primers, complementary to the template DNA near to the point from where to extend the known sequence, can be used. This approach may be good for sequencing small DNA fragment. However, it is costly and time-consuming. (2) Progressive deletions of the insert can be made from one end, in such a way as to bring different regions of DNA close to the primer extension site (Figure 2.25) (Henikoff, 1984).

In both of these approaches, the entire DNA fragment is to be cloned into M13 or plasmid. Inserts much greater than 2 kb are usually unstable in the M13 vector. There are problems of secondary structure in DNA sequencing, therefore sequencing of the template in both orientations is usually necessary. There are some other problems with ExoIII, such as dependence upon specific restriction enzyme sites.

To overcome the above problems, the DNA fragment to be sequenced is broken down into fragments of reasonable size that can be easily cloned into M13. Sequencing of these subfragments leads us to determine the sequence of the entire fragment. If the fragment is very small (i.e., approximately 1 kb) and its accurate restriction map is available, the easiest way of breaking DNA fragment is by using restriction enzymes.

A more random approach is needed to generate subfragment from a larger DNA fragment. There are two methods for generating random fragments.

(1) Enzymatically with DNAaseI (Anderson, 1981)

(2) Mechanically by sonication (Deininger, 1983)

A sequencing procedure in which random subfragments are generated mechanically by sonication is described (Bankier et al., 1987). A fragment of up to 20 kb can be sequenced by this procedure. In the sonication method, the sequence of each base in a DNA fragment is read an average of six to eight times before the sequence of the entire fragment is known. This will take care of sequencing errors. The rate of sequencing can be greatly enhanced by using a microtiter plate in place of microfuge tubes for sequencing reaction.

## 2.10.1 Enzymatic Sequencing Reactions Utilize Three Basic Reagents

(1) Template: Utilizes a single-stranded DNA or double-stranded DNA that has been denatured, usually by alkali. However, single-stranded DNA templates isolated from recombinant M13 give the best results.

(2) Primer: A synthetic oligonucleotide complementary to a specific sequence on the template is used for priming of DNA synthesis. Usually the ''universal'' primer which is complementary to the vector sequences, that flank the target DNA, is used for DNA synthesis. Universal primers used for the sequencing of M13 clones are usually 17 to 20 nucleotides long, and complementary to the sequences immediately adjacent to the HindIII site in the polyclonal region of M13mp18.

(3) DNA polymerases: For the dideoxynucleotide-mediated sequencing, the commonly used enzymes are as follows:

- Klenow fragment of *E. coli* DNA polymerase (Sanger et al., 1977)
- bacteriophage T7 DNA polymerases that have been chemically modified to stop $3' \rightarrow 5'$ exonuclease activity (e.g., sequenase) (Tabor and Richardson, 1987)
- thermostable DNA polymerase (e.g., Taq DNA polymerase) (Innis et al., 1988)
- reverse transcriptase (Mierendorf and Pfeffer, 1987)

## 2.10.2 Random Generation of Overlapping Clones

The following methods are summarized from Bankier et al. (1987) (with permission from Applied Science Pub., New York).

### 2.10.2.1 Isolation of the Target DNA Fragment

(1) Digest a sufficient amount of recombinant DNA (to yield 5 to 10 $\mu$g of target DNA) with suitable restriction enzyme(s) to excise the target DNA. Use the suppliers' recommended buffer and reaction conditions.

(2) Prepare 0.8% LGT agarose gel mixture using 50 mL of 1 × TBE and 0.4 g LGT agarose (ultrapure low-melting-point agarose). Dissolve at 100°C or in a microwave. Add 2 $\mu$L of 10 mg/mL ethidium bromide just before pouring. Pour the gel in a 10 × 10 cm minigel plate, put in the slot former and leave the gel to solidify.

(3) When the digestion of recombinant vector DNA is complete, add 1/10 volume of loading mix.

(4) Load the DNA onto the minigel and run at 100 V until the required band has separated clear enough from the other band(s).

(5) Cut out the band with a scalpel blade, put the excised gel slice into an Eppendorf tube and melt at 70°C.

(6) Extract three times with TE-saturated phenol.

(7) Add 1/10 volume of 3 M sodium acetate (pH 4.8), 2.5 volumes of ethanol and precipitate at −20°C for 20 min.

(8) Centrifuge in a microfuge for 5 min and pour off the supernatant.

(9) Wash the DNA pellet with 96% ethanol.

(10) Dissolve the dried DNA pellet in 20 to 25 $\mu$L of TE.

### 2.10.2.2 Self-Ligation of the Target DNA

For self-ligation of the target DNA, it is best to use restriction enzymes that generate compatible termini. If the target DNA is isolated using restriction enzymes that generate incompatible protruding termini, self-ligation will generate chains of fragments in alternating orientations. In such cases, the junctions are represented by large inverted repeats. The cloning of such junction fragments into M13 vector will be a problem. To resolve this problem, save 1/3 of the target DNA from ligation and mix the ligated DNA with the unligated DNA before sonication.

**Ligation Reaction**

In a 0.5 mL microfuge tube add:
5 $\mu$g purified target DNA
3.0 $\mu$L 10 × ligation buffer
2 units T4 DNA ligase
Water to make 30 $\mu$L
Incubate the mixture at 16°C for 16 h

## 2.10.2.3 Fragmentation of the Target DNA by Sonication Using W-375 Cup Horn Sonicator from Heart Systems Ultrasonics

(1) Fill the cup horn of the sonicator with ice-cold water to a depth of about 3 cm and clamp the microfuge tube containing the self-ligated DNA just above the probe.

(2) Sonicate at maximum output for a total of two 80 s bursts. After each burst, centrifuge the microfuge tube and change the ice-cold water of the cup horn to keep it cool.

(3) After sonication, analyze a 1 to 2 $\mu$L aliquot on a 1.5% agarose gel to find out the sizes of fragments generated. Use pBR322 DNA cut with Sau3A as marker.

(4) Peak size distribution should be around 300 to 600 bp.

## 2.10.2.4 End Repair and Size Selection of DNA

The DNA subfragments resulting after sonication cannot be directly cloned into a vector. Before cloning such subfragments, their ends need to be repaired. It is achieved by using enzymes to produce blunt ends. There are a number of enzymes that can be used to repair ends, such as S1 nuclease, mung bean nuclease, and DNA polymerases. However, T4 DNA polymerase and Klenow fragment of *E. coli* DNA polymerase are used, in combination, to yield good results.

**For end repair mix**

Sonicated DNA 22 $\mu$L
50 mM MgCl$_2$ 3 $\mu$L

0.5 mM dATP 1 $\mu$L
0.5 mM dGTP 1 $\mu$L
0.5 mM dCTP 1 $\mu$L
0.5 mM dTTP 1 $\mu$L
10 units of Klenow fragment of *E. coli* DNA polymerase
10 units of T4 DNA polymerase

(1) Incubate at room temperature for 30 min.

(2) Fractionate the required DNA fragments on a 1.2% agarose gel and cut out bands of desired size (300 to 600 bp) for sub-cloning.

(3) Recover the DNA from the agarose gel as described earlier.

## 2.10.2.5 Vector Preparation

For cloning DNA fragments of 300 to 600 bp size range, a M13 vector is cut with suitable restriction enzyme to yield blunt ends; and to prevent self-ligation of the vector DNA, alkaline phosphatase treatment is usually preferred.

(1) To yield blunt ends and removal of 5′ phosphates from the termini in a single step, to a 0.5 mL microfuge tube add:

- 5 $\mu$g of M13 mp RF (replicating form) DNA
- 5 $\mu$L of 10× SmaI buffer
- 15 units of SmaI
- 20 units of calf alkaline phosphatase
- water to make 50 $\mu$L

(2) Incubate at 37°C for 1 h.

(3) Digestion should be complete. It can be checked on an agarose gel.

(4) Extract twice with equal volume of TE-saturated phenol.

(5) Precipitate DNA with 2.5 volumes of ethanol after adding 1/10 volume of 3 M sodium acetate.

(6) Leave at −20°C for 15 min.

(7) Centrifuge in a microfuge centrifuge for 5 min.

(8) Pour off the supernatant, wash the DNA pellet with 96% ethanol and recentrifuge to recover the pellet.

(9) Redissolve the dried pellet in TE and store at −20°C.

(10) To determine the effectiveness of the alkaline phosphatase, it is better to carry out test ligations and transfect competent *E. coli* of the appropriate strain. Add IPTG and X-gal in the medium containing transfected bacteria, put top agar and plate on the 2 TY agar plates.

TABLE 2.5. *Ligation of small subfragments to vector DNA. Set up ligation reaction in 1.5 mL microfuge tubes as follows.*

| | Tube Number | | | | | | |
|---|---|---|---|---|---|---|---|
| | **1** | **2** | **3** | **4** | **5** | **6** | **7** |
| Vector DNA (SmaI cut and dephosphorylated) | 40 ng | 40 ng | 40 ng | 40 ng | 40 ng | 40 ng | 40 ng |
| End repaired subfragments | 20 ng | 40 ng | 60 ng | 80 ng | – | – | – |
| λ DNA fragments (AluI cut) | – | – | – | – | 10 ng | – | – |
| 10× ligase buffer | 1 μL | 1 μL | 1 μL | 1 μL | · 1 μL | 1 μL | 1 μL |
| T4 DNA ligase | 0.5 μL | 0.5 μL | 0.5 μL | 0.5 μL | 0.5 μL | 0.5 μL | – |

Sterile distilled water to make 10 μL.
Incubate at 16 °C overnight.

(11) Incubate at 37°C overnight and count the number of blue and colorless plaques. Then, if the results of the test ligations are satisfactory, continue with the actual ligation reaction (Table 2.5).

## 2.10.2.6 Bacterial Transformation

Make bacterial cells (strain JM 101) competent by Hanahan's method.

(1) Mix 200 μL of competent cells with 10 μL of subfragment/vector ligation mix in a sterile glass tube on ice.

(2) Incubate on ice for 1 h.

(3) Heat shock competent cells/DNA mixture at 42°C for 2 min.

(4) Add 4 mL of molten top agar in a sterile glass tube, keep it at 48°C in a water bath, and add 25 μL of X-gal and 25 μL of IPTG.

**Top agar**
1% tryptone
0.8% sodium chloride
0.7% agar

**X-gal**
20 mg/mL 5-bromo-4-chloro-3-indolyl-β-D-galactoside in dimethylformamide

**IPTG**

25 mg/mL isopropyl-$\beta$-D-thio-galactopyranoside in water

(5) Add molten agar to the competent cells/DNA mix.

(6) Mix and pour onto a 2 TY agar plate.

(7) Spread evenly and keep at room temperature for 15 min for the agar to set.

(8) Incubate at 37°C overnight.

(9) The vector containing insert will form white (colorless) plaques.

## 2.10.2.7 Phage Growth and Single-Stranded Template Preparation

(1) Pick up a single colony from a plate of JM cells and grow in 2 TY at 37°C overnight.

(2) Dilute the overnight culture 1/100 with 2 TY.

(3) Add 1.5 mL of the diluted culture to sterile tubes.

(4) Pick up a number of single white plaques with sterile toothpicks and shake into tubes containing the diluted culture.

(5) Grow the bacterial cultures at 37°C for 4.5 to 5.5 h with vigorous shaking.

(6) Centrifuge 1.5 mL cultures in a microfuge for 5 min.

(7) Transfer 1.0 mL of the supernatant into a clean microfuge tube.

(8) Add 200 $\mu$L of PEG solution, vortex and leave at room temperature for 10 min.

**PEG solution**

20 g polyethylene glycol 8000
14.6 g sodium chloride
Water to bring volume to 100 mL

(9) Centrifuge the mixture in a microfuge centrifuge for 10 min and suck off the supernatant.

(10) Recentrifuge and remove all residual PEG solutions.

(11) Resuspend the pellet in 100 $\mu$L TE (10 mM Tris, 0.1 mM EDTA; pH 8.0).

(12) Add an equal volume of TE-saturated phenol, vortex well and leave for 5 min.

(13) Recover the aqueous phase after centrifugation, add 1/10 volume of 3 M sodium acetate (pH 4.5) and precipitate with 2.5 volume of ethanol in a dry ice/isopropanol bath for 20 min.

(14) Centrifuge in a microfuge for 5 min and discard the supernatant.

(15) Dry the pellet, dissolve in 30 $\mu$L of TE and store at $-20°C$.

## 2.10.3 Sequencing Reaction

Buffers and solutions for sequencing:

**TM**

100 mM Tris-HCl (pH 8.5)
50 mM MgCl$_2$

**Formamide dyes mix**

100 mL de-ionized formamide
0.1 g xylene cyanol FF
0.1 g bromophenol blue
2 mL 0.5 M EDTA (pH 8.0)

**10 × TBE**

216 g Tris
110 g boric acid
18.6 g EDTA

Add distilled water to make 2 L. Adjust pH to 8.3.

**40% Acrylamide**

380 g acrylamide
20 g bis-acrylamide

TABLE 2.6. *dNTP:ddNTP mixes (Bankier et al., 1987).*

| | T (in $\mu$L) | C (in $\mu$L) | G (in $\mu$L) | A (in $\mu$L) |
|---|---|---|---|---|
| 0.5 mM dTTP | 25 | 500 | 500 | 500 |
| 0.5 mM dCTP | 500 | 25 | 500 | 500 |
| 0.5 mM dGTP | 500 | 500 | 25 | 500 |
| 10 mM ddTTP | 50 | – | – | – |
| 10 mM ddCTP | – | 8 | – | – |
| 10 mM ddGTP | – | – | 16 | – |
| 10 mM ddATP | – | – | – | 1 (3 if using $^{32}$p-dATP) |
| TE | 1000 | 1000 | 1000 | 500 |

Add distilled water to make 1 L. Stir with 20 g amberlite NB-1, and filter.

### 2.5 TBE mix

150 mL 40% acrylamide
250 mL 10× TBE
460 g urea
50 mg bromophenol blue

Add distilled water to make 1 L.

### 0.5 TBE mix

150 mL 40% acrylamide
50 mL 10× TBE
460 g urea

Add distilled water to make 1 L.

### Ammonium persulphate solution

25% ammonium persulphate (w/v)

### Klenow mix (sufficient for 10 clones)

4 $\mu$L (20 units) Klenow
6.25 $\mu$L (62.5 $\mu$Ci) $^{35}$S-dATP
12.5 $\mu$L 0.1 M DTT
77 $\mu$L distilled water

### Chase mix

0.25 mM dTTP
0.25 mM dATP
0.25 mM dCTP
0.25 mM dGTP

The appropriate dNTP and ddNTP mixes are given in Table 2.6.

(1) For an annealing reaction, to a 0.5 mL microfuge tube, add:

- 5 $\mu$L template
- 1 $\mu$L primer (0.2 pmol)
- 1 $\mu$L TM
- 3 $\mu$L distilled water

Incubate at 60°C for 1 h.

(2) Aliquot 2 $\mu$L each of the primed template into four wells marked as T, C, G, and A of a multi-well plate.
(3) Add 2 $\mu$L of dNTP: ddNTP (T, C, G, and A) mixes into the marked wells (T, C, G, and A), respectively.
(4) Dispense 2 $\mu$L of Klenow mix to each well.
(5) Centrifuge briefly and incubate at room temperature for 20 min.

(6) Add 2 $\mu$L of chase mix to each well.

(7) Centrifuge and continue incubation at room temperature for 15 min.

(8) Terminate the reaction by adding 4 $\mu$L of formamide dyes mix to each well.

(9) Keep the plate at 80°C for 5 min.

(10) Cool the plate on ice. The samples are ready to be loaded onto a sequencing gel.

## 2.10.4 Buffered-Gradient Polyacrylamide Gel

There is a logarithmic relationship between the DNA fragment length and its mobility on a polyacrylamide gel, therefore the sequence ladder at the bottom of the gel is largely spaced. However, the spacing of adjacent bands starts to decrease towards the top of the gel. To decrease the spacing between the consecutive bands, increased salt concentration of the buffer in the polyacrylamide gel mix for the bottom part of the gel is used, which sets up a potential difference gradient with lower voltage drop in the bottom part of the gel. This buffered gradient retards the mobility of fast-moving fragments at the bottom part of the gel so that consecutive bands are evenly spaced. For setting up a buffered-gradient polyacrylamide gel for sequencing:

(1) Clean a pair of sequencing plates.

(2) Silanize the inner surface of the smaller or notched plate with 2% dimethyl-dichlorosilane in trichloroethane.

(3) Clean the silanized plate first with water and then with 96% ethanol.

(4) Assemble the glass plates using 0.35 mm side spacers and a wide, good-quality adhesive tape.

(5) For a 40 × 20 cm gel, set up acrylamide polymerization reaction into two separate beakers.

|  | **Beaker 1** | **Beaker 2** |
|---|---|---|
| 2.5× TBE mix | 7 mL | — |
| 0.5× TBE mix | — | 40 mL |
| 25% Amm. persulphate | 14 $\mu$L | 70 $\mu$L |
| TEMED *N,N,N',N'*-tetramethyl-ethylenediamine) | 14 $\mu$L | 70 $\mu$L |

(6) From Beaker 2, take up about 35 mL of 0.5× TBE gel mix into a 50 mL plastic syringe.

(7) From Beaker 2, first take up 3 mL of 0.5× TBE gel mix, and then place 7 mL of 2.5× TBE gel mix from Beaker 1 into a 10 mL pipet. Allow 3 to 4 air bubbles to pass up the pipet.

(8) Hold the assembled plates at an angle of about 45° and pour the gradient gel from the pipet slowly. After adding gel mix from the pipet, lower the plates to horizontal to stop the flow. Then lift the plate to an angle and start adding the gel mix from the syringe and continue pouring until the plates are filled.

(9) Put the comb in place and clamp the edges of the gel plates with clips.

(10) Allow the gel to polymerize for 1 h.

Note: There are a number of electrophoresis apparatuses commercially available. The length, width, and arrangement of the glass plates vary from manufacturer to manufacturer.

## 2.10.4.1 Loading and Running of Sequencing Gel

(1) Remove any dried urea/polyacrylamide present on the outside of the gel plates with a damp paper towel.

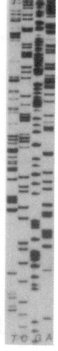

**FIGURE 2.26.** Nucleotide sequencing by dideoxynucleotide chain termination technique. Lanes from the left to right are: T = thymine, C = cytosine, G = guanine, and A = adenine. Reproduced with permission from Mittal, 1989.

(2) Carefully remove the comb, and tape from the bottom of the gel plates, and wash the slots with water to remove the unpolymerized acrylamide/dried urea.

(3) Put 2.5× TBE buffer to the bottom tanks of the electrophoresis apparatus, and clamp the gel to the electrophoresis apparatus.

(4) Fill the top tank with 0.5× TBE.

(5) Flush the slots using a Pasteur pipet.

(6) Load 2 μL of each sample onto adjacent slots of the gel using a finely drawn out pipet tip. Keep the same order for each reaction set.

(7) When all the samples are loaded, connect the apparatus to a power pack (positive to the bottom tank) and run at constant power (35 to 40 W for a 40 × 20 cm gel) until the bromophenol blue dye is about to leave the bottom of the gel.

(8) Switch off the power supply and take the gel plates from the electrophoresis apparatus.

## 2.10.4.2 Autoradiography of Sequencing Gel

(1) Remove the tap from the gel plate. Carefully take off the small plate with the help of a spatula.

(2) Submerge the gel (stuck to one plate) in a solution of 10% acetic acid in water for 30 min.

(3) Remove the plate along with the gel and drain off the acetic acid.

(4) Place a piece of 3 MM Whatman paper (slightly bigger than the size of the gel) over the gel. Apply gentle pressure all over the gel and carefully peel the gel/paper off the glass plate.

(5) Cover the gel with a piece of Clin film and dry at 80°C under vacuum.

(6) Remove the film from the gel, place the gel in contact with an X-ray film in a film cassette and leave for 24 h at room temperature.

(7) Develop the X-ray film to obtain an autoradiographic image of the gel (Figure 2.26).

(8) Record the nucleotide sequence from the autoradiograph starting from the bottom.

## 2.11 **ANALYSIS OF PROTEINS BY SDS-POLYACRYLAMIDE GEL ELECTROPHORESIS (SDS-PAGE)**

Within a cell a large number of different proteins are present, which perform a variety of cellular functions. Some of the proteins are enzymes that are required in a number of metabolic activities and DNA replication. Many proteins are associated with the cell membrane.

The mRNA are translated at ribosomes into proteins. A number of proteins undergo secondary modifications, such as glycosylation (addition of sugar moieties), phosphorylation (addition of phosphate groups), sulphation (addition of sulphur groups), and proteolytic cleavage (cleavage of complex proteins). Proteins that undergo secondary modifications are fully functional only after secondary modifications.

Different cellular proteins can be separated on SDS-polyacrylamide gels by electrophoresis, visualized by staining with Coomassie blue or silver stain. Proteins can be labeled *in vitro* with [$^{35}$S] methionine or [$^{35}$S] cysteine before being analyzed by SDS-polyacrylamide gel electrophoresis.

## 2.11.1 Radiolabeling of Mammalian Cells with [$^{35}$S] Methionine

Mammalian cells grown in monolayers can easily be labeled with [$^{35}$S] methionine. The cells at G1 phase (rapidly growing) incorporate more label than cells at S phase (static).

(1) Grow cells in monolayers using appropriate growth medium.

(2) Wash the monolayers with prewarmed (37°C) medium deficient in methionine and serum, and then incubate in a medium deficient in methionine and serum for 20 to 30 min at 37°C to deplete the intracellular pool of methionine. Replace the medium with medium deficient in methionine and serum but containing an appropriate amount of [$^{35}$S] methionine.

(3) Incubate at 37°C for a fixed period depending on the experiment.

(4) Remove the medium and resuspend cells in a minimum volume of 2 × sample buffer with the help of a cell scraper.

If the interest is in secretory proteins, the medium should also be saved.

### 2 × Sample buffer

125 mM Tris-HCl (pH 6.8)
4% SDS
20% glycerol
0.20% bromophenol blue

Protease inhibitors [such as PMSF (phenylmethylsulfonyl fluoride) may be added in the 2 × sample buffer to protect the target proteins from degradation due to release of intracellular proteases.

(5) Shear the chromosomal DNA by sonication, centrifuge for 10 min in a microfuge and save the supernatant.

(8) Depending upon the size of the well, 15 to 40 $\mu$L of the sample can be added into each well. Add molecular weight marker into one of the wells. Load 1 × gel-loading buffer into unused wells, if there are any.

(9) Connect the electrophoresis equipment to a power supply. The polypeptides migrate towards the anode. Run at 5 to 10 V/cm of the gel. Continue electrophoresis until the bromophenol blue dye reaches the bottom of the gel.

(10) Remove the gel cassette from the electrophoresis equipment, and take off one of the glass plates by applying gentle pressure with the help of a spatula.

(11) If the samples are radiolabeled, fix the gel with gel fixer for 30 min and dry onto a filter paper using a gel drier under vacuum. Expose an autoradiographic image of the gel to an X-ray film for various lengths of time at 70°C, and develop the film at the end of the exposure period.

The samples that are not labeled, the polypeptide bands, can only be visible if the gel is stained with Coomassie Brilliant blue or silver stain.

## 2.11.2.2 Coomassie Brilliant Blue Staining of the Gel

(1) Prepare a 0.25% solution of Coomassie Brilliant blue in methanol:acetic acid:water (45:10:45). Filter the solution to remove the undissolved dye, if any.

(2) Stain the gel in the Coomassie Brilliant blue solution for 4 to 5 h at room temperature with a constant gentle shaking.

(3) Destain the gel in methanol:acetic acid:water (30:10:60) for 5 to 10 h at room temperature with constant gentle shaking.

(4) After destaining, photograph the gel or dry onto a filter paper using a gel drier under vacuum.

## 2.11.2.3 Drying the Gel with Gel Drier

(1) Place the fixed gel on a piece of 3 MM Whatman paper. Cover the gel with a piece of Clin film slightly larger than the gel.

(2) Place a 3 MM Whatman paper larger than the gel on the surface of the gel drier and then place the gel covered with Clin film.

(3) Cover the gel with the rubber lining of the gel drier and switch on the suction pump to apply suction. The rubber lining will make a tight seal around the gel. Dry the gel at 60°C for 2 to 3 h.

(4) Pull the rubber lining to break the suction, remove the dried gel attached to Whatman 3 MM paper.

(5) [$^{35}$S] emits low energy $\beta$ particles, therefore remove the Clin film to obtain an autoradiographic image of the gel on an X-ray film.

To detect the target antigens in a mixture of proteins, there are two methods that are widely used.

## 2.11.3 Immunoprecipitation

A mixture of radiolabeled proteins containing the target antigens is treated with the specific antiserum or monoclonal antibodies to form specific immune complexes. The mixture is treated with protein A coupled to Sepharose. Immune complexes attached to Sepharose beads are purified, resuspended in gel loading buffer, boiled to release the protein associated to the antibodies, and electrophoresed on SDS polyacrylamide gel.

## 2.11.4 Western Blotting

The transfer of polypeptides from SDS-polyacrylamide gels to solid supports (such as nitrocellulose filter) is popularly known as Western blotting (Towbin et al., 1979; Burnette, 1981). In this technique, unlabeled proteins are separated on SDS-polyacrylamide gels and transferred to nitrocellulose filters. The filters are incubated with the specific antiserum or monoclonal antibodies and washed to remove unbound antibodies. When these filters are incubated with radiolabeled or enzyme-linked antibodies against specific antibodies (known as anti-antibodies), complexes are formed with specific antibodies bound to the target antigens. The filters are washed to remove unbound radiolabeled or enzyme-linked anti-antibodies. If radiolabeled anti-antibodies have been used, put the filter with an X-ray film in a cassette to obtain the autoradiographic image of the target antigens. Otherwise, use suitable substrate for visual detection of the target antigens.

## 2.12 **CLONING IN YEAST**

Yeasts have an important role in the food industry as they are being used in brewing, bread-making, and in a number of other foods. Yeasts are single-cell eukaryotic organisms. Out of more than 350 species of yeast, only *Saccharomyces cerevisiae* has been well-characterized genetically and used extensively for introduction of the foreign DNA into yeast cells. There are, in fact, several features that make yeast an attractive organism for the large-scale production of biologically active proteins. Yeast can be produced in large amounts using the technology of brewing and fermentation.

Initially the transformation of yeast cells was not very efficient because they have a thick cell wall composed of polysaccharides, lipids, and protein components. Now, the transformation of yeasts can easily be achieved by removing the cell wall with the mixtures of $\beta$-lucanases to release spheroplasts, which are later exposed to $Ca^{++}$ and DNA. The final mixture is treated with polyethylene glycol (Hinnen et al., 1978; Beggs, 1978).

With the other transformation procedure, in which there is no need to digest the cell wall, yeast cells are treated with $Li^+$, DNA and polyethylene glycol (Ito et al., 1983).

The DNA sequences to be cloned and analyzed in yeast first need to be isolated as recombinant DNA in *E. coli*. The cloned DNA can be easily manipulated in two hosts by taking advantage of each host. The vectors used for gene cloning in yeast contain sequences of bacterial plasmid or bacteriophage DNA and yeast DNA. There are mainly two types of cloning vectors used for gene cloning in yeast: integrating vectors and independently replicating vectors. Integrating vectors have sequences from a bacterial cloning vector and sequences of a yeast gene for the selection of transformants. On transformation of yeast cells, integration of the vector DNA sequences takes place into the yeast genome by the process of recombination between yeast genome and the yeast DNA sequences in the vector. However, independently replicating vectors replicate extrachromosomally, and are known as yeast episomal plasmids which contain autonomously replicating sequences.

A number of genes from bacteria, viruses, and mammals have been successfully expressed in yeasts. However, since yeasts do not recognize mammalian promoters for transcription, they are unable to process mammalian introns, and fail to carry many of the secondary modifications of the translated proteins of mammalian origin. To overcome many of these problems, instead of mammalian genes, mammalian cDNA is cloned in the yeast under the control of a yeast promoter. Many foreign gene products for commercial use are under production in yeasts, including bovine growth hormone, interferon, and human hepatitis B antigen.

## 2.13 **Ti PLASMID-ASSISTED CLONING IN PLANTS**

*Agrobacterium tumefaciens*, a soil microbe, possesses plasmids known as tumor-inducing (Ti) plasmids. On infection, agrobacterium transfers a segment of the Ti plasmid DNA, known as T-DNA, to plant cells. There is integration of the T-DNA into the plant DNA genome and the T-DNA expresses itself within plant cells. For introduction of the foreign DNA into plant cells, Ti plasmids have been exploited. The Ti plasmid-based vectors can be used to introduce the foreign DNA in most of dicotyledenous plants. The large pieces of DNA can easily be cloned into the Ti plasmid-based vectors. Due to expression of the T-DNA integrated into the plant DNA genome, the transformed tissue grows and forms a mass of tissue known as crown gall tumor. This gall tumor is not desirable when plant tissues are transformed with the Ti plasmid containing the foreign DNA, because tumor tissues are difficult to regenerate into normal plants. The deletion in the tumor-producing region of the Ti plasmid-based vector has retained the T-DNA transfer ability but eliminated its oncogenicity. For easy selection of the transformants, these vectors contain a dominant selection marker such as the gene coding for resistance to kanamycin or methotrexate (Lichtenstein and Draper, 1985).

Genetically engineered plants are being exploited to study the mechanisms of gene regulation in plants. Cotton, soybean and alfalfa are the major target crops to be improved by genetic engineering in the near future. Work is underway to develop disease-resistant and insect-resistant plants.

## 2.14 **GENE TRANSFER INTO MAMMALIAN CELLS**

Post-translational modifications of many eukaryotic proteins in bacterial or other prokaryotic hosts do not take place correctly. Therefore, in order to express mammalian genes for which correct post-translational modifications are essential, mammalian cells are the best choice.

Genes introduced into mammalian cells either express transiently or stably. In transient gene expression, the DNA introduced into mammalian cells does not need to be integrated into the cellular genome. However, in stable gene expression, the DNA introduced into the cells gets integrated into the host cell genome to form transformed cell lines.

The DNA fragments can be introduced into mammalian cells by a number of physical methods, which include:

(1) Calcium phosphate precipitation (Graham and van der Eb, 1973): The DNA is dissolved in a buffer containing phosphate and then precipitated with $CaCl_2$. The calcium phosphate-DNA coprecipitates are taken up by the cells.

(2) DEAE-dextran method (Vaheri and Pagano, 1965; McCutchan and Pagano, 1968; Warden and Thorne, 1968): The DNA is treated with DEAE-dextran and then cells are exposed to the DEAE-dextran/DNA mixture.

(3) Formation of complexes with polycations or lipids (Felgner et al., 1987; Mannino and Gould-Fogerite, 1988): The DNA is encapsulated within artificial membrane vesicles (liposomes). The DNA is taken up by the cell following fusion of the liposomes with the cell membrane.

(4) Electroporation (Neumann et al., 1982; Zimmermann, 1982): Cells are

exposed to a quick pulse of high-voltage current, during which the DNA will be taken up by cells.

(5) Micro-injection (Capecchi, 1980): The DNA is introduced into the cell by puncturing it with a micro-needle.

(6) Particle gun (Klein et al., 1987): The DNA is coated onto tungsten micro-particles, which are then introduced into cells with a high-velocity particle gun.

In the DNA transfer experiments, only a small proportion of the mammalian cells stably express the foreign gene. For the easy isolation of the transformed cells, the foreign DNA is attached to a gene coding for a selective marker such as thymidine kinase (Wigler et al., 1977), xanthine-guanine phosphoribosyl transferase (Mulligen and Berg, 1980), aminoglycoside phosphotransferase (Colbere-Garapin, 1981) and dihydrofolate reductase (Subramani et al., 1981).

## 2.14.1 Virus Vectors for Delivery of DNA into Mammalian Cells

Viral vectors are capable of infecting every target cell more efficiently than physical methods of gene transfer. Many of the viral vectors are tumor viruses which, on infection, integrate into the host cell genome and express foreign genes. Simian virus 40 (Elder et al., 1981), adenoviruses (Solnick, 1981; Thummel et al., 1981) and retroviruses (Dick et al., 1986; Gilboa et al., 1986) have been studied extensively for the delivery of the foreign DNA into the mammalian cells for stable expression.

Other viruses such as vaccinia virus (Mackett et al., 1984; Moss and Flexner, 1987), papilloma-viruses (Campo, 1985), and herpes viruses (Shih et al., 1984; Geller and Breakfield, 1988) can also be exploited as expression vectors in mammalian cells.

## 2.15 **AUTOMATION OF DNA SYNTHESIS**

Three chemical protocols have been used in the solid phase synthesis of DNA. These are (1) the triester method developed by Gait and Itakura, (2) the methyl-phosphoramidite method developed by Carruthers, and (3) the cyanoethyl-phosphoramidite (CED) method. The main components of a modern automatic DNA synthesizer are a series of airtight reagent bottles or reservoirs; a reagent delivery system; a reactor chamber; and a computer control system. The operating principles of a fully automated DNA synthesizer were exemplified by the model BT-8500 (Biotech Instruments, England) (Newton, 1989).

The applications are discussed in Chapter 11.

# CHAPTER 3

# Plant Tissue/Cell Culture Techniques

## 3.1 **INTRODUCTION**

*Plant* tissue culture, i.e., the propagation of plant tissue in aseptic nutrient media, has many applications in food processing including biosynthesis of high-value natural products such as flavors, colors, preservatives, and nutritional supplements (Wasserman et al., 1988). One-fourth of all prescription drugs in the United States are of plant origin. These pharmaceuticals include ajmalicine, atropine, codeine, diosgenin, digoxin, hyoscyamine, quinine, morphine, serpentine, scopolamine, and vincristine (Payne et al., 1987). Many food colors, flavors and fragrances, and agricultural chemicals are produced from plant products. Plant tissue culture allows one to bypass growing whole plants to get desired products.

Plant tissue culture techniques are being used to improve stress, salt, cold, and chemical tolerance; disease, pest and pesticide resistance; and to increase the nutrient content in various crops (Teutonico and Knorr, 1985).

Plant cells are up to 100 times larger than microbial cells, affecting the mixing and gassing regimes, nutrient transfer, etc., in a reactor. The metabolic rate of plant cells is significantly lower than that of microbial cells. Plant cells are surrounded by a semi-rigid, cellulose-based cell wall that is quite different in microbial cells (Fowler et al., 1987). The biotechnological techniques of clonal propagation, somaclonal variation, gametoclonal variation, and protoplast fusion are used to develop new plant varieties.

### 3.1.1 Clonal Propagation

Clonal propagation provides for large-scale production of a genetically uniform plant population. Many plant species can be clonally propagated from plant tissues using suitable culture media. The clones are generated at both the organ and the cellular levels. For micropropagation or meristem-cloning, the apical meristem, the growing tip of the shoot, is removed and induced to develop several additional shoots. Each of the developing shoots can be separately rooted to recover clonal copies (Morris, 1986; Evans and Sharp, 1986).

To establish these cultures, meristems are cut from an elite stock plant and sterilized to remove contaminating organisms. With the appropriate balance of plant growth regulators, temperature and light, the meristem will develop into a rooted plantlet. A major advantage of micropropagation is the avoidance of viruses and other pathogens in commercial stock (McKersie et al., 1990).

In another method, donor plant cells can be induced to multiply in culture, and then the plants can be regenerated. With this approach, many cells or plants can be produced in a shorter span of time. From these cells, plants are produced by (1) organogenesis: induction of shoots or roots from roots and shoots, respectively, and (2) somatic embryogenesis: induction of embryos that directly develop into plants. Somatic embryos develop from somatic cells without the involvement of gametic fusion and have shoot and root axes.

In summary, tissue explants including pieces of leaves, shoots, roots, or immature fruits are transferred to a solid medium containing nutrients and various growth regulators. The cells proliferate to form a non-differentiated mass called callus. Embryos are formed by transferring the callus tissue to a growth medium. Selective agents can be incorporated into the growth medium to select mutant varieties that demonstrate resistance to the agent used. Callus tissue can be transferred into a liquid culture medium and grown as a suspension culture in bioreactors for the production of secondary metabolites (Wasserman et al., 1988) (see Chapter 6 for details).

### 3.1.2 Somaclonal Variation

Regeneration of plants from callus, leaf tissue explants, or plant protoplasts (wall-less cells) can create somaclonal variants. This permits modification of genes in cultured plant cells. This is used to develop new breeding lines with improved characteristics (Evans and Sharp, 1986; Morris, 1986).

In corn, a new fully functional electrophoresis variant at the alcohol dehydrogenase locus has been found among somaclones and subsequently characterized as resulting from a single nucleotide substitution (Scowcroft and Larkin, 1988) (more details in Chapter 2). In wheat, variants have been analyzed which affect traits such as height and alcohol dehydrogenase synthesis, as well as complex gene loci involved in the synthesis of grain amylases. Somaclonal analysis in corn and alfalfa has shown that cell culture greatly enhances the activation of transposable elements.

### 3.1.3 Gametoclonal Variation

Gametoclonal variations are created by regenerating plants from cultured microspore or pollen cells or pollen still contained within the anther. This genetic variation results from both recombination during cell meiosis and mutation induced by tissue culture (Evans and Sharp, 1986; Morris, 1986).

## 3.1.4 Protoplast Fusion

Protoplast or plant cell fusion combines similar or different plant cells to produce hybrid plants. These, in some cases, cannot be produced by conventional breeding methods. Potential applications are in tomato, potato, canola, lettuce and alfalfa.

The contents of a plant cell are protected by the cell wall, which can be enzymatically digested to produce easily manipulatable wall-less cells referred to as protoplasts. Protoplasts from cells of two-parent plants can be fused together with a multi-step chemical treatment using polyethylene glycol. Fused protoplasts can then be grown in a culture medium suitable for callus formation and plant generation. This is a suitable breeding technique for the species that can be regenerated from protoplasts.

New genes can also be introduced in protoplast. After introducing the required genes, the protoplasts are grown in tissue culture (Wasserman et al., 1988). To modify gene expression, sequences are first identified which regulate gene expression, and then it is established exactly which changes in nucleotide sequences are required to increase or decrease the level of expression. These sequences are generally present in the DNA upstream (5′) to the coding region (Shewry et al., 1987) (see Chapter 5 for detailed techniques).

## 3.2 **TISSUE CULTURE TECHNIQUES**

When a piece of a plant (explant) is placed in a test tube, it will produce tiny replicas of its single parent, in the absence of microorganisms and in the presence of a balanced diet of chemicals. Plants multiplied by tissue culture generally have identical genetic makeups, and thus will prove to be true clones. Tissue culture techniques can be used for most plants, if the right chemical solution and processes are developed. The diseases transmitted from parent to offspring can often be eliminated through tissue culture procedures. This also provides savings of time and space, and can be less costly than other means of vegetative propagation.

Meristematic cells are located at the tips of stems and roots, in leaf axils, and in stems as cambium, on leaf margins, and in callus tissue. Meristematic tissue usually initiates new growth. These will differentiate into leaves, stems, roots, and other organs and tissues. However, this will only occur under the influences of light, temperature, nutrients, hormones, etc.

### 3.2.1 Sterilization

All instruments, glassware, equipment and nutrient media must be sterilized. The sterilization of glassware and many instruments can be performed in dry heat. A minimum exposure of 3 h at 160 to 180°C will be sufficient. In plastic lab-ware, only polypropylene, polymethylpentene, polyallomer™,

Teflon™, and Tefzel™ can be autoclaved at 121°C (122 kPa gauge). Other material such as polystyrene, polyvinylchloride, polyethylene, and styrene-acrylonitrile are not autoclavable. Investigators' hands should be relatively aseptic, which can be accomplished by washing with an antibacterial detergent followed by spraying with 40 to 70% ethanol. Plant material can be sterilized by a variety of sterilizers.

## 3.2.2 Materials and Equipment

Glassware, instruments, equipment, etc., required for tissue culture should also be sterilized. Cultures are incubated under controlled temperature, relative humidity, and light conditions. For this purpose, controlled environment chambers are suitable.

## 3.2.3 Nutrition

The nutrients required in tissue culture consist of inorganic salts, vitamins, carbon and energy sources, and phytohormones. Other important optional components are organic acids, organic nitrogen compounds, and complex substances.

- Inorganic salts: Salts of N, K, P, Ca, S and Mg are required in millimole quantities. The optimum concentration of each nutrient for maximum tissue growth rates differs considerably. For most purposes, the medium should contain at least 25 mM nitrate and potassium. A concentration of 1 to 3 mm calcium, sulfate and magnesium is usually adequate (Gamborg, 1984).
- Carbon and energy source: Sucrose or glucose is the standard source. However, other carbohydrates such as maltose were also found to be suitable in certain studies.
- Vitamins: Vitamins are required for growth and development. Thiamine is an absolute requirement. Other vitamins required included ascorbic acid, choline chloride, folate, *p*-amino-benzoic acid, and riboflavin.
- Phytohormones: Plant tissue culture also requires cytokinin and auxin classes of compounds.
- Organic nitrogen: Organic nitrogen is available from amino acids, adenine, asparagine and glutamine.
- Organic acids: Plant cells cannot utilize organic acids as carbon sources. Supply of acids such as citrate, fumarate, malate or succinate allows plant cell growth on ammonium as the sole nitrogen source.

- Complex substances: These include malt extracts, protein hydrolyzates, yeast extracts, and many other plant preparations such as coconut milk and potato starch.

## 3.2.4 Media Formulas

Many chemical combinations (formulas) have been developed for plant growth and multiplication in culture. Standard media formulas provide optimum nutrients and growth regulators for specific cultivars.

Murashige and his associates (Kyte, 1983) developed specific media for each variety and three stages of culture:

- stage I: explant establishment
- stage II: multiplication stage
- stage III: culture stage in which rooting is induced

Two methods are available for multiplication: (1) direct shoot route, which promotes lateral and adventitious shoots directly, beginning with the explant; and (2) callus route, which first induces the explant to produce callus (unorganized) tissue, then multiplies these to produce embryoids or shoots.

Several tissue culture supply companies sell premixed powdered media. These premixes are available with or without sugar, agar or hormones. The premixes are added to water, heated to dissolve agar, supplemented and pH-adjusted where necessary, and dispensed.

Table 3.1 shows three typical media compositions used for plant tissue culture (Brodelius, 1985).

## 3.2.5 Media Preparation

The required chemicals are mixed with demineralized water, then desired stock solutions are added, and finally required pH is adjusted. The medium is autoclaved at 121°C for 15 min, cooled at room temperature, and stored at 10°C.

The following is a typical composition of a medium for Blackberry multiplication (Zimmerman and Broome, 1980). This can be prepared directly or by mixing stock solutions beforehand:

- $NH_4NO_3$ (1650 mg)
- $KNO_3$ (1900 mg)
- $CaCl_2 \cdot 2H_2O$ (440 mg)
- $MgSO_4 \cdot 7H_2O$ (370 mg)
- $KH_2PO_4$ (170 mg)

TABLE 3.1. *Composition of three typical plant tissue culture media (numbers are in mg/L).*

| Ingredient | MS | B5 | SH |
|---|---|---|---|
| Inorganic macronutrients | | | |
| $KNO_3$ | 1900 | 2500 | 2500 |
| $NH_4 \cdot NO_3$ | 1650 | – | – |
| $NaH_2 \cdot PO_4$ | – | 150 | – |
| $KH_2 \cdot PO_4$ | 170 | – | – |
| $NH_4 \cdot H_2PO_4$ | – | – | 300 |
| $CaCl_2 \cdot 2H_2O$ | 440 | 150 | 200 |
| $MgSO_4 \cdot 7H_2O$ | 370 | 250 | 400 |
| $(NH_4)2SO_4$ | – | 134 | – |
| Inorganic micronutrients | | | |
| $Na_2MoO_4 \cdot 2H_2O$ | 0.25 | 0.25 | 0.1 |
| $CuSO_4 \cdot 5H_2O$ | 0.025 | 0.025 | 0.2 |
| $MnSO_4 \cdot H_2$ | 16.9 | 10.0 | 10.0 |
| $ZnSO_4 \cdot 7H_2O$ | 8.6 | 2.0 | 1.0 |
| $CoCl_2 \cdot 6H_2O$ | 0.025 | 0.025 | 0.01 |
| $H_3BO_3$ | 6.2 | 3.0 | 5.0 |
| KI | 0.83 | 0.75 | 0.1 |
| $FeSO_4 \cdot 7H_2O$ | 27.8 | 27.8 | 15.0 |
| $Na_2 \cdot EDTA$ | 37.3 | 37.3 | 20.0 |
| Organic nutrients | | | |
| Thiamine-HCl | 0.1 | 10.0 | 5.0 |
| Pyridoxine-HCl | 0.5 | 1.0 | 0.5 |
| Nicotinic acid | 0.5 | 1.0 | 5.0 |
| Glycine | 2.0 | – | – |
| Inositol | 100 | 100 | 100 |
| Sucrose | 30,000 | 20,000 | 30,000 |
| Hormones | | | |
| 2,4-D | – | 1.0 | 0.5 |
| 1AA | 1.0 | – | – |
| Kinetin | 0.1 | 0.1 | 0.1 |
| CPA | – | – | 2.0 |
| pH | 6.0 | 5.8 | 6.0 |

MS, B5 and SH are tissue culture media names.

Reprinted by permission from Brodelius, P. 1985. "Immobilized Plant Cells," in *Enzymes and Immobilized Cells in Biotechnology*, A. I. Laskin, ed., Menlo Park, CA: The Benjamin/Cummings Pub. Co.

- $MnSO_4 \cdot H_2O$ (16.9 mg)
- $ZnSO_4 \cdot 7H_2O$ (8.6 mg)
- $H_3BO_3$ (6.2 mg)
- KI (0.83 mg)
- $Na_2MoO_4 \cdot 2H_2O$ (0.025 mg)
- $CoCl \cdot 6H_2O$ (0.025 mg)
- $CuSO_4 \cdot 5H_2O$ (0.025 mg)
- $FeSO_4 \cdot 7H_2O$ (27.8 mg)
- $Na_2 \cdot EDTA$ (37.2 mg)
- inositol (myoinositol) (100 mg)
- thiamine HCl (0.4 mg)
- $N_6$-benzyladenin (BA) (1.0 mg)
- indole-3-butyric acid (IBA) (0.1 mg)
- gibberellic acid ($GA_3$) (0.5 mg)
- sucrose (30 g)
- agar (5 g)

Mix all of these except agar in 1 liter of distilled water and adjust the pH to 5.2 by adding 1 M NaOH or 2 M HCl. Add the agar, and heat the medium until it boils vigorously. Put 15 mL medium in each test tube and cap them. Place the test tubes in racks in the autoclave and heat for 15 min at 105 kPa (gauge) pressure. After autoclaving, remove tubes and place on an angle so that the medium will solidify on a slant. Place the cooled, labelled tubes of medium in a cool and clean place. In place of tubes, jars, petri dish, etc., are also used.

Some formulas prescribe the use of a liquid medium for one or more stages of growth. Usually liquid media are agitated on a shaker or rotator. To avoid agitation, a filter paper (7.6 × 2.0 cm) can be inserted in the test tube, to form a bridge, to support the culture in the liquid medium. The paper supports the culture and acts as a wick carrying nutrients to the culture.

### 3.2.6 Explant Material

Cells from any plant species can be cultured aseptically on or in a nutrient medium. According to Gamborg et al. (1981), the materials in order of priority are sections of seedlings, swelling buds, stems, and leaves.

Sterilized seedlings are prepared by germinating sterilized seeds under sterile conditions. First, seeds are soaked in 70% ethanol for 2 min. After removing the alcohol, 20% commercial bleach containing 5% hypochlorite or 1 to 2% sodium hypochloride is added and stirred for 15 to 20 min. Remove floating seeds, and repeat second bleaching treatment, if required. Rinse well the bleached seeds in sterilized distilled water, and place them on double layers of sterilized filter paper in petri dishes. Add sterilized distilled water and seal the dishes with Parafilm™. Place in the dark at 25 to 28°C for seed germination. Any part of the seedling can be used as explant for callus formation.

Bud, leaf, and stem sections of plants can be sterilized by dipping in 70% ethanol for 1 to 3 min followed by 2 to 3 washings in sterilized distilled water. Similarly, bulbs, roots, and tuber sections are sterilized in 70% ethanol followed by washings in 20% commercial bleach (5% sodium hypochlorite) solution.

## 3.3 **CELL CULTURES**

Plant cultures are classified into callus, cell, organ, meristem, and protoplast. To initialize the cell culture, 3 to 5 pieces of about 0.5 cm each from explant materials are placed into nutrient medium containing 2,4-D and 0.6 to 0.8% Difco Bacto agar in glass jars or flasks. Generally, plant cells in culture require an acidic pH (5.5 to 5.8). A temperature between 26 to 28°C is considered optimum, and proper aeration by shaking (100 to 150 rpm) is needed for optimum growth. Light is generally not required for growing cell cultures. Thus, the sections should be incubated in the dark at 26 to 28°C for 3 to 4 weeks.

The callus from 2 to 3 containers is then transferred into 20 mL of liquid medium in a flask, and incubated at 26°C in the presence of light with continuous shaking at 150 rpm. This cell suspension is used to regenerate plants.

### 3.3.1 Plant Regeneration from Cell Suspension Cultures

5 to 10 mL of cell suspension are washed 2 to 4 times in hormone-free medium; place them on hormone-free agar plates and incubate at 25 to 27°C in light. Plantlets are then placed in hormone-free agar medium in tubes or jars, and incubated at 25°C and 16:8 h photoperiod. Rooted plants are transferred to soil or peat-vermiculite and provided with nutrient solution. Plants are then grown in a growth room.

Hong et al. (1989) investigated strawberry tissue culture. Immature strawberry fruits were sterilized for 30 s in 70% ethanol, dipped in 1% solution of sodium hypochlorite for 4 min, and washed 3 times with sterile distilled water. Tissue explants were cut into 4 mm cubes and placed into MS medium with 4% sucrose, 0.6% agar, and different concentrations of growth regulators.

The pH was adjusted to 5.5 and autoclaved at 121°C for 20 min. The callus tissues were transferred into 20 mL MS medium in 50 mL flasks, and incubated on a shaker (120 rpm). The suspension cultures were transferred into fresh medium every 2 to 3 weeks.

The predicted growth index was 1.3 for a culture with 4.5 mg/L of both 2,4-D and kinetin. For a culture with 1 mg/L of 2,4-D and no kinetin, the growth index was approximately 2.0. The strawberry liquid suspension cultures grew faster at the lower levels of added hormones. Medium composition, types and levels of phyto-hormones, and inoculation size influenced callus formation and growth of cell in liquid suspension cultures.

### 3.3.2 Embryogenesis and Organogenesis

Somatic embryogenesis is the formation from somatic cells of embryo-like structures capable of growing into new plants. It can be induced artificially in tissue cultures (Tisserat, 1985). Embryogenesis can occur in either callus or suspension cultures, but the cells should be closer to the zygotic condition and show negligible differentiation. In organogenesis, roots or shoots can be induced directly on differentiated explants by culturing them on a medium having low concentrations of auxin and cytokinin in the suitable ratio (Goldsworthy, 1988).

Techniques such as protoplast fusion, mutagenesis, or DNA uptake can be used to genetically modify a cell with clonal propagation used to recover genetically altered regenerated plants (Evans et al., 1981). Callus growth establishment with subsequent organogenesis or embryogenesis were obtained from many species of plant cells. Many plant cells will undergo mitosis *in vitro*.

Cell division may provide genetic variability. The plant cell division is classified as meiotic and mitotic. The meiotic division is related to sexually reproductive cells. Mitotic (somatic or vegetative) division provides two cells with the same number of chromosomes as the original cells. If chromosomal changes, known as mutations, occur during mitosis, they will be carried in all further cell divisions.

Somatic embryogenesis can be used to produce artificial seeds. Somatic tissue is cultured on a growth medium, usually containing the synthetic plant growth regulator, 2,4-D. The cells pass through a period of rapid division and growth leading to the formation of a callus. In this, some cells develop like fertilized egg cells into embryos, which are termed "somatic" embryos because they originate from vegetative, not reproductive tissue. Under appropriate physical environments, with proper nutrients and growth regulators, these embryos can be separated from the callus mass and will develop like zygotic (fertilized) embryos (McKersie et al., 1990).

McKersie et al. (1990) have refined techniques to modify the development of somatic embryos to prevent early germination, by treating the embryos at the early cotyledonary stage of development with the medium with 6 to 9% sucrose. This blocked the early germination and the somatic embryos increased in weight and acquired tolerance to desiccation. Somatic embryogenesis has been reported for a range of tropical plants including papaya, lemon, coffee, sweet potato, mango, cassava, and sugarcane. Artificial seed would be a viable alternative for the propagation, exchange, and germplasm preservation of these species.

According to Evans et al. (1981), somatic embryogenesis of 40% crop species used embryo or hypocotyl explants during primary culture. About 70% of explants were cultured on MS medium or a modification of it. They reviewed the compositions of media for various crop species. The method used for each crop species has been described. The MS medium is also used in 75% of somatic organogenesis, and was also used in 87.5% species of direct shoot formation, 58% of crop species capable of callus-mediated regeneration, and 85% of graminaceous species. About 46% of crop species had no growth regulator requirement.

Many morphological variants can be developed using somaclonal variation. These variants contain novel leaf, flower, and fruit colors, which are useful in ornamental varieties. Evans (1988. "Applications of Somaclonal Variation," in *Advances in Biotechnological Processes*, A. Mizrahi, ed., New York: Alan R. Liss, Inc., 9:204−225) regenerated plants of *Nicotiana alata* using somaclonal variation in the following way:

(1) Leaves were sterilized in 8% Clorox™ for 8 min, then rinsed 3 times in sterile distilled water. All other operations were conducted under absolute sterilized conditions. Leaves were dried, cut into 2.3 × 2.5 cm pieces and placed onto agar solidified media consisting of MS macro- and micronutrients and vitamin $B_5$. This medium (MMS) was further supplemented with 0, 0.5, 1.0, 5.0, or 10 $\mu$M of 6-benzyladenine (BA) to determine plant regeneration capacity.

(2) Leaf sections were cultured under 16 h of light at 30°C up to 3 cm shoot height, excised, and placed onto MMS with 0.1 mM 3-aminopyridine (rooting medium). After root development, shoots were transplanted to Jiffy pots™, later to 10 cm pots and, when secondary leaves developed, the mature plants were transferred to a greenhouse.

(3) Shoots were regenerated from explants cultured on either 1$\mu$M or 5$\mu$M BA within 28 days.

A great deal of variability was observed in these plants regenerated from leaf explants. This somaclonal variability included variation for morphological traits (flower shape, leaf shape, plant height, etc.), pollen viability, and chromosome number.

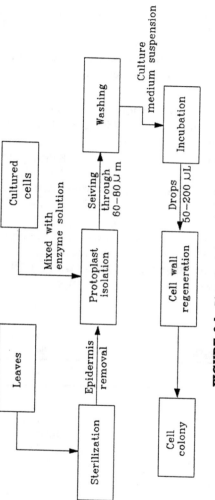

**FIGURE 3.1.** Plant protoplasts — isolation and culture.

## 3.4 **PROTOPLAST FUSION**

Somatic hybridization by protoplast fusion is a potential method for producing hybrids between closely related and unrelated plants. The cell fusion is a means of developing required crosses in plant breeding that are infeasible by other methods. The protoplast isolation and culturing can be performed by the following method (Gamborg et al., 1981).

Protoplast isolation and culture from leaf is shown in Figure 3.1. Aseptic conditions in a laminar airflow cabinet should be maintained for all operations.

(1) Sterilize leaves in 70% ethanol for 1 min and place in 15% Clorox™ bleach and Tween-80™ (1 drop/25 mL) for 15 min in a petri dish. Wash with sterile distilled water (3×).

(2) Peel off the lower epidermis with a suitable forceps, section the remaining portion (1 cm² surface area) and incubate in a 1:1 mixture of enzyme solution (E1, Table 3.2) and protoplast culture medium (Table 3.3) in a petri dish. After sealing the dish with parafilm, incubate for 5 to 15 h at 22 to 24°C with shaking (50 rpm).

(3) Protoplast release can be observed with an inverted microscope. Using a Pasteur pipette, remove protoplast suspension and sieve through 60 to 80 $\mu$m mesh in a centrifuge tube. Centrifuge at 50 g's for 6 min, and remove the supernatant with a Pasteur pipette.

TABLE 3.2. *Solutions used in protoplast isolation, culture and fusion (from Gamborg et al., 1981, reprinted by permission of Academic Press, Orlando, FL, U.S.A.), pH = 5.8.*

| Chemical | Molarity | Amount/100 mL | Enzyme | mg/100 mL |
|---|---|---|---|---|
| **S1** | | | **E1** | |
| CaCl$_2$·2H$_2$O | 6 mM | 90 mg | Hemicellulose, Rhozyme HP150 | 100 |
| KH$_2$PO$_4$·H$_2$O | 0.7 mM | 10 mg | Onozuka R10 | 200 |
| Glucose | 0.5 M | 9.1 g | Pectinase | 100 |
| **Polyethylene Glycol (PEG) Solution*** | | | | |
| Chemical | Molarity | Amount/5 mL | | |
| Glucose | 0.2 M | 0.18 g | | |
| CaCl$_2$·2H$_2$O | 10 mM | 7.35 mg | | |
| KH$_2$PO$_4$ | 0.7 mM | 0.48 mg | | |

*Mix 25 g PEG per 5 mL of the solution and filter.

TABLE 3.3. *Protoplast culture medium (from Gamborg et al. 1981, reprinted by permission of Academic Press, Orlando, FL, U.S.A.).*

| Chemical | Amount/100 mL |
|---|---|
| B₅ salts and vitamins (10×) | 10 mL |
| Glucose, 0.38 M | 6.84 g |
| Ribose | 25 mg |
| Sucrose | 25 mg |
| L-Glutamine, 4 mM | 58 mg |
| Casamino acids | 25 mg |
| NH₄NO₃, 3 mM | 25 mg |
| CaCl₂·2H₂O, 4 mM | 60 mg |
| Coconut milk* | 2 mL |
| Multivitamins (stock solution A) | 0.2 mL |
| Multivitamins (stock solution B) | 0.1 mL |
| Naphthaleneacetic acid | 1 μm |
| 2,4-Dichlorophenoxyacetic acid (2,4-D) | 1 μm |
| Zeatin or benzyladenine | 1 μm |
| Adjust to pH 5.8 and filter sterilize | |
| **Multivitamin Solution A** | **mg/100 mL Water** |
| Calcium pantothenate | 50 |
| Ascorbic acid | 100 |
| Choline chloride | 50 |
| *p*-Aminobenzoic acid | 1 |
| Folic acid | 20 |
| Riboflavin | 10 |
| Biotin | 1 |
| **Multivitamin Solution B (in 70% Ethanol)** | **mg/100 mL** |
| Vitamin A | 1 |
| Vitamin D₃ | 1 |
| Vitamin B₁₂ | 2 |

*Drain mature nuts, sterilize at 80°C, filter and store frozen.

(4) Immerse the protoplasts in 5 to 10 mL of S1 solution (Table 3.2), centrifuge at 50 g's for 5 min. Repeat washing once for fusion and twice for culturing.

(5) The protoplast concentration can be measured with a coulter counter. Rinse about 10 protoplasts in 1 mL of S1 solution (Table 3.2) for fusion or in protoplast culture medium for culturing.

(6) For culturing place 150 μL (6 to 8 droplets) protoplast suspension in each petri dish, and seal with parafilm. Incubate at 25°C in diffused light in a box humidified by a blotting paper in a 1% CuSO₄ solution.

If suspension cell culture is available for protoplast fusion, then mix equal amounts of cell culture and enzyme solution (E1) (Table 3.2) in a petri dish. After sealing the dish with parafilm™, incubate at 25°C for 5 to 15 h with shaking at 50 rpm. Then follow steps 3 to 6.

Another method of protoplast fusion is discussed here and shown in Figure 3.2.

(1) Place a 3 μL drop of sterile silicone-200 fluid in a petri dish and put a cover slip on the drop.

(2) Mix cell culture protoplasts with an equal volume of mesophyll protoplasts, and put 150 μL of the mixture on the cover slip. This may settle onto the slip in about 5 min.

(3) Drop by drop, add 450 μL (about 6 drops) of polyethylene glycol solution (Table 3.2) to the protoplast suspension. Protoplast agglutination or adhesion can be observed on an inverted microscope. Incubate this for 5 to 25 min at the room temperature.

(4) Slowly add 0.5 mL of eluting solution (Table 3.4) to the mixture, and add 1 mL more solution after about 10 min.

(5) Leave 500 μL of culture medium on the cover slip, and put 0.5 to 1 mL of medium on the dish to maintain humidity. Seal the dish and incubate

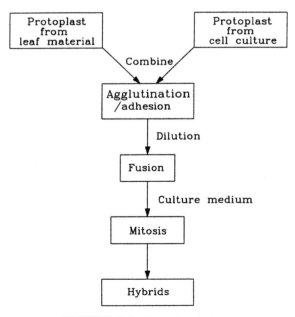

**FIGURE 3.2.** Protoplast fusion.

TABLE 3.4. *Eluting solutions (from Gamborg et al., 1981, reprinted by permission of Academic Press, Orlando, FL, U.S.A.).*

| Chemical | Molarity | Amount/100 mL |
|----------|----------|---------------|
| Glycine | 100.0 mM | 750.0 mg |
| Glucose | 0.3 M | 5.4 g |
| Adjust to pH 10.5 with NaOH, filter, sterilize and store at 4°C in the dark. | | |
| $CaCl_2 \cdot 2H_2O$ | 100.0 mM | 1.47 g |
| Glucose | 0.3 M | 5.40 g |
| Filter sterilize. Mix both in equal volume at the fusion time. | | |

at 25°C in diffused light in a plastic box with a blotting paper in 1% $CuSO_4$ solution (Gamborg et al., 1981).

In the presence of inorganic salts, high molecular weight dextrans favor protoplast clustering and fusion (Kameya, 1979). It has also been reported that higher frequency protoplast fusion can also be achieved by electric impulse (Zimmermann and Scheurich, 1981).

## 3.5 PLANT CELL SYNTHESIS

The development of complex polysaccharide gel matrices can provide effective immobilization systems for plant cells, and provide continuous recovery of secondary plant metabolites. The increase of glucan synthase activity in plants by chitosan could assist in the recovery and purification of callose synthase (Knorr et al., 1990). The product yields from dill cultures could be more than doubled when chitosan treatment was used in combination with a precursor treatment. This approach combined with other ways of product yield improvement such as cell culture, immobilization, medium selection, and effective removal and recovery of the required materials may provide optimum process development for the production of food ingredients from plant cells.

The immobilization of plant cells is discussed in Chapter 6.

## 3.6 CULTURING PLANT CELLS FOR FLAVOR

Typically, a high-yielding, good quality line of fruit or vegetable is selected. Sterilized leaf, stem, bud, etc., are segmented, and segmented sections are placed in a petri dish on a solid agar medium containing solution (vitamins, nutrients, hormones, etc.) for callus growth. A callus, selected to express particular properties, is subcultured to amplify its enzymatic activity. The selected callus is then aseptically transferred and dispersed in a liquid medium containing nutrients, vitamins, micronutrients, hormones, minerals, etc. A particular hormone, determined empirically, will induce expression of the flavor component (Dziezak, 1986).

The plant suspension cells are multiplied in a bioreactor, generally an air lift or sparging type. These cells are grown until they reach the maximum cell density, and obtain the optimum level of enzymatic activity to form desired products such as flavor. The time required for the cells to obtain optimum enzymatic activity may be 5 to 20 days, based on the nutrients, cell density and other process conditions such as aeration. The desired product either leaks into the medium or remains within the cells. Special techniques are needed for further separation and purification.

To decrease the cost of production, immobilized cells are used as biocatalysts in continuous production of desired product or flavor. For this, metabolically active cells are encapsulated in a calcium alginate or polyamide matrix. These immobilized cells are packed into a column reactor with a medium for cell maintenance. The cells produce the desired flavor compound and release it into the medium, from which it is separated. For this, other types of reactors such as hollow fiber bioreactors can also be used (Dziezak, 1986).

This technique has been used to produce vanilla and other fruit flavors. Many fruit plants have been successfully cultured using the leaf, shoot, bud, root, petiole or embryo portions of the plant. Details are given by Dziezak (1986).

Somaclonal variation can be used to induce genetic variations in plants that produce flavor compounds, because somaclonal variants have been observed with modified levels of flavor compounds. Protoplast fusion technology has also been used to develop genetically new plant cells for the production of chemicals from plants.

Cell aggregate cloning has also been used successfully to select cell cultures that overproduce chemical compounds such as alkaloids, pigments, and vitamins (Whitaker and Evans, 1987).

Many essential oils are used as flavoring agents in food and beverages. Production of essential oils and related products by plant cell and tissue cultures has been reviewed (Mulder-Krieger et al., 1988). Many volatiles, different from those occurring in the plant, have been isolated from cell cultures. They focused the review on monoterpenes, $C_9-C_{11}$ compounds, phthalides, pyrethrins, valepotriates, cannabinoids, furanoterpenes, aromas and their derivatives. The occurrence of these compounds in dicotyledonae, monocotyledonae, coniferae and bryophyta are also summarized.

The synthesis and yield of essential oils can be stimulated by (Mulder-Krieger et al., 1988): (1) inducting morphological differentiation into tissue cultures, (2) creating artificial accumulation sites for these volatiles, (3) changing the culture medium composition, (4) changing physical conditions for the culture growth, such as temperature, light, gaseous environment composition, and (5) inducting polyploid cells.

More basic research related with the regulation of the primary and secondary metabolism of plant cells cultured *in vitro* is required.

CHAPTER 4

# Microbial Synthesis and Production

## 4.1 INTRODUCTION

*Microorganisms* have been used since ancient times for the production of special foods. Many food ingredients, particularly colors and flavor compounds such as lactones, esters, acetone, pyrazines, L-menthol, diacetyl, terpenes, and volatile fatty acids, are derived from microbial cells. Many current food processing techniques use fermentation to produce various organic compounds for providing unique flavor and color to many foods. Well known fermented foods are beer, bread, yogurt, cheese, sauerkraut, sausages, etc. Less well-known fermented products are tea, coffee, soy sauce, pickled cucumber, etc. (Hill, 1987). Table 4.1 is a summary of microbial production of food ingredients (Wasserman et al., 1988). These ingredients could be utilized to replace the presently used synthetic ingredients. These products could be food-stabilizing agents, emulsifiers, vitamins, amino acids, single-cell proteins, water-binding agents, foaming agents, etc. (Kosaric, 1984).

## 4.2 FLAVOR

Microorganisms and enzymes produce flavor compounds in many foodstuffs. The sources of enzymes in foods are (1) those inherent in the food, (2) those of added microorganisms, (3) those arising from microbial contaminants, and (4) those that are intentionally added to foods. Examples of flavors produced using enzymes and/or via microbial fermentation are enzyme-modified cheeses, lipolyzed milk fats, and the blue cheese flavor (Gatfield, 1988).

Diacetyl (2,3-butane dione) is a metabolic end product of lactic acid bacteria that is synthesized from the intermediary metabolite pyruvate (Daeschel, 1989). This is an essential ingredient in butter flavor formulations. *Leuconostoc dextranicus, Lactobacillus citrovorum* and *Streptococcus lactis* var. *diacetylactis* produce butter flavor. The dairy industry uses diacetyl-producing streptococci to impart butter flavor to yogurt and other fermented products (Trivedi, 1986).

**105**

TABLE 4.1. *Microbial production of food ingredients (Wasserman et al., 1988, used with permission from Inst. of Food Technol., Chicago).*

| Ingredient | Function | Microorganism |
|---|---|---|
| Acetic acid | Acidulant | *Acetobacter pasteurianus* |
| N-Acetyl tripeptide | Immune enhancer | *Bacillus cereus* |
| D-Arabitol | Sugar | *Candida diddensii* |
| Beta-carotene | Pigment | *Blakeslea trispora* |
| Chrysogenin | Pigment | *Penicillium chrysogenum* |
| Citric acid | Acidulant | *Aspergillus niger* |
| Citronellol | Fruity flavor | *Ceratocystis* spp. |
| Curulan | Thickener | *Alcaligenes faecalis* |
| Diacetyl | Buttery flavor | *Leuconostoc cremoris*, *Streptococcus lactis* |
| Dextrans | Thickeners | *Leuconostoc mesenteroides* |
| Emulsifier | Emulsification | *Candida lipolytica* |
| Fatty acid esters | Fruity fragrances | *Pseudomonas* spp. |
| Gamma-decalactone | Peach fragrance | *Sporobolimyces odorus* |
| Geranoil | Roselike fragrance | *Kluyveromyces lactis* |
| Glycerol | Humectant | *Bacillus licheniformis* |
| Glutamic acid | Flavor enhancer | *Corynebacterium glutamicum* |
| Lactic acid | Acidulant | *Streptococci* and *lactobacilli* |
| Leucine | Amino acid | *Brevibacterium lactofermentum* |
| Lysine | Amino acid | *Corynebacterium glutamicum* |
| Mannitol | Sugar | *Torulopsis mannitofaciens* |
| Methanol | Flavor | *Pseudomonas putida* |
| 3-Methoxy-isopropyl-pyrazine | Potato odor | *Pseudomonas perolens* |
| Methylbutanol | Malt flavor | *Streptococcus lactis* var. *maltigenes* |
| 3-Methylbutylacetate | Banana fragrance | *Ceratocystis moniliformis* |
| Monascin | Pigment | *Monascus perpureus* |
| Nisin | Antimicrobial | *Streptococcus lactis* |
| 5′-Nucleotides | Flavor enhancers | *Corynebacterium glutamicum* |
| 6-Pentyl-2-pyrone | Coconut fragrance | *Trichoderma viride* |
| L-Phenylalanine | Aspartame precursor | *Bacillus polymyxa* |
| Proline | Amino acid | *Serratia marcescens* |
| Sesquiterpenes | Fruity fragrance | *Lentinus lepideus* |
| Surfactant | Wettability | *Bacillus licheniformis* |
| Tetramethylpyrazine | Nutty flavor | *Bacillus subtilis*, *Corynebacterium glutamicum* |
| Thermogelable polysaccharides | Thickeners | *Agrobacterium radiobacter* |
| Vitamin $B_{12}$ | Vitamin | *Propionibacterium* |
| Xanthan gum | Thickener | *Xanthomonas campestris* |
| Xylitol | Sweetener | *Torulopsis candida* |

Pyrazines are heterocyclic, nitrogen-containing compounds that provide a roasted or nutty flavor upon heating.

Lactones provide fruity, coconut-like, buttery sweet, and nut-like flavors. At present, these are chemically synthesized.

Monosodium glutamate is the principal flavor component of soy sauce. It is used in many foods to provide a spicy flavor. Many other flavors such as vanilla, mushroom, mint, onion, etc., can be produced using microorganisms.

The popular glutamic acid producing bacteria is *Corynebacterium glutamicum*, which is capable of producing large amounts of L-glutamic acid.

Terpenes are hydrocarbons, and are of great interest to the food industry since they produce flavors of oils and grape aroma. These include essential oils, vitamin A, and certain plant pigments.

Presently, the flavorants inosinic acid and guanylic acid are produced by extraction of yeast. rDNA techniques (see Chapter 2) will amplify the production of these compounds in bacteria. Other flavorings, spices, and compounds can be economically produced by bacteria or plant cell cultures (for details see Chapter 3).

Fragrance production technologies can be improved by using microbial rather than chemical routes. This can be achieved by cloning the plant genes in microorganisms (Schindler and Schmid, 1982).

## 4.3 **COLOR**

*Phaffia rhodozyma*, a yeast, and *Monascus purpureus*, a fungus, produce red pigment (Trivedi, 1986). *Candida utilis* yeast was used to ferment red beet juice, which is used as a colorant after concentration (Von Elbe and Amundson, 1977).

rDNA techniques can also be applied to increase the activity of the biosynthetic paths of microorganisms to produce natural pigments (e.g., carotenoids) using fermentation.

## 4.4 **VITAMINS**

Pure riboflavin is produced by microbiological synthesis, mainly by the plant pathogen *Eremothecium ashbyii*, a yeast-like fungus. Vitamin $B_{12}$ is produced by various bacteria, but *Streptomyces griseus* is the producer of the compound (Brown et al., 1987).

Vitamins such as $B_2$, $B_{12}$, C and D, which are produced by microbial processes, can be more efficiently synthesized by using rDNA techniques. By characterization and alteration of the microbial biosynthetic pathway for vitamin E, it can be produced by fermentation (Haas, 1984).

## 4.5 **SINGLE-CELL PROTEINS (SCP)**

SCP can be produced from agricultural residue anaerobically in yields of about 20% by using a mixed culture of rumen bacteria. Removal of acidic end products by cyclic microfiltration into a methanogenic fermenter increased the cell yields (Finn and Ercoli, 1986). Cassava and other starch-containing products can be hydrolyzed to give a glucose syrup. If available in excess, their conversion to SCP will be economical and appropriate. Similarly, clarified cane juice is also a suitable fermentation substrate for SCP (Vasey, 1984). A number of microorganisms (such as algae, bacteria, and fungi) and substrates are used to produce various single-cell proteins (Table 4.2). Canning and food processing wastes can be used to grow yeast. The yeast can be grown at low pH (4.5 to 5.5) to minimize bacterial competition. The use of fungal mycelium (such as *Fasarium*) produces protein that can be processed to a meat-like texture (Brown et al., 1987).

For human food, SCP must be a dry, soluble powder, with minimal color and odor, no pathogens and low nucleic acid, no viable cell count and toxic materials. Thus, for human consumption, protein extraction and purification, cell lysis, and nucleic acid hydrolysis are required.

The torula yeast is used as (1) a flavor carrier in processed foods, (2) a protein supplement for human food products and animal feeds, and (3) a source of vitamins and nucleic acids in pharmaceuticals (Joglekar et al., 1983).

SCP production generally requires aerobic fermentation, cell separation, and cell drying. Cells can be separated in batch operations by conventional vacuum filters or a solid-bowl centrifuge. Rotary vacuum filters are used for continuous operation. Pasteurization can be performed by using the high temperature (60 to 76°C) and short time (0.75 to 2.5 min) technique. For drying of SCP, spray dryers are preferred over drum dryers to avoid contamination. The process can be made efficient and economical by minimizing the power required for aeration and agitation, e.g., air lift fermenter, and by improved product separation through the use of membranes.

The Waterloo SCP bioconversion process has been viewed as the most technically and economically feasible for SCP production. It was developed at the University of Waterloo, Waterloo, Ontario, Canada (Joglekar et al., 1983; Wasserman et al., 1988). The process uses a cellulolytic fungus (*Ch. cellulolyticum*) pretreated at an alkaline pH, sterilized (at 150°C for 3 min or 120°C for 15 min and cooled to 37°C), and supplemented with nutrients (manure, nitrogen, potassium and phosphorus salts) before aerobic fermentation (pH 5.5 to 6 for 4 to 12 h depending on the substrate). If animal manure is used as a nutrient source, then methane is produced by anaerobic fermentation and used to produce energy. The fermenter is maintained at 37°C through the use of a cooling coil, and aeration is achieved with a combination

TABLE 4.2. *Single-cell protein production using organic wastes (modified after Joglekar et al., 1983).*

| Substrate | Microorganism |
|---|---|
| **Dairy Wastes** | |
| Lactose | *Kluyveromyces fragilis* |
| | *Lactobacillus bulgaricus* |
| | *Saccharomyces fragilis* |
| | *Trichospora cutaneum* |
| | *Candida curvata* |
| **Agriculture Wastes** | |
| Cellulose and hemicellulose | *Chaetominum cellulolyticum* |
| **Cereal and Sugar Wastes** | |
| Mixed carbohydrates | *Candida quilliermondi* |
| | *Debaryomycas kloeckeri* |
| | *Hansenula anomalc* |
| | *S. cerevisiae* |
| Sucrose or glucose | *C. utilis* |
| Saccharose, glucose, fructose | |
| and raffinose | *S. cerevisiae* |
| | *C. utilis* |
| Starch | species of: |
| | *Aspergillus, Cephalosporium,* |
| | *Rhizopus, Penicillium,* |
| | *Trichoderma, Gliocledium* |
| **Fruit and Vegetable Wastes** | |
| Mixed carbohydrates and lactic acid | *S. cerevisiae, C. utilis, K. fragilis* |
| Glucose, fructose and sucrose | *Pichia spartina* |
| Glucose and fructose | *S. cerevisiae, C. utilis, S. rouxii* |
| Starch | *Aspergillus foetidus* |
| | *L. bulgaricus, thermophilous* |
| | *L. acidophilus* |
| | *Endomycopsis fibuliger, T. virdis* |
| | *C. utilis, Gliocladium deliquescens* |
| **Meat Wastes** | |
| Collagen | *Bacillus megaterium* |
| **Brewery Wastes** | |
| Reducing sugars | *Aspergillus niger, C. utilis, S. cerevisiae,* |
| | *A. bisporus, M. esculenta, C. utilis* |
| | *Rhodotorula glutinis* |

of bubbled air, mechanical aeration, and mixing. The fermentation broth is passed through a belt press, vacuum drum filter, or centrifuge, and the fungal biomass is dried and milled.

For human consumption, further treatment is necessary to remove nucleic acids. Mogren et al. (1974) and Lindblom and Mogren (1976) have described a process to remove RNA from SCP. In this process, the cell wall is degraded by mechanical disintegration, in which shearing is provided in a homogenizer with milling particles. Then, suspended cells are incubated at 30 to 70°C and pH 5 to 9 in the presence of a salt (chlorides of sodium, potassium or calcium, etc.) which activates cellular enzymes (RNAse). The addition of 0.1 wt% salt reduces 75 wt% RNA. The fungal SCP obtained from the Waterloo process is composed of 45% protein, 35% carbohydrate, 10% fat, 5% nucleic acids and 5% minerals and ash.

Batt and Sinskey (1984) reviewed some of the developments in SCP production and strategies to improve its production via genetic engineering. *Saccharomyces cerevisiae* is suitable for strain improvement via classical and recombinant DNA manipulation to yield a better SCP. Presently, *S. cerevisiae* does not ferment a large range of substrates. This yeast can be improved to utilize less expensive substrates in order to decrease the cost of production of the SCP. For example, the whey can be utilized if a lactose-catabolizing strain of *S. cerevisiae* is available. Similarly, development of a yeast strain to overproduce amino acids may decrease or eliminate the need to add amino acids to SCP products (Batt and Sinskey, 1984). A protein can be enriched for a given amino acid by insertion of a synthetic polynucleotide sequence using genetic engineering techniques. Kangas et al. (1982) used this approach by modifying the ampicillinase protein of pBR322 by the insertion of poly G-C sequences to increase the proline content of the ampicillinase protein in *E. coli*. Chemical mutagenesis can also be used to improve the yeast strain. By modifying the cell wall structure of the yeast, the functional attributes can be manipulated to produce the SCP of desired properties.

The structure-function relationship of SCP can be controlled by knowing the polymer biosynthesis genetics and manipulating the properties by recombinant DNA techniques (Batt and Sinskey, 1984).

Phillips Petroleum Co., Bartlesville, Oklahoma, developed a high-cell-density direct-dry process for the production of yeasts from methanol or other feed stocks (Shay and Wegner, 1985). The process used mechanical agitation and foam breakers, and high heat and oxygen transfer rates. It utilized low-cost carbohydrates such as molasses. High cell density allowed product recovery by direct-drying, which simplified recovery, produced minimum waste streams, and eliminated contamination during product recovery.

Presently, SCP contains high levels of nucleic acids. By introducing the genes into the cells for the appropriate nucleic acid-degrading enzymes and timing their expression, the DNA and RNA breakdown could be induced before harvesting of the cells (Haas, 1984).

## 4.6 **ANTIMICROBIAL SUBSTANCES**

Bacteriocins are antibacterial substances produced by many bacterial species such as colicins (*Escherichia coli*). From the group *N-Streptococci*, the bacteriocins nisin and diplococcic have been characterized. The nisin is currently used as a food preservative in many countries. It has shown to be effective in inhibiting certain gram-positive species but not gram-negative bacteria, yeasts, or fungi. It is also effective in preventing the outgrowth of *Clostridium botulinum* spores (Daeschel, 1989).

The lactic acid bacteria, comprising the genera *Lactobacillus*, *Lactococcus* (*N-Streptococci*), *Leuconostoc* and *Pediococcus*, are involved in the preservation of certain foods including milk, meat, fruits, and vegetables. These are capable of producing inhibitory substances that are antagonistic toward other microorganisms. These also have the ability to generate hydro-peroxide during growth. Reuterin is a low molecular weight, non-proteinaceous, highly soluble, pH-neutral product produced by the *Lactobacillus reuterii*. It is a broad spectrum antimicrobial agent, active against certain gram-negative and gram-positive bacteria, yeasts, fungi, and protozoa (Daeschel, 1989).

## 4.7 **POLYSACCHARIDES AND BIOPOLYMERS**

Many polysaccharides used by the food industry are plant products. To produce these by bacteria, the genes coding for their synthesis may be transferred to microorganisms. Presently, xanthan gum is produced by *Xanthomonas campestris*. Its production may be increased by using genetic engineering techniques.

Polysaccharides are used in food processing as coagulants, dispersants, stabilizers, surfactants and thickeners, and also as matrix materials for separating, chromatographic, and immobilized-enzyme technologies. Major sources of new polysaccharides are the extracellular polysaccharides secreted by mucoid microorganisms. Table 4.3 summarizes food-related microbial polysaccharides (Sinskey et al., 1986).

Alginates are the salts of alginic acid, which is a structural component of the cell walls of brown algae. They are heteropolysaccharides containing D-mannuronic acid and L-guluronic acid that are $\beta$-linked. Curdlan is a homopolymer of (1-3) linked $\beta$-D-glucose. The following procedure to isolate curdlan was recommended by Morris (1987). The fermentation broth is treated with alkali. Under alkaline conditions curdlan is soluble, and the bacterial cells are removed by centrifugation. The supernatant was then neutralized and the insoluble polymer removed by centrifugation, washed, and dried.

Gellangum is the generic name for the extracellular polysaccharide secreted by microorganisms, and it is produced by aerobic submerged fermentation. Pullulan is non-toxic and non-digestible by human enzymes, and possible food uses include preparing food films as emergency rations.

Scleroglucan has the calorific equivalence of starch, and was shown to lower cholesterol levels in chicks and rats (Morris, 1987). Xanthan gum

TABLE 4.3. *Food-grade polysaccharides from microorganisms (modified after Sinskey et al., 1986; Morris, 1987).*

| Polysaccharides | Organism | Application |
|---|---|---|
| Alginate | *Azotobacter vinelandii*<br>*Pseudomonas aeruginosa* | – – |
| Baker's yeast–glucan | *Saccharomyces cerevisiae* | To simulate mouthfeel of fats and oils |
| Biozan | *Alcaligenes* | – – |
| Cellulose | *Acetobacter agrobacterium*<br>*Alcaligenes* | – – |
| Chitosan | *Mucorale* | – – |
| Curdlan | *Alcaligenes faecalis*<br>*Agrobacterium* | – – |
| Dextran | *Lactobacillus*<br>*Leuconostoc mesenteroides*<br>*Streptococcus* | – – |
| Emulsan | *Acinetobacter calcoaceticus* | Emulsifier |
| Gellan gum | *Pseudomonas elodea* | – – |
| Pullulan | *Aureobasidium pullulans* | Substitute for biodegradable plastic and edible packaging |
| S-194 | *Alcaligenes* | – – |
| Scleroglucan | *Sclerotinum rolfsii* | Suspending, coating and gelling agent |
| Xanthan gum | *Xanthomonas campestris* | Thickening agent |
| XM-6 | *Enterobacter* | – – |

consists of the extracellular polysaccharides secreted by bacteria, and is produced by aerobic fermentation in batch culture. XM-6 is also a secreted extracellular polysaccharide. This may be isolated by alcohol precipitation and clarified by dispersion in water, centrifugation, dialysis against water, and freeze-drying.

The microorganisms *Bacillus polymyxa* and *Rhizobium meliloti* produce polysaccharides that gel. The microorganisms *Arthrobacter viscus*, *Arthrobacter stabilis*, *Azotobacter indicus* var. *myxogenes*, *Erwina tahitica*, *Alcaligenes* and *Acetobacter* produce polysaccharides with interesting rheological properties (Morris, 1987).

## 4.7.1 Cloning Polysaccharide Genes

Both classical and recombinant DNA techniques offer opportunities to alter polymer structures *in vivo* by altering the genetic structure of the organism. The isolation and characterization of exopolysaccharide biosynthetic genes can be used to control polysaccharide structures and functions. This could provide a number of novel biopolymers with diverse properties and many potential applications.

A *Z. ramigera* (I-16-M), cosmid gene library was transduced into *E. coli* (HB101), and screened directly for exopolysaccharide production on the basis of increased fluorescence in the presence of cellufluor dye (Sinskey et al., 1986).

Easson et al. (1986) (quoted by Sinskey et al., 1986) showed the strategy for the cloning of polysaccharide genes in *Z. ramigera* or a related organism. The genes for polysaccharide synthesis must be ligated onto a plasmid, which can then be introduced into a *Z. ramigera*, expressed, and identified by some screening technique. The strategy for cloning in *Z. ramigera* involved the conjugal transfer of a broad host range cloning vector (pcp13) from *E. coli* to *Z. ramigera*.

## 4.7.2 Cloning for β-Glucan

DNA isolated from wild-type yeasts was cloned into *E. coli*:yeast shuttle vectors, and then transformed into the glucan mutant strain. Transformants containing the appropriate cloned gene were selected by reversion from the mutant to the wild-type phenotype. Alternatively, DNA isolated from the mutant strains with increased resistance to a particular β-glucanase, for example, was cloned into *E. coli*:yeast shuttle vectors and transformed into a wild-type strain. Transformants were selected by increased resistance to the β-glucanase relative to the wild-type level (Sinskey et al., 1986; Botstein and Davis, 1982).

## 4.8 **FAT AND OIL**

The process of microbial oil production, i.e., single-cell oil (SCO), is similar to that of single-cell protein (SCP) production. The potential of bacteria, yeasts, and molds to produce oils was reviewed by Ratledge (1984). Table 4.4 lists yeast and fungi species and their lipid contents.

The mycobacteria, corynebacteria, and nocardia comprise high lipid contents, but these lipids tend to be associated with toxic and allergic factors (Ratledge, 1984). Arthrobacter (AK-19), a soil bacterium that can utilize lower hydrocarbons, accumulates up to 80% of lipid when grown on glucose. The lipid contained ca. 90% triacyglycerols, and ca. 50% of their fatty acids were unsaturated. The best lipid yield was 12.5 to 14 g/100 g glucose; under these conditions the biomass yield was 20 g/100 g glucose (Wayman et al., 1984).

The oils from yeasts and molds contain 80 to 90% triacylglycerols. Triglycerides derived from the yeast *Candida curvata* have a melting range

TABLE 4.4. *Oleaginous yeasts and fungi containing* $\geq 45\%$ *lipid (modified after Ratledge, 1982).*

| Yeasts | |
|---|---|
| *Candida curvata* (58%) | *Cryptococcus terricolus* (65%) |
| *Endomycopsis vernalis* (65%) | *Lipomyces lipofer* (63%) |
| *Lipomyces starkeyi* (63%) | *Lipomyces tetrasporus* (64%) |
| *Rhodosporidium toruloides* (66%) | *Rhodotorula glutinis* (71%) |
| *Trichosporon cutaneum* (45%) | *Trichosporon pullulans* (65%) |
| **Fungi** | |
| *Entomophthora coronata* (45%) | *Cunninghamella enchinulata* (45%) |
| *Cunninghamella elegans* (56%) | *Mortierella vinacea* (66%) |
| *Mucor albo-ater* (45%) | *Mucor circincelloides* (65%) |
| *Mucor muceus* (51%) | *Mucor ramannianus* (56%) |
| *Mucor spinosus* (47%) | *Rhizopus arrhizus* (49%) |
| *Pythium ultimum* (49%) | *Aspergillus fischeri* (53%) |
| *Aspergillus nidulans* (51%) | *Aspergillus ochraceus* (48%) |
| *Aspergillus terreus* (57%) | *Chaetonium globosum* (54%) |
| *Fusarium bulbigenum* (50%) | *Gibberella fujikuroi* (48%) |
| *Penicillium lilacinum* (56%) | *Geotrichum candidum* (45%) |
| *Sclerotinum bataticola* (46%) | *Tricholoma nudum* (48%) |
| *Ustilogo zeae* (51%) | — — |

similar to that of cocoa butter (Anon, 1989). The predominant fatty acids in order of their usual abundance were oleic acid, palmitic acid, linoleic acid, stearic acid and palmitoleic acid (Ratledge, 1984). The major fatty acids of most species are in the saturated and unsaturated groups, which are similar to vegetable oils (Shifrin, 1984). Yeast oil is close in composition and properties to palm oleon, a palm oil fraction with a lower melting point. Yeast oil can be easily fractionated to produce non-hydrogenated, non-lauric confectionery fats (Anon, 1989).

Compared to real yields from soybeans, microalgae could produce up to 30 times more oil. Moreover, cell generation times are shorter compared to oil plants (Shifrin, 1984). An organism cultivated for its oil content could be grown to 2 to 2.5 times the density of the same organism being grown for its protein content (Ratledge, 1984). The cost of substrate will be the largest input to any product cost. By utilizing cheap substrates, the oil can be produced at reasonable cost (Wayman et al., 1984). Total lipid content ranged from 12.5 to 46% of dry weight for the 30 algae species. Green algae averaged $17.1 \pm 44.4\%$ total lipids ($\pm$ 1 s.d. for 17 species of green algae and 11 species of diatoms), whereas diatoms averaged $24.5 \pm 4.7\%$ (Shifrin, 1984).

The production of fats and oils from food processing and agricultural wastes by the yeast *Candida curvata* was investigated by Glatzt et al. (1984).

To induce mutations, suspensions of *C. curvata*-D in 0.1 M phosphate buffer at pH 7 were exposed to ultraviolet light for 4 min so that 90 to 99% of the population was killed. The cells were transferred to the medium supplemented with 0.1% yeast extract and incubated for 6.5 h in the dark at 25°C to allow the survivors to divide once. Then the cells were selected by plating the culture on agar and selecting the surviving colonies. By remutagenizing this strain, a mutant was found that survived at 38°C on plates.

### 4.8.1 Oil Modification

The possibility of modifying fats by fermentation with *Candida lipolytica* was explored by Glatzt et al. (1984). *C. lipolytica* grows on fats and oils as its primary carbon source, and at the same time accumulates oils. The fatty acid composition of the oil that was accumulated resembled that of the medium. Linolenate tended to be reduced, and palmitoleic acid was produced in modest amounts. Thus, oils of particular properties can be obtained by feeding different oils, such as corn and coconut oils, together and synthesized them into non-random mixture by *C. lipolytica*. Cholesterol might be removed from animal fats by digestion and recovery from *C. lipolytica*.

### 4.8.2 De-Emulsification/Emulsification

Bacterial surfaces induced de-emulsification of both oil-in-water and water-in-oil emulsions. Hydrophobicity is important to break the oil-in-water emulsion, but important for de-emulsification of both types to ensure that the bacteria are interfacially active as opposed to being dispersed in the bulk phases of the emulsion. Poly-$\beta$-hydroxy-butyric acid was found to be an active de-emulsifier, especially in the presence of the monomer, hydroxybutyric acid (Cairns et al., 1984).

Emulsifiers are classified according to the size and strength of the hydrophilic and lipophilic portions of the surface-active molecule. The ratio of these two groups is known as hydrophile-lipophile balance (HLB). Thus, a lipophilic emulsifier will have a lower HLB number than a hydrophilic one.

Microbial cells possess emulsification properties due to their complex outer surface, which shows both hydrophilic and lipophilic moieties. The cell membrane contains proteins, phospholipids and lipo-polysaccharides; receptors for bacteriophages and surface attachment sites; and it acts as a diffusion barrier against some compounds (Zajic and Seffens, 1984). The microorganisms *Zymomonas mobilis*, *Bacillus subtilis* and *Corynebacterium fascians* had HLB value of 22.5, 13.2 and 4, respectively measured by blending method.

Microorganisms produce a variety of biosurfactants under specific physiological conditions. These allow dispersion of oil in water by reducing the surface tension of the aqueous phase and by reducing the oil/water interfacial tension. Surfactant molecules concentrate at the oil/water interface with concomitant emulsion stabilization. Table 4.5 summarizes typical microbial surfactants (Finnerty, 1984).

In food processing plants, surfactants are also used for cleaning and sanitizing, and for the removal of pesticides and wax coating from fruits and vegetables. They prevent spattering in cooking fats and oils, solubilize flavor oils and retard staling in the bakery, and reduce processing time in sugar crystallization (Kachholz and Schlingmann, 1987). Biosurfactants are also biodegradable. Food surfactants are classified as solubilizers, emulsifiers, detergents, crystallization modifiers, and as foaming, wetting, and lubricating agents (Kosaric et al., 1987). New and better surfactants can be biologically produced by biosynthesis of microorganisms. Biotechnological techniques will be used in the future to produce new biosurfactants for the food industry.

### 4.8.3 Oxidation

Microorganisms oxidize various hydrocarbons or hydrocarbon derivatives to characteristic and specific lipids. Paraffinic hydrocarbons ranging from methane to kerosene are oxidized to acids, alcohols, ketones, dibasic acids, aldehydes, wax esters and glycerides. Aromatic hydrocarbons are oxidized to 1,2-diols, dibasic acids, phenols, acids, alcohols, aldehydes, and hydroxy acids. Symmetrical ethers ranging in chain length from $C_{14}$ to $C_{20}$ are converted to alkoxyacetic acids and dibasic acids through biotransformation (Finnerty, 1984).

TABLE 4.5. *Microbial biosurfactants (modified after Finnerty, 1984).*

| Biosurfactant | Microorganism |
|---|---|
| Acidic polysaccharide-protein complex | *Acinetobacter* sp. H01-N |
| Corynemycolic acids | *Corynebacterium lepus* |
| Fatty acids, diglyceride, lipoprotein complex | *Acinetobacter* sp. H01-N |
| Fatty acid — mannan complex | *Candida tropicalis* |
| Glycolipid | H-13A; *Rhodococcus erythropolis* |
| Lipopeptide | *Candida petrophilum* |
| Polysaccharide-lipid-protein complex | *Corynebacterium hydrocarbonoclastus* |
| Rhamnolipid | *Pseudomonas aeruginosa* |
| Sophorose lipid | *Torulopsis bombicola* |
| Surfactin (lipoprotein) | *Bacillus subtilis* |
| Trehalose lipid | *Arthrobacter paraffineus* |

## 4.8.4 **Wax-Ester**

Wax-ester mixtures such as sperm whale oil and jojoba oil are of interest as high-temperature and high-pressure lubricants, and the latter has increased usage in various health care products. The wax-ester compositions produced by the bacterium *Acinetobacter* sp. H01-N grown on *n*-hexadecane ($C_{16}$) and *n*-eicosane ($C_{20}$) at growth temperatures of 17, 24, and 30°C were investigated (Ervin et al., 1984). Average chain length of the wax-esters was proportional to the chain length of the *n*-alkane substrate, but was independent of the growth temperature. Average unsaturation level was a function of both the *n*-alkane substrate and the growth temperature.

# CHAPTER 5

# Mutagenesis and Protein Engineering Techniques

## 5.1 INTRODUCTION

*The* manipulation of the side chains of a purified protein and X-ray diffraction studies of purified protein crystals provide information on the structural and functional relationships of a protein.

To identify the regions of functional importance within cloned DNA sequences, a number of mutagenesis methods have been developed to introduce mutations within any DNA sequence. The expression of mutated DNA sequences is studied under appropriate conditions to know the effect of each mutation, individually. Once the importance of functional regions within the target sequence is known, it may be possible to construct novel proteins. The effect of mutation within the upstream regulatory elements is studied by linking the mutated upstream regulatory element to a reporter gene whose expression can be monitored.

The easiest way of making mutations is by simply deleting a restriction fragment of the foreign DNA cloned in a vector by the use of a restriction enzyme and T4 DNA ligase. Similarly, a restriction fragment can be introduced at a specific site within the cloned DNA sequence. If there is no unique restriction site within the foreign DNA, it can be created by oligonucleotide-mediated mutagenesis.

The deletion mutagenesis can also be performed by partial digestion of the target sequence with a restriction enzyme that cleaves DNA frequently (e.g., HaeIII, AluI). For this type of mutagenesis, the size of the target sequence should be in the range of 0.5 to 2.0 kb.

The nucleotides from one or both ends of a sequence can be progressively deleted. This results in a nested set of deletion mutants. Such deletions are usually generated by using Bal-31 or exonuclease III.

## 5.2 GENERATION OF DELETION MUTANTS BY BAL-31

Bal-31 enzyme has two functions: (1) By exonuclease activity, it degrades a double-stranded linear DNA molecule by progressively removing

**119**

nucleotides from the 3' ends. (2) By weak endonuclease activity, it degrades single-stranded DNA or RNA molecules. The procedure is described below (see Figure 5.1).

(1) Digest the recombinant double-stranded DNA vector containing the target sequence, with a restriction enzyme that makes a cut at one end of the target sequence.

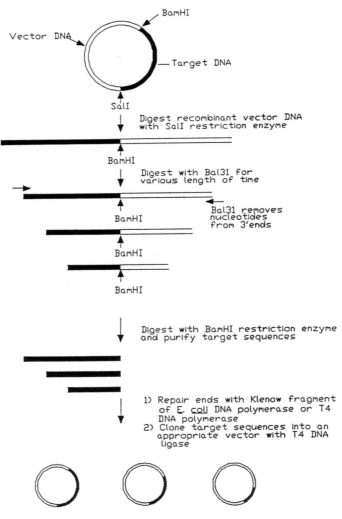

**FIGURE 5.1.** Generation of deletion mutants with Bal-31.

(2) Digest the double-stranded linear DNA with Bal-31 for various lengths of time.

(3) Digest the linear DNA with a suitable restriction enzyme to cleave the target sequences from the vector.

(4) Separate the target DNA from the vector DNA by agarose gel electrophoresis and purify the target DNA from the agarose gel.

(5) Repair the ends with a Klenow fragment of *E. coli* DNA polymerase or bacteriophage T4 DNA polymerase.

(6) Digest a vector with a restriction enzyme that generates blunt ends, and clone the target DNA by blunt-end ligation with bacteriophage T4 DNA ligase.

Exonuclease III enzyme removes nucleotides from 5′ overhang or blunt ends. However, 3′ overhangs are completely resistant. The generation of a nested set of deletions with exonuclease III is described elsewhere (Figure 2.25).

## 5.3 **OLIGONUCLEOTIDE-MEDIATED MUTAGENESIS**

By oligonucleotide-mediated *in vitro* mutagenesis, a specific site within a target DNA—whose sequence is known—can be altered in order to study regions of functional importance (Hutchison et al., 1978; Zoller and Smith, 1987; Shewry et al., 1987). The procedure is as follows ( Figure 5.2):

### 5.3.1 Cloning of the Target Sequence in M13 Vector

The target sequence is isolated after digesting with appropriate restriction enzymes and cloned into a M13-based vector. The isolation of reasonably pure single-stranded circular DNA from single-stranded phage such as M13 is relatively easy. The single-stranded target DNA is required for oligo-nucleotide-directed mutagenesis.

### 5.3.2 Synthesis of the Oligonucleotide to Introduce Mutation

To ensure a good binding of the synthetic oligonucleotide to the single-stranded circular target DNA, the oligonucleotide is carefully designed to keep the mismatch in the center of the oligonucleotide. The preferable size of the synthetic oligonucleotide is 15 to 20 nucleotides long. Such oligo-nucleotides are more likely to bind to the unique target sequence rather than to any other site. Moreover, even after the mismatch their binding would be sufficient to allow them to serve as primers for DNA synthesis at room temperature.

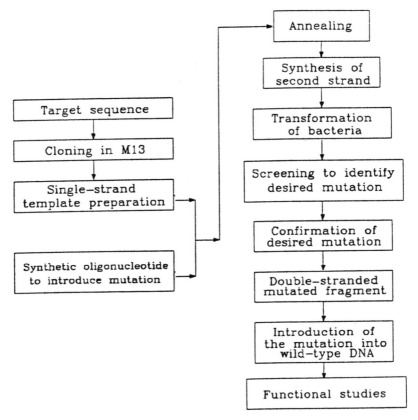

**FIGURE 5.2.**   Steps involved in oligonucleotide-mediated mutagenesis.

### 5.3.3 Annealing of the Synthetic Oligonucleotide to the Single-Stranded DNA Template and Primer Extension

The mutagenic oligonucleotide in 20−50 molar excess is annealed to single-stranded DNA template usually at 50 to 55°C. The synthesis of the second strand is achieved by primer extension using a Klenow fragment of *E. coli* DNA polymerase, and ligating with T4 DNA ligase to form double-stranded closed circular DNA (Figure 5.3). After primer extension and ligation there will be a mixture of double-stranded closed circular DNA and heteroduplex (partially double-stranded) DNA due to inefficient conversion of single-stranded template DNA into double-stranded closed circular DNA. If the bacteria are transformed with this mixture, a substantial population of plaques will be of the wild type, and thus a large number of plaques will have to be screened to isolate the mutant plaque (recombinant M13 in which the target DNA sequence has been mutated). To avoid having to screen a large number of plaques in order to isolate the recombinant M13 containing the desired mutation, double-stranded closed circular DNA can be separated from the partially double-stranded DNA by alkaline sucrose density gradient centrifugation.

The purification of double-stranded closed circular DNA molecules can be avoided if two oligonucleotides (mutant oligonucleotide and oligonucleotide complementary to M13 sequences such as universal primer) are used for primer extension.

Transform the appropriate strain of *E. coli* and place the transformed bacteria onto agar plates. Incubate the plates at 37°C overnight for the development of plaques.

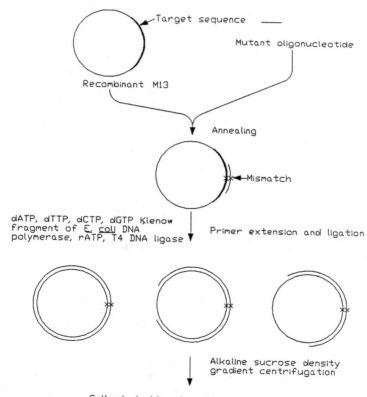

**FIGURE 5.3.** Synthesis of second strand in oligonucleotide-mediated mutagenesis.

## 5.3.4 Screening of Recombinant M13 Plaques to Isolate Mutant

The plaques from the agar plates are transferred to a nitrocellulose filter or nylon membrane just by placing a circular membrane over the plaques. To hybridize such membranes using a $^{32}$P-labeled mutant oligonucleotide as probe, the M13 recombinant plaque containing the desired mutation within the target sequence is identified by progressively increasing the stringency conditions.

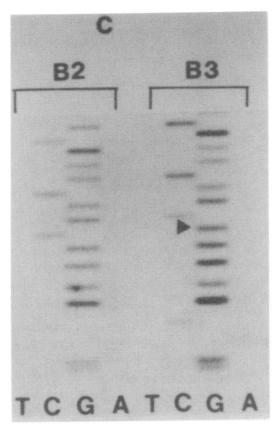

**FIGURE 5.4.** Confirmation of the point mutation by dideoxynucleotide chain termination sequencing technique. The nucleotide sequence of two (B2 and B3) thymidine kinase-deficient mutants of bovine herpesvirus-1 at the region in the *tk* gene where the point mutation was observed. The arrowhead points to the additional "G" inserted in the mutant B3, as compared to the mutant B2. T = thymine, A = adenine, G = guanine, C = cytosine. (Reproduced with permission from Mittal, 1989.)

The putative mutant clone should be plaque-purified and rescreened by hybridization to isolate homogeneous population of the mutant. The desired mutation is confirmed by sequencing single-stranded DNA template prepared from the purified mutant plaque (Figure 5.4). However, the sequencing of the entire mutant target sequence should be completed to rule out the possibility of mutations at any other sites.

### 5.3.5 Purification of the Mutated Target Sequence

(1) Recover the double-stranded replicative form of the recombinant M13 containing the mutated target sequence.
(2) Digest with restriction enzymes to release the mutated target sequence from bacteriophage M13 DNA and separate the DNA fragments onto an agarose gel by electrophoresis.
(3) Cut the appropriate DNA band containing the mutated target sequence and purify the DNA from the gel.

### 5.3.6 Introduction of the Mutated Target Sequence into the Wild-Type DNA

(1) To replace the homologous fragment of the wild-type DNA with the mutated target sequence, digest the wild-type DNA with appropriate restriction enzymes; and gel-purify the wild-type DNA fragment lacking the DNA fragment homologous (except for the mutation) to the mutated target sequence.
(2) Ligate the wild-type DNA fragment and the mutated target sequence with T4 DNA ligase.
(3) Transform bacteria and recover the recombinant clone containing the desired mutation. Quantitate the effect of the mutation by appropriate assays.

## 5.4 **SITE-DIRECTED MUTAGENESIS WITHOUT PHENOTYPIC SELECTION**

The site-directed *in vitro* mutagenesis technique described above has been improved to select the recombinant bacteriophage M13 containing the desired mutation without phenotypic selection (Kunkel, 1985; Kunkel et al., 1987). In this technique the wild-type DNA templates are subsequently degraded on transformation, therefore most of the plaques generated onto lawns of a suitable bacterial strain contain the desired mutation.

The Kunkel method of site-directed mutagenesis utilizes *dut⁻* and *ung⁻*. mutants of *E. coli*. The *E. coli dut⁻* mutant is deficient in the enzyme UTPase

(Konrad and Lehman, 1975; Hochhauser and Weiss, 1978) which converts dUTP to dUMP. In the absence of UTPase, the intracellular pool of dUTP is comparatively high and, therefore, competes with dTTP for incorporation into the DNA. This means that some of the thymine residues of DNA are replaced by uracil. The *E. coli ung⁻* mutant is deficient in the enzyme uracil-*N*-glycosylase, which removes uracil residues if they get incorporated into the DNA (Lindahl, 1974; Duncan et al., 1978). Thus, when the recombinant M13 is grown in the *E. coli dut⁻ ung⁻* mutant, at some places in the target sequence, thymine residues are replaced by uracil and remain in the DNA.

The site-directed mutagenesis is performed by using a DNA template in which a small number of thymine residues are replaced by uracil.

The steps are as follows:

(1) Clone the target sequence, in which a particular mutation needs to be introduced, into bacteriophage M13.

(2) Prepare the single-stranded DNA template containing uracil by growing the recombinant bacteriophage M13 in the *E. coli dut⁻ ung⁻* mutant.

(3) Anneal the single-stranded DNA template containing uracil with mutant oligonucleotide.

(4) Synthesize the second strand by using a Klenow fragment of *E. coli* DNA polymerase and T4 DNA ligase. This will produce a heteroduplex DNA molecule, in which some uracil residues are present in the template strand but not in the strand synthesized by DNA polymerase.

(5) Transfect a wild-type (*ung⁺*) *E. coli* strain to degrade the template strand containing uracil. The enzyme uracil-*N*-glycosylase present in the wild-type *E. coli* removes uracil from the template strand, and thus generates sites that block DNA synthesis and are cleaved by endonucleases. Most of the template strands will be degraded and, therefore, the majority of plaques arise from the complementary strand containing the desired mutation.

(6) Prepare single-stranded template DNA from the recombinant M13 plaques, and sequence them to identify the one containing the desired mutation.

## 5.5 **RANDOM MUTAGENESIS**

The oligonucleotide-directed mutagenesis is particularly useful when one wishes to mutate only a few specific sites in the target sequence. These specific sites are identified by a number of biochemical and X-ray crystallographic studies. However, when a large number of mutations are needed in a DNA fragment in which the information regarding any other specific functional analysis is not available, chemical mutagenesis is a good choice. Sodium bisulphite (Pine and Huang, 1987), hydroxylamine (Kadonaga and Knowles, 1985) or methoxylamine (Tobian et al., 1985) are used for random mutagenesis to introduce mutations at various sites in the target sequence.

The procedure of random mutagenesis using sodium bisulphite as mutagen is as follows.

(1) Prepare single-stranded recombinant M13 DNA containing the target sequence.

(2) Bacteriophage M13 genes are essential for the replication of the phage, therefore they must be protected from random mutagenesis. To protect M13 gene from undergoing mutation, digest wild-type replicative form of M13 DNA with the same enzymes as are used for cloning the target sequence.

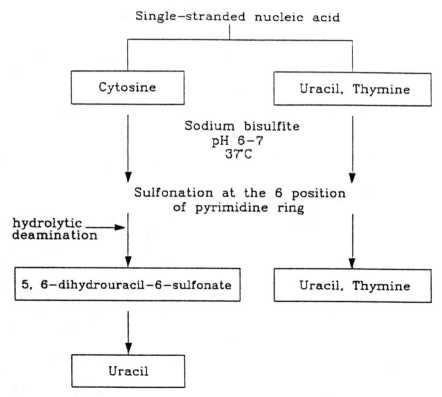

**FIGURE 5.5.** Conversion by cytosine into uracil by sodium bisulfite.

(3) Anneal the single-stranded recombinant M13 DNA with the denatured linearized wild-type replicative form of M13 to form a heteroduplex. Thus only the target sequence will be susceptible to sodium bisulphite.

(4) Treat the heteroduplex DNA molecule with sodium bisulphite to convert some of the cytosine residues of the single-stranded target sequence to uracil (Figure 5.5), and dialyze under appropriate conditions to complete the conversion of cytosine residues to uracil.

(5) Transfect *E. coli ung⁻* (lacks uracil-*N*-glycosylase) strain with heteroduplex DNA having mutations in the single-stranded target DNA. When recombinant M13 DNA containing uracil is introduced into an *E. coli ung⁻* strain, the uracil residues will not be removed from the DNA.

(6) Pick up the plaques and grow them on normal bacterial cells.

(7) Pick up individual plaques, prepare single-stranded template DNA, and confirm the mutation by sequencing.

## 5.6 MUTAGENESIS USING DEGENERATE OLIGONUCLEOTIDES

In this method, a number of point mutations can be generated within a specific region of the target sequence, with degenerate oligonucleotides (a mixture of related oligonucleotides) produced by including low percentage of the three non-wild-type nucleotides during oligonucleotide synthesis (Hill et al., 1987).

## 5.7 PROTEIN ENGINEERING

### 5.7.1 Introduction

The amino acid sequence of a protein can be altered by changing the nucleotide sequence of the gene encoding it using protein engineering techniques. This involves mapping amino acids in the specific site on the protein molecule to improve the protein. This will change the functional properties of the protein. Inherent structural limitations in some food proteins are responsible for the shortcomings in requisite physicochemical characteristics. Recombinant DNA technology can be used systematically to change their functional characteristics (Jimenez-Flores and Richardson, 1987). Various techniques are available to delete, substitute or reorganize the primary amino-acid sequence of a protein by altering its DNA coding sequence.

Protein engineering can change a microbial version of the enzyme so that it retains its primary, desired activity but loses secondary activities. Limited applicability of some enzymes, due to stability to heat, pH, or shear forces; inadequate kinetic parameters, etc., can be improved through protein engineering (Wetzel, 1986). Chemical synthesis is too costly for proteins containing more than 20 to 30 amino acids. For enzymes, therefore, it is easier to clone and express synthetic genes in bacteria, or possibly to use a combination of the two techniques. Potential applications include engineering chymosin in microbial rennin to retain its primary activity of cleaving a specific peptide bond in casein without generating peptides that cause off-flavors in cheese. Splicing the calf chymosin gene into microorganisms to produce low-cost rennin is another approach (Morris, 1986).

In protein engineering, the gene of interest is first isolated, cloned, and characterized. Then a number of different strategies are used to introduce desired changes. The enzyme sequence can be coded by a single gene, and each amino acid can be coded by a three-nucleotide segment of DNA. If one nucleotide within the sequence is changed, this results in a different amino acid being incorporated, and therefore a modified protein being produced. Active sites can be identified by antibody probing of the natural molecules, NMR, X-ray crystallography, and molecular modelling to characterize the three-dimensional configuration of these sites. Computer graphics, DNA synthesis, and gene splicing can create new and more versatile proteins and enzymes. Computerized molecular modelling can provide a three-dimensional image of structures, which allows their manipulation in real time and visualization of molecule parts (Anon, 1986b).

For example, a West African fruit *Thaumatococcus daniellii Benth* contains thaumatin protein, which has 100,000 times the sweetness of sucrose. The gene for thaumatin has been cloned and expressed, and its three-dimensional structure has been studied (Wasserman et al., 1988). The protein engineering of the cloned gene may become a valuable tool for studying the mechanism of sweetness perception.

Casein cDNAs (DNAs complementary to the mRNA) derived from various species have been cloned into *E. coli*. Bovine cDNAs coding for $d_{s1}$-casein and $x$-casein have been cloned, and resultant plasmids were used to transform *E. coli*. This opens the way for the systematic structure-function studies of the caseins. After incorporating an appropriate expression vector of cDNA into a host cell such as *E. coli*, the primary sequences of the caseins will be produced by microorganisms. It may be possible to inject the structural gene along with its controlling elements into the bovine embryo and have the gene stably integrated into the bovine genome for expression in the adult under appropriate circumstances. Once the gene is stably integrated into the genomic DNA, it will be transmitted to the progeny via the germ cells (Jimenez-Flores and Richardson, 1987). Similarly, additional bonds with enhanced chymosin sensitivity to accelerate the rate of textural development in cheese may be inserted into the caseins.

## 5.7.2 Protein Synthesis

Protein engineering techniques are also being used to engineer improved properties into the proteinaceous constituents of food sources. This generally requires the gene for the engineered protein to be transferred into the genome of the plant or animal that is utilized as a food source. The new gene should support the expression in a way comparable to the wild-type protein, and it may be necessary to remove or otherwise inactivate the wild-type gene of the protein (Wetzel, 1986). For example, the gene for a poorly-behaved lysine-enriched protein may improve through random mutagenesis.

Protein synthesis is related to the cell genetic material because the primary sequence is described by the transcription of the genetic material composition. As shown in Figure 2.1 (Chapter 2), the DNA is first transcribed into mRNA, which is then translated into a protein by ribosomes, involving the activation of amino acids by an activating enzyme. There are 20 amino acids used for the synthesis of proteins, of which 19 (except histidine) are derived from a few metabolites (Evans and Sharp, 1986).

In the transcription process, RNA polymerase binds to a specific site on the gene (promoter) and synthesizes messenger RNA (mRNA) complementary to a strand of the DNA. Since the rate of chain elongation is dependent on the temperature, transcription is regulated at the RNA polymerase binding site, progressing from the promoter into the open-reading frame of gene and termination. In *E. coli*, RNA polymerase occurs in a single form consisting of an enzyme of four polypeptides providing mRNA synthesis, and two other polypeptides responsible for initiating and terminating transcription at the correct sites (Brown et al., 1987). Transcription can also be controlled by placing a regulatory protein between the promoter and structural genes. Termed attenuation involves the premature termination of transcription between the operator and structural genes providing a truncated mRNA. This offers fine control over protein synthesis.

Transcription and translation are generally coupled in prokaryotes. Therefore, translation is more important in eukaryotes (Brown et al., 1987).

Protein secretion across the cytoplasmic membrane of bacteria or the membrane of the endoplasmic reticulum of eukaryotes is an important aspect of this process. Secreted proteins are synthesized with an amino-terminal peptide extension of $20-30$ amino acids, the signal or leader peptide (Brown et al., 1987).

## 5.7.3 Mutagenesis

Computer graphics analysis and site-directed mutagenesis approaches rely on the knowledge of protein's three-dimensional structure and primary sequence. A model for the relation between protein structure and func-

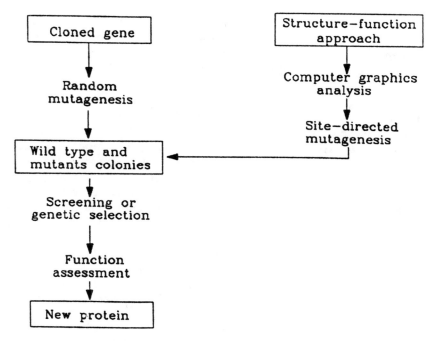

**FIGURE 5.6.** Protein engineering techniques.

tionality is needed for proper modification (Figure 5.6). Based on the sensitivity of the amino acid sequence to oxidation or hydrolysis, a desired protein structure model can be developed for the required functionality. Then, computer graphics analysis is used to design amino acid replacements. Deletions or insertions of the sequence are carried out by altering the DNA sequence of the cloned gene by site-directed mutagenesis.

A structure-function hypothesis may connect an amino acid sequence to a particular function or property. When this approach fails to predict the changes in properties due to the amino acid replacement, the nucleotide sequence for the region of interest is applied in several ways for region-specific mutagenesis (random mutagenesis of a short stretch of oligonucleotides) or site-saturation (replacement of an amino acid with all 19 possible alternatives) (Wetzel, 1986). A structure-function based approach can provide starting material for random mutagenesis for further modification. Similarly, materials of the random mutagenesis approach also serve as starting points for the directed approach.

The site-directed mutagenic approach includes mechanisms of protein folding, protein−protein interaction, nucleated polymerization, and cooperative protein−ligand interactions. This modifies the genetic sequence so a gene makes a different protein. The altered DNA is then cloned and expressed in a fermentation process to produce modified protein. Computer graphics designs new proteins by predicting how specific changes in amino acid sequence can alter three-dimensional structures (Wasserman et al., 1988).

Substitution, insertion or deletion of single or small numbers of bases can be achieved by oligonucleotide or oligodeoxynucleotide site-directed mutagenesis (described in Section 5.3). For more changes, whole sequences are deleted from proteins by specific digestion with restriction endonucleases, followed by the joining of the ends. Specific restriction sites should be inserted at the correct positions using oligonucleotide-directed mutagenesis.

According to Jimenez-Flores and Richardson (1987), a protein can be restructured by (1) relevant gene cloning into an appropriate vector that is capable of autonomous replication (e.g., plasmids), (2) determining cloned gene expression and inserting DNA sequence, (3) making the mutagenized site available in a single-stranded form, (4) using the appropriate oligodeoxynucleotide, containing base mismatches for altering the clone, to sequence for specifically annealing to the single-stranded site, (5) identifying and screening mutant colonies, and isolating and characterizing the mutant DNA.

## 5.7.4 Computer-Aided Protein Engineering

The new protein is modeled with the help of computer graphics, which identify possible sites for amino acid exchange. The local changes are applied to the three-dimensional structure. The loop orientations are predicted without global energy minimization. Then, the three-dimensional structure of the modified protein is predicted by force field calculations including energy minimization and protein dynamics (Shomburg, 1990). The new protein is evaluated by applying the knowledge of structure-stability and structure-function relationships. The new protein is then synthesized or remodeled. For this purpose, a number of data banks are helpful, including the Cambridge Crystallographic, the Brookhaven protein sequence, and others.

Local geometry of the modified protein should be estimated accurately prior to overall energy optimization and protein dynamics calculation. During modification of torsion angles, accurate and detailed information is needed regarding the intramolecular potential of intermolecular interactions (Shomburg, 1990). Finally, energy minimization of the modified protein dynamics calculations is required to remove any local close contacts and to localize the new energy minimum.

# CHAPTER 6

# *Enzyme Engineering and Immobilization Techniques for Enzymes and Cells*

## 6.1 **INTRODUCTION**

*Enzymes* are protein catalysts produced by living systems. They are generally composed of about 200 to 1000 amino acid residues covalently linked in a sequence (Anon, 1985b). It has been estimated that 25,000 different enzymes exist. Of these, some 2100 have been officially recognized by the International Union of Biochemistry (Neidleman, 1986). The system devised by the International Commission of Enzymes divides enzymes into six major classes based on the type of reactions they catalyze: hydrolases, transferases, oxidoreductases, isomerases, lyases, and ligases. Each class is further divided into subclasses based on the reaction involved, the need for coenzymes and cofactors, the nature of the isomerization, and the type of bonds involved.

Although over 2000 enzymes have been isolated from microorganisms, plants, and animals, fewer than 20 are used on a large scale (Smith, 1985). The majority of the enzymes are hydrolases such as amylases, cellulases, pectinases, and proteases, which are used in the baking, dairy, brewing, and fruit juice industries. The food processing industry is currently the largest consumer of industrial enzymes, making up about 40% of the market (Newell and Gordon, 1986). The use of enzymes in the food industry involves the production of food components such as flavors and fragrances, and the control of color, texture, appearance and nutritive value (Table 6.1). Their use for the production of new products could increase dramatically over the next decade.

Enzymes can perform the following tasks in food processing (Boyce, 1986):

(1) Hydrolysis causes viscosity reduction.
(2) Degrading the pectin and cellulose causes an increase in the yield of extracted apple juice.
(3) Separation: Renneting enzymes separate milk protein and fat from the water and milk sugar in cheese making.
(4) Functionality change: Soy protein whipping agents can be prepared

**133**

TABLE 6.1.  *Traditional enzyme-based food processes and products.*

| Food | Enzyme System | Enzyme Function |
|------|---------------|-----------------|
| Bread | Yeast, malt | 1, 2, 3 |
| Beer | Yeast, malt | 1, 2, 4, 5, 6 |
| Cheese | Calf stomach extract | 7, 8, 9 |
| Distilled beverages | Yeast, malt, cultures | 1, 4 to 6, 8, 9 |
| Meats | Papaya, pineapple, cultures | 8, 9, 10 |
| Pickles/sauerkraut | Cultures | 8, 9, 10 |
| Soy sauce | Cultures | 5, 8, 9 |
| Wine | Yeast | 6, 8 |
| Vinegar | Cultures | 6, 8 |

1) fermentation, 2) carbon dioxide production, 3) gluten modification, 4) starch liquefaction, 5) protein hydrolysis, 6) ethanol production, 7) coagulation, 8) flavor development, 9) acid production, and 10) texture modification.

through controlled hydrolysis with proteases. Enzymes keep the bread texture soft for a longer time when incorporated into the dough.

(5) Flavor modifier: Lipases added in the process cause peculiar flavor in certain Italian cheeses.

(6) Some enzymes can be used to synthesize compounds.

(7) Cleaning improvements: Amylases and protease enzymes in detergents improve cleaning and stain removal.

## 6.2 ENZYMES IN FOOD PROCESSING

### 6.2.1 Proteases

These are enzymes that hydrolyze peptide bonds in protein. Amino acids can be joined together to form peptide chains. When the peptide chain becomes longer than 10 amino acids, the structure is termed a polypeptide. The susceptibility of a protein to protease attack is related to its structure. Proteases split peptide bonds with water. Endoprotease enzymes cleave interior peptide bonds, while exoproteases attack peptide bonds on the interior of the protein chain.

The degree of specificity for a protease is inversely related to the number of cleavages it can make on a peptide chain. To select proteases for a particular application, one should consider their relative degrees of specificity as well as the desired degree of hydrolysis and desired end products. The degree of hydrolysis is the extent to which a protease acts on a protein.

Some of the proteases used in food processing are (Boyce, 1986):

• rennin (chymosin): an endoprotease used to clot milk in cheese making

- pepsin: an endoprotease that will hydrolyze a broad range of synthetic peptides
- fungal acid proteases: hydrolyze a wide range of peptide bonds
- trypsin: cleaves bonds next to arginine and lysine residues
- papain: hydrolyzes a wide range of peptide bonds
- bromelain/ficin: hydrolyzes a wide range of peptide bonds similar to papain

Proteases control viscosity, elasticity, cohesion, emulsification, foam stability and whipability; develop flavor; modify texture; maintain nutritional quality; and increase solubility, digestibility, and extractability. These are used for meat flavor development and tenderization; for cracker and cookie gluten modification; in the brewing industry for nitrogen concentration, malt supplementation, and chill-proofing; and for hydrolysis of protein gels to lower viscosity for concentration or filtration.

## 6.2.2 Amylases

Amylases are enzymes that catalyze the hydrolysis of starch. Common amylases are:

- $\alpha$-Amylases: random cleavage of interior bonds of starch polymer to yield dextrins; these have a pH optimum of 6 to 7.
- $\beta$-Amylases cleave maltose units from the ends of starch chains, and bypass 1-6 bonds.
- Fungal amylases are $\alpha$-amylases that hydrolyze starches, yielding mostly maltose and some oligomers.
- Amyloglucosidases or glucoamylases catalyze the hydrolysis of 1-4 linkages in starch; these are used in the corn syrup industry to break down dextrins in the production of glucose syrups.
- Pullulanases hydrolyze the 1-6 bonds in amylopectin molecules.

## 6.2.3 Cellulases

These are enzymes that hydrolyze fibers (cellulose). Cellulose is a linear glucose polymer coupled by $\beta(1-4)$ bonds. Cellulose fibers swell after absorbing water. The number of bonds available for enzyme action depends on the degree of swelling of the cellulose. The degree of swelling can be increased by mechanical treatments such as milling and steam treatment. Common cellulases include the following:

- endocellulases: capable of hydrolyzing the $\beta(1-4)$ bonds randomly along the cellulose chain, yielding oligosaccharides

- exocellulases: cleave off glucose molecules from one end of the cellulose strand
- exo-cellobiohydrolases: hydrolyze $\beta$(1-4) bonds in cellulose to release cellobiose from the non-reducing ends of the chains
- cellobiases: hydrolyze the $\beta$(1-4) bonds in cellobiose, giving two molecules of glucose

Hemicelluloses are hetero-polymers and may have up to 5 or 6 different sugar components. Enzymes from a wide variety of sources hydrolyze hemicelluloses.

- endohemicellulases: random cleavage of interior bonds
- exohemicellulases: systematic hydrolysis from the non-reducing end of the chain

Pectins are nearly unique among common carbohydrates, in that the major subunit in pectin is galacturonic acid.

- endogalacturonases: random cleavage of interior glucoside bonds
- exogalacturonases: systematic cleavage of disaccharides from non-reducing end
- pectinestarases: hydrolysis of methylesters of galacturonic acid
- transeliminases: nonhydrolytic cleavage of glycoside bond to form unsaturated sugars

$\beta$-glucans are similar to cellulose in that they are linear polymers of glucose.

- endo-$\beta$-glucanases: random cleavage of interior 1,3 bonds in laminarin, and random cleavage of interior 1,4 bonds in Lichenan

### 6.2.4 Isomerases

These enzymes catalyze isomer conversion reactions. Glucose isomerases catalyze the conversion of glucose to fructose.

### 6.2.5 Lipases/Phospholipases

These enzymes hydrolyze fats and oils by attacking the ester bonds. An ester is the combination of a carboxylic acid and an alcohol. Lipases act on triglycerides, while phospholipases act on phospholipids. Sometimes lipases also hydrolyze specific types of fatty acids at the C-1 and/or C-3 position on the triglyceride. Animal lipases are added to milk with the renneting enzyme to ensure the development of Romano or Parmesan flavors (Boyce, 1986).

Pancreatic phospholipase A2 can be used to convert lecithin into lysolecithin (Pardun, 1972; Van Dam, 1978). This is an emulsifier for making mayonnaise or for baking bread.

## 6.2.6 Redox Enzymes

These catalyze chemical reductions and oxidations, and are involved in the breakdown or synthesis of many biochemicals. Many redox enzymes require cofactors. Cofactors are substrates or cosubstrates involved in reactions.

In food processing, these enzymes prevent undesirable Maillard browning reactions, and act as oxygen scavengers, to prevent off-flavors in juices.

Redox enzymes are classified as follows.

- Glucose oxidases oxidize glucose to gluconic acid while reducing molecular oxygen to hydrogen peroxide.
- Catalases oxidize one molecule of $H_2O_2$ to molecular oxygen while reducing another molecule of $H_2O_2$ to water.
- Lipoxidases oxidize polyunsaturated fatty acids to their corresponding hydroperoxide forms.

## 6.2.7 Other Enzymes

Pectins are used in jellies, ice cream, salad dressings, frozen pies, and films and coatings. These are tailor-made for slow or rapid gelling, emulsifying, and confectionery.

Pectinases are used in grape juice and wine processing to increase yield, color, and stability; to reduce pressing; and to provide better clarity.

Microbial enzymes are also used in dairies, pickling, juice clarification, and production of sweeteners. They may be cheaper, and they do not remain in food after processing (Kosaric, 1984). Two very potent flavor groups are aldehydes and esters. Aldehydes can be produced from alcohols by the action of alcohol dehydrogenase, and esters can be synthesized in organic media using lipases (West, 1988).

Lactase ($\beta$-galactosidase) is used to hydrolyze lactose in ice cream to prevent crystallization and provide sweetness. Enzymic hydrolysis of the lactose into glucose and galactose using an immobilized lactase with simultaneous protein recovery was reported by Brown et al. (1987). The enzyme immunoassay (EIA) technique allows detection of salmonellae directly from pre-enrichment, resulting in the reduction of total samples requiring analysis, and a substantial savings in time (Swaminathan and Minnich, 1985).

Oxido-reductases are enzymes that catalyze oxidation and reduction reac-

Enzyme origin refers to whether it is indigenous in the food stuff or exogenous to the food stuff either through a conscious addition of an enzyme or enzyme-producing organism or through accidental contamination, particularly by microorganisms (Neidleman, 1986).

### 6.3.6 Process Development

At the optimum conditions (pH, temperature, substrate concentration), the enzyme works at its maximum rate. This may or may not be desired for a particular process, and so the desired rate must be taken into consideration when setting process conditions.

Heat, pH change or inhibitor addition are used to inactivate the enzyme at the end of the reaction. Thermally labile enzymes can be inactivated by heat. To inactivate thermostable enzymes, pH can be adjusted beyond activity range. This may raise the salt content to an unacceptable high level in foods. Enzyme inhibitors are used when the heating and pH adjustment methods are inappropriate. However, these have limited applications in foods because many inhibitors are toxic.

## 6.4 ENZYME IMMOBILIZATION

### 6.4.1 Introduction

Immobilization is the conversion of enzymes from a water-soluble, mobile state to a water-insoluble, immobile state. It prevents enzyme diffusion in the reaction mixtures and facilitates their recovery from the product stream by solid/liquid separation techniques. The advantages of immobilization are (1) multiple and repetitive use of a single batch of enzymes; (2) creation of a buffer by the support against changes in pH, temperature and ionic strength in the bulk solvent, as well as protection from shear forces; (3) no contamination of processed solution with the enzyme; and (4) analytical considerations, especially with respect to long half-life for activity and predictable decay rates. Generally, an enzyme is attached to a solid support material so that substrate can be continually converted to product. Thus, enzymes can be recycled and used many times. The goals are to increase the enzyme's stability, to increase the ability to recycle the enzyme, and to separate the enzyme easily from the product (Shoemaker, 1986).

Chibata (1978) reviewed enzyme immobilization techniques; while Johnson (1979) provided information on various techniques for immobilizing enzymes, taken from the U.S. patents issued between 1974 and 1979. Techniques such as adsorption; ionic bonding; entrapment; and covalent bonding to inorganic supports, carbohydrates, proteins, cellular materials, and synthetic polymers were discussed in detail.

## 6.4.2 **Immobilization Techniques**

Immobilization techniques include adsorption, covalent attachment, cross-linking, entrapment, and encapsulation. Many immobilization requirements could be met by the use of a porous inorganic support that combines high strength with a structure containing pores of appropriate dimensions. The Biofix range (English China Clays International) consisted of four support materials, two for cell support and two for immobilization of enzymes (Adams et al., 1988). All four products were derived from the clay mineral kaolinite. Supports exist in variations such as natural or synthetic, organic or inorganic with differences in shape, size, density, and porosity. They are used in various forms as sheets, tubes, fibers, cylinders and spheres to suit the need for a specific reactor design (Cheetham, 1985). These procedures are used to manufacture immobilized glucose isomerase, and are used in batch or continuous reactors for the production of high fructose syrups. Other immobilized enzymes include glucoamylase, lactase, and aminoacylase. Indirect applications of enzyme immobilization have been in the development of enzyme probes and enzyme-linked immunosorbent assays (Brown et al., 1987).

Figure 6.1 shows the schematic representation of different methods (Smith, 1985; Poulsen, 1984; Powell, 1984; Cheryan and Mehaia, 1986).

## 6.4.2.1 Adsorption

This method is based on the physical adsorption of enzymes onto the surfaces of solid matrices, and is accomplished by allowing an aqueous solution of enzyme to contact the carrier by several means. The adsorption success is dependent on the pH, solvent type, ionic strength, quality of enzyme, and adsorbence time and temperature (Kennedy, 1985). Some commonly used adsorbents include alumina for acylase and amylase, cellulose for cellulase, clays for catalase, glass for urease, hydroxylapatite for NAD pyrophosphorylase, carbon, and various siliceous materials. An enzyme solution is added to the support, mixed, and surplus enzymes are then removed by washing. The process can be carried out by contacting enzyme and support in an agitated reactor or by passing the enzyme through a packed bed, column, or membrane formed from the support material. Adsorption on the support can be improved by the attachment of cofactors such as pyridoxal phosphate or by the use of hydrophilic side chains (Powell, 1984). Enzymes can be attached to ion-exchange resins by electrostatic attractions.

This method of immobilization can be very weak, and the enzyme may be easily desorbed and lost. Important factors in adsorption are the surface area to volume ratio, the particle size, and the ratio of hydrophilic to hydrophobic groups.

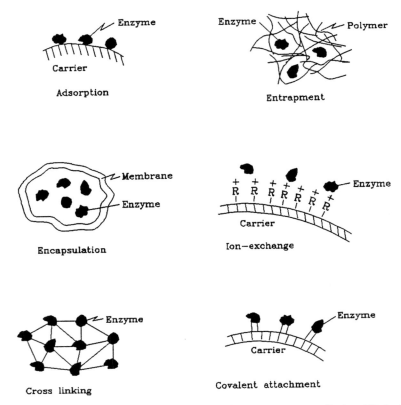

**FIGURE 6.1.** Typical methods for immobilizing enzymes or cells (modified after Weetal, 1974).

Ichijo et al. (1985) used adsorption on poly(vinyl alcohol) superfine filaments to immobilize invertase, which was used to convert sucrose to fructose. Immobilization was carried out by circulating invertase solution through a reactor packed with either fiber or knitted fabric. The rate of adsorption on the fibers was higher because the fiber packing induced turbulent flow, which made diffusion of enzymes to the active sites easier. Upon immobilization onto the knitted fibers, the Michaelis-Manten constant, $K_m$, was found to decrease with increasing flow rate due to the reduction in mass transfer resistance in laminar flow.

## 6.4.2.2 Covalent Bonding

This is based on the covalent attachment of enzymes to water-insoluble matrices. Attachment must involve functional groups on the enzyme not responsible for catalytic action, and thus attachment reagents must be selected carefully. The following methods have been developed: (1) covalent attachment in the presence of a competitive inhibitor or substrate; (2) a reversible covalently linked enzyme inhibitor complex; (3) a chemically modified soluble enzyme whose covalent linkage involves new active residues; and (4) a zymogen precursor (Kennedy, 1985). Common support includes porous glass and ceramics for amyloglucosidase, cellulose for amylase, nylon for urease, alumina for glucose oxidase, and polymers for trypsin. This method involves support activation and enzyme attachment. Chemical bonds are formed by amino, hydroxyl, carboxyl, etc., groups. One of the most widely used covalent attachment techniques is the cyanogen bromide (CN.Br) activation of Sephadex and Sepharose. CN.Br-activated Sephadex (G-200) was chosen because it gave an enzyme immobilization with good activity, coupling yield, stability, filterability, and economical cost (Powell, 1984).

The lactase enzyme was immobilized by covalently bonding to a controlled pore silica carrier using the silane glutaraldehyde technique (Shoemaker, 1986). Corning Glass Works immobilized lactase (*A. niger*) on silica beads, while Finnish Co-op Dairies immobilized it on adsorption resin, and Snam Progetti immobilized yeast lactase on cellulose acetate fibers. All of these multi-layer enzyme immobilizations, consisting of the enzyme immobilized within consecutive protein gel layers, were produced by gelation due to a concentration polarization phenomenon (Greco et al., 1980).

Manjon et al. (1985) immobilized narginase, an enzyme that removes the bitter taste in fruit juices, by covalent binding to glycophase-coated controlled pore glass. The controlled pore glass was activated by periodic acid oxidation. The enzymes reacted with the ethylenediamine, $\beta$-mercaptoethylamine or $p$-phenylenediamine. Upon immobilization, the optimum pH was unchanged, the stability to a wider range in temperatures increased, the activation energy decreased, and $K_m$ was reduced.

## 6.4.2.3 Cross-Linking

This is based on covalent bonding between enzyme molecules using bi- or multi-functional reagents, producing network aggregates of enzymes that are completely insoluble in water but do not require the use of water-insoluble carriers. Some of the most commonly used cross-linkers are aliphatic diamines, dimethyl adipimate, and glutaraldehyde. Cross-linking involves adsorption on a cellophane membrane and cross-linking with glutaraldehyde for amylase, catalase, and glucose oxidase. Polymerization with bifunctional agents is the basis.

Chitin was selected as a suitable support for lactase and invertase immobilization with glutaraldehyde. At optimum conditions, immobilization efficiency was 46% and 71%, and catalyst operational half-life was 10 and 26 days for lactase and invertase, respectively (Illanes et al., 1988). Similarly, chymosin (rennet enzyme) was immobilized onto nylon filters of varying mesh using different chemical pathways. The carbodiimide condensation technique gives the best result, both in terms of residual activity and stability. Addition of 0.5 M NaCl increased the half-life (Amourache and Vijayalakshmi, 1986).

Nakanishi et al. (1985) used adsorption and cross-linking to immobilize thermolysin on Amberlite XAD-7. Thermolysin was used in the continuous production of a precursor of aspartame. The enzyme was first shaken with the Amberlite XAD-7. Amberlite XAD-7 was coated with polyacrylic ester and had hydrophilic sites. Once the enzyme was adsorbed, the particles were mixed with glutaraldehyde. The stability was found to be similar to that of the free enzyme while activity was 50%. The reaction was tested in aqueous buffer, a biphasic system, and in organic solvent. Organic solvent was found to be best for continuous production because the need for an extraction step was eliminated.

Similarly, $\beta$-galactosidase was produced by means of solid state fermentation with a strain of *Scopulariopsis* sp. The crude enzyme extract from solid culture medium was immobilized on DEAE-cellulose and phenol formaldehyde resin (Duolite), respectively. The enzyme immobilized on duolite was superior to the others with respect to its practical application. After one passage of whey through the enzyme packed column (2 × 11.5 cm), this enzyme reduced the lactose in the whey by 50% (Park and Pastore, 1988).

Illanes et al. (1988) immobilized lactose and invertase on cross-linked chitin. Immobilization of invertase has been studied over a wide variety of supports such as activated clay, bentonite, cellulose fibers, ion exchangers, krill chitin, polyacrylamide gel, and porous glass. Chitin flakes were obtained from waste shrimp shells. Immobilization was carried out by using four different methods. Three of them involved glutaraldehyde as a cross-linking agent. After immobilization, the solid catalysts were washed with distilled water, 0.1 M acetate buffer pH 4.5 or 4.0, and 2 M NaCl to eliminate any unbound protein.

Park and Pastore (1988) summarized recent progress in the immobilization of lactase and its application. Friend and Shahani (1982) immobilized *Aspergillus oryzae* lactase on regenerable affinity chromatography support (RACS) attached to Sepharose. RACS was placed in twice the volumes of dioxane and dicyclohexyl-carbodiimide, and 0.1 M hydroxysuccinamide. After 90 min of reaction at room temperature, the activated resin was washed thoroughly with dioxane, absolute methanol, and dioxane. The gel was placed in a 0.067 M phosphate buffer (pH 6.5) containing 40 mg lactase/mL gel, and the reaction proceeded overnight at 4°C. After filtering the resin, it was placed in fresh buffer containing 0.2 M glycine. The resin was washed after 2 h with 10 volumes of buffer, and stored at 4°C.

Nakanishi et al. (1983) immobilized β-galactosidase from *Bacillus circulans* on Duolite ES-762, in the presence of glutaraldehyde. The enzyme immobilized on the Duolite showed a significantly higher activity than those on other supports, such as Dowex MWA-1 and sintered alumina. Leonil et al. (1984) immobilized *E. coli* β-galactosidase on photoactivable chitosan. The chitosan was dried after successively washing with 1 M ammonia and water. The suspension of 1 g chitosan in 20 mL dimethyltormamide mixed with 0.8 mL triethylamine and 2 g 4-azido-3,5-dichloro-2,6-difluoro-pyridine. This reaction proceeded at 80°C for 72 h. The resin was obtained after filtration and successive washing with methanol and methylene chloride. This was then dried under vacuum. 100 mg photoactivable chitosan was washed with 0.1 M sodium phosphate buffer (pH 7.3) and then mixed with 1 mg β-galactosidase in 2 mL 0.01 M phosphate buffer (pH 7.3).

Kaul et al. (1984) immobilized *E. coli* lactase in 5 mL poultry egg white using 2% glutaraldehyde. 5 mL egg white was mixed with the 5 mg enzyme in 0.5 mL water, and treated with glutaraldehyde to a 2% final concentration. The mixture was stirred for 2.5 h at 25°C. It was washed with water, and stored at 4°C. Cowan et al. (1984) immobilized Thermus strain 4-1 A lactase on 1.5 g of controlled pore glass, which was silanated with λ-aminopropyl-triethoxysilane and then treated with 10% glutaraldehyde. 50 mL β-galactosidase solution (2.86 mg/mL in 0.04 Tris-HCl, pH 9) was added, and the mixture was incubated at 40°C while stirring for 72 h. This was washed at the end.

## 6.4.2.4 Entrapment

Enzymes are entrapped in gels such as silica gel, starch, collagen, alginates, *x*-carrageenan, and polyacrylamides. Gel was formed either by temperature changes or by adding a gel-inducing chemical. Polyacrylamides have been used for asparaginase, glucose isomerase, peroxidase, etc. The technique involves the occlusion of enzyme within the lattice structure of a polymer matrix, which is usually effected by mixing enzyme with suitable monomers and initiating a polymerization reaction (Powell, 1984).

Chang et al. (1984) coimmobilized oxidase and catalase on calcium alginate beads to convert glucose and oxygen to gluconic acid. A rotating packed disc reactor was developed to carry out the three-phase biochemical reaction. The discs were packed with the immobilized enzymes and half submerged in glucose solution and were rotated at constant speed.

## 6.4.2.5 Encapsulation

Enzymes are encapsulated in semi-permeable membranes that are impermeable to enzymes but permeable to low molecular weight substrates and products. Collodion membranes have been used for catalase and L-asparaginase, cellulose derivatives for lipase, and nylon for urease and trypsin.

A polymer (such as cellulose nitrate) can separate out around enzyme microdroplets formed by agitating an aqueous dispersion of enzyme in a water-immiscible solvent, e.g., ethyl cellulose, nitrocellulose, polystyrene, polyethylene, polyvinyl acetate, polymethylmethacrylate, and poly-isobutylene (Chibata, 1978). This can be accomplished by interfacial polymerization, in which an aqueous solution of enzyme and monomer can be dispersed in a water-immiscible solvent. The second, hydrophobic monomer is then added, dissolved in the solvent, and polymer then formed by chemical reaction at the interface of the microdroplets (Powell, 1984).

Kumakura and Kaetsu (1984) developed a technique of encapsulation by radiation polymerization, using cellulase and encapsulating it in polymers of 2-hydroxyethylmethacrylate and tetraethyleneglycol diacrylate. Cellulase was used to break down cellobiose into glucose. The enzymes were placed in a solution with buffer and frozen at $-78°C$. The frozen solution was ground and mixed with monomer at $-24°C$. The mixture was stirred until the enzyme was coated with monomer. The particles were then irradiated at $-78°C$ by gamma rays from a $60°C$ source. The temperature was allowed to rise to $25°C$. There was an optimum capsule thickness since leakage occurred when the capsule was thin while diffusion resistance was too high when the capsule was thick. The capsule was porous due to the presence of ice crystals during polymerization.

## 6.4.2.6 Other Examples

$\alpha$-Amylase has been immobilized by adsorption on membranes, collagen, dextran, cationite gels, and other ion exchange carriers; entrapment into membranes, and organic gels; and covalent bonding on solid supports (Reilly, 1980).

Glucoamylase has been immobilized in gels and ultrafiltration reactors and attached to solid matrices by adsorption, ion-exchange and covalent bonding. The enzyme has been entrapped in many gels with poor retention of activity. It has also been immobilized to many inorganic and a few organic adsorbents, with different degrees of success (Reilly, 1980). It was also adsorbed by ion exchange to many carriers such as CM-Sephadex, CM-cellulose, Amberlite, Duolite, Dower, etc. A large number of matrices and binding agents were employed to attach this enzyme by covalent bonding. These include organic and inorganic matrices with different binding agents including glutaral-dehyde, metal chlorides, and nitrous acid.

Similarly, these methods were also used to immobilize $\beta$-amylase, pullulanase, invertase, and $\alpha$-galactosidase. Reilly (1980) provided a review of these methods with their limitations.

### 6.4.3 Technique Selection

The immobilization technique should be mild, cheap, safe, versatile, and suitable for scale-up. It should provide easy control of the type and amount of enzymes, minimize enzyme leakage from the support, and prevent partitioning between substrate, product, and support (Kennedy, 1985). Enzyme purity prior to immobilization affects the stability of the immobilized complex. A crude enzyme preparation can withstand more denaturing and inactivation mechanisms inherent in immobilization techniques. Economic considerations play a large part in determining if immobilization is feasible and beneficial to the particular application. Immobilization is generally not required for reactions that can be performed satisfactorily with inexpensive, crude enzyme preparations, or where large quantities of products are not required.

Although any enzyme can be immobilized, its use, desired end result, economics, and substrate constraints should be considered. Each enzyme has a unique surface chemistry and operational stability. Consequently, no one immobilization method is generally applicable (Adams et al., 1988). Substrate diffusion is critical in immobilized enzymes, since it must penetrate the immobilized matrix to get to the enzyme.

### 6.4.4 Kinetics

The kinetics of immobilized enzymes can be studied by one of the two approaches: (1) the diffusion of substrate to the enzyme surface as the only transport mechanism, and (2) electrostatic forces as well as diffusion.

#### 6.4.4.1 Diffusional Approach

By combining diffusion due to concentration gradient and Michaelis-Menten kinetics:

$$V = k_s (So - S) = V_{max} \cdot S/(K_m + S)$$

Where $V$ = reaction rate, $V_{max}$ = maximum reaction rate, $K_m$ = Michaelis-Manten constant, $k_s$ = mass transfer coefficient, $S$ = concentration of substrate at enzyme surface, and $So$ = concentration of substrate in bulk solution. Damkohler number $(Da) = V_{max}/(k_s \cdot So)$. When $Da \ll 1$, the enzyme kinetics is in the reaction limited regime, and when $Da \gg 1$, it is in the diffusion limited regime.

## 6.4.4.2 Diffusional Resistance and Electrostatic Forces

In this approach the movement of substrate is governed by diffusion and by electrostatic forces. Combining steady state conditions and Michaelis-Menten kinetics:

$$V = V_s \cdot a = V_{max} \cdot S/(K_m + S)$$

Where $V_s$ [5] reaction rate per unit surface area of the immobilized enzyme, and $a$ [5] surface area per unit volume of the immobilized enzyme.

The reaction rate is, therefore, affected by the ionic strength and pH of the substrate solution as well as the temperature.

## 6.5 **MICROBIAL AND ANIMAL CELL IMMOBILIZATION**

It is possible to immobilize almost any cell structure and keep the cell viable. Immobilized cells of *Bacillus coagulans* produced a major proportion of high fructose syrups, and aspartic and malic acids (Smith, 1985).

### 6.5.1 Cell Entrapment

Suspended cells can be immobilized by entrapment of cells within a porous matrix. It includes porous supports such as ceramics and gels, which can be either preformed or formed around the cells. Several reviews of methodology exist (Mosbach, 1983). The basic principle is shown in Figure 6.2.

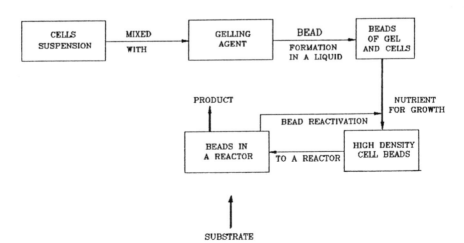

**FIGURE 6.2.** Entrapment of cells—basic principles.

The matrix in which cells are entrapped is, generally, known as gel. The materials used in this process are:

- agar
- agarose
- $x$-carrageenan
- proteins
- alginates
- chitosan
- polyacrylamides
- celluloses
- alginate-gelatin systems

We will look at each of these individually.

## 6.5.1.1 Agar

This is a natural polymeric complex polysaccharide extracted from various species of marine macro-algae. It consists of $\beta$-1,4-D-galactopyranoside and $\alpha$-1,3,3,6-anhydro-$\alpha$-galactopyranoside. For cell immobilization, the agar is diluted (2 to 4% w/v) in a suitable medium depending on the type of cell. This is done at high temperatures much above the settling point of the agar. Generally, a 1:1 ratio of cell suspension to agar solution is used.

Banerjee et al. (1982) immobilized *Saccharomyces anamensis*, containing $\beta$-galactosidase, by entrapping the cells in agar-agar. 1 mL cell suspension in phosphate buffer was placed in 5 mL agar prepared by melting 125 mg of agar in 5 mL of distilled water, and then cooled to 45 to 50°C. After cooling, it was shaken thoroughly. The mixture was cast into bead shape by injecting it into an ice-cold toluene-chloroform (3:1) mixture. These were then washed with phosphate buffer (Park and Pastore, 1988).

Vuillemard and Amiot (1988) immobilized *Serratia marcescens* for the hydrolysis of milk proteins. Culture was maintained at 4°C on tryptic soy agar. After centrifugation, for 15 min at 15,000 g's, about 2 g cells (100 mL broth) were placed in 5 mL saline and mixed to 100 mL of a 2% (w/v) sodium alginate solution. The mixture was extruded through a syringe to a gently stirred 3% (w/v) $CaCl_2$ solution, and hardened in the solution for 30 min. The alginate gel beads (4 mm average dia.) were washed with saline water.

## 6.5.1.2 Agarose

This is separated from agar and purified. Two types of agarose (Sigma type VII and FMC Sea Plague) were suitable for the entrapment of animal cells (Nilsson et al., 1987). The required amount of agarose was mixed with PBS and autoclaved. After sterilization it was stored at room temperature. Before use it was remelted by heating to 70°C and maintained at either 37 or 45°C depending on the type being used. A convenient way to immobilize cells is to prepare the agarose solution at twice the final concentration and to mix it with an equal volume of suspended cells. After mixing the cell suspension and agarose, the mixture was dispersed in the paraffin oil, and when the desired bead size was reached, the dispersion was cooled to at least 10°C below the gelling temperature. Medium was added, and after sedimentation of the beads, the oil phase and most of the washing medium were aspirated. The beads were further washed with medium until they were essentially free from oil (Nilsson et al., 1987). Cells have been entrapped in agarose at concentrations between 0.1 to 20 million cells per mL beads.

### 6.5.1.3 $x$-Carrageenan

This is a natural polymeric complex polysaccharide extracted from species of marine algae, and available in a purified food grade form. It consists of $\beta$-D-galactose sulphate and 3,6 anhydro-galactose. Cells are mixed with it at 37 to 50°C, and at a concentration of 4% (w/v). Beads (3.5 to 5.5 mm dia.) are formed by adding the mixture dropwise into a cold liquid. Gel beads are hardened by dipping in diisocyanate, carbodiimide or isothiocyanate at a concentration of 0.01 to 1 g/mL and 30°C. According to Wang and Hettwer (1982), mixing of 5% (w/v) tricalcium phosphate in the gel provided better growth of yeast. Cell leakage was reduced from the beads by increasing the potassium chloride concentration to 4% (w/v). Mattiasson (1983) listed 18 species covering 11 products that have been immobilized in $x$-carrageenan. Microbial cells immobilized into the carrageenan showed higher activity than those attached to the acrylamide matrix.

Microbial cells were immobilized in $x$-carrageenan by Tosa et al. (1979). $x$-carrageenan was dissolved in physiological saline previously warmed at 70 to 80°C at a concentration of 3.4% (w/v) and the solution was kept at 40°C. Gelation occurred when the solution was cooled below 10°C; contacted with 0.1 M metal salt solution; contacted with 0.5 M ammonium chloride dissolved in 0.5 M phosphate buffer (pH 7.0); contacted with 0.5 M diamine hydrochloride dissolved in 0.5 M phosphate buffer (pH 7); and contacted with water-miscible organic solvents. Cubic-type matrices were produced by using the following procedure: 100 mg enzyme or 16 g (wet) microbial cells were dissolved or suspended in 32 or 16 mL physiological saline at 25 to 50°C, respectively, and 3.4 g carrageenan was dissolved in 68 mL of the physiological saline at 38 to 60°C. The two were mixed, and the mixture was cooled at 10°C for 30 min. To increase the gel strength, the obtained gel was soaked in a cold 0.3 M potassium chloride solution. After this treatment, the resulting stiff gel was formed to a cubic gel of 3 × 3 × 3 mm. The bead-type support matrix was formed as follows: 50 mL of the mixture of carrageenan and enzyme or microbial cells were dropped into a solution containing one of the gel inducing reagents through a special nozzle having an orifice of 1 mm in diameter at a constant speed. Bead-type gels of 3 mm in diameter were obtained.

Using x-carrageenan, the immobilization can be performed under very mild conditions without the use of chemicals, e.g., *Corynebacterium simplex* cells were immobilized with collagen as follows (Chibata, 1978): *C. simplex* cells (20 g dry weight) were suspended in 100 mL of water and the suspension was gently added to a 2% collagen solution. The mixture was adjusted to pH 8.5 with 0.1 N NaOH. To this mixture 3 mL of 50% glutaraldehyde was added, and the mixture was allowed to thicken for 5 to 10 min before being cast as a membrane on a mylar sheet. The membrane was dried at room temperature, hardened by dipping in alkaline glutaraldehyde solution, and the hardened cell-collagen membrane was cut into small pieces.

Ethanol can be continuously and efficiently produced by using yeast cells immobilized with x-carrageenan. Glycerol can be produced using living cells of *S. cerevisiae* immobilized with x-carrageenan by adding sodium sulfite to an ethanol production medium (Chibata, 1978).

## 6.5.1.4  Proteins

Any easily available protein can be used for the cell entrapment. Glutaraldehyde is used as a cross-linking agent (Figure 6.3). Collagen is hydrophilic and swells in the presence of water, and thus can be used for the immobilization of cells. Cells in suspension (0.5 to 3%) are mixed with a collagen

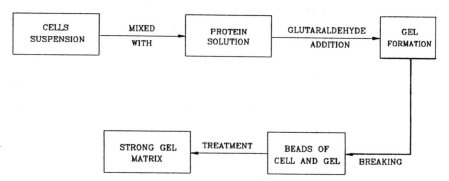

**FIGURE 6.3.** Principle of glutaraldehyde cross-linking of protein gels.

dispersion at pH 6.5. After flocculation, the pH of the mixer is slowly raised to 11.2. The dispersion is casted, dried at room temperature, and tanned by dipping in glutaraldehyde solution. Collagen has been successfully used as a support for the immobilization of various cells for primary and secondary metabolite production.

High cell densities were achieved by immobilizing the cells inside porous microbeads made from a collagen-sponge matrix. The porous microbeads formed a thick slurry, which was fluidized in the bioreactor and operated under continuous culture conditions for extended periods of time (Venkat, 1986).

## 6.5.1.5 Alginates

These are available as water-soluble sodium salts in many grades, and extracted from different species of marine algae. These are available in purified food grades. Cells are mixed with sodium alginate solution and added dropwise to dilute aqueous calcium chloride. Rochefort et al. (1986) found a two-fold increase in the gel strength of beads using 0.1 M aluminum nitrate. Rehg et al. (1986) provided a method for the large-scale production of alginate beads, by using a dual fluid atomizer in which sodium alginate beads were sheared from the tips of hypodermic needles by an air stream into calcium chloride solutions.

Conidia of *Aspergillus niger* TMB 2022 were immobilized in calcium alginate for the production of citric acid (Cayle et al., 1986). One mL of conidia suspension containing approximately $2.32 \times 10^8$ conidia was entrapped into sodium alginate solution to prepare 3% calcium alginate (w/v) gel bead. Immobilized conidia was inoculated into a productive medium containing 14% sucrose, 0.25% $(NH_4)_2 \cdot CO_3$, 0.25% $KH_2 \cdot PO_4$ and 0.025% $Mg \cdot SO_4 \cdot 7H_2O$ with addition of 0.06 mg/L $CuSO_4 \cdot 5H_2O$, 0.25 mg/L $ZnCl_2$, 1.3 mg/L $FeCl_3 \cdot 6H_2O$, pH 3.8, and incubated at 35°C for 13 days by surface culture to produce 61.53 g/L anhydrous citric acid.

Immobilized conidia was prepared under sterile conditions using calcium-alginate as the carrier (Tsay and To, 1987). A total of 10 g of 3% calcium-alginate gel beads (mean diameter of 4 mm) entrapped with conidia were formed. Immobilized conidia beads (10 g) were transferred into a sterile 500 mL Erlenmeyer flask containing 100 mL productive medium. The flask was incubated at 30°C in an incubator for 13 days.

Suspended mammalian cells entrapped in beads of calcium-alginate were encapsulated within a semipermeable membrane by a procedure in which the bead surface was stabilized with polysine and the bead interior was sub-sequently liquified (Nilsson, 1987). These cells can also be entrapped in agarose beads by a two-phase method in which cells suspended in a solution of agarose can be dispersed in paraffin oil. The formed droplets can then be subsequently solidified by cooling.

A *Pseudomonas* sp. was immobilized in alginate beads, for the purpose of producing protease (Anon, 1988). The beads were packed in an upflow column reactor. Similarly, skim milk proteins were hydrolyzed in a bioreactor with immobilized cells. The bacteria used were *Serratia marcescens* or *Myxococcus xanthus* immobilized in calcium alginate beads ( Vuillemard and Amiot, 1988).

Linko (1980) immobilized *Lactobacillus lactis* cells into calcium alginate gel beads for lactic acid production. The strain *L. casei* ssp. *rhamnosis* was reported to be the best producer of lactic acid on a glucose medium. The organism was grown on MRS-medium (deMan et al., 1960): glucose (or lactose) (20 g), bacto-protease-peptone (10 g), bacto-meat extract (10 g), yeast extract (5 g), Na-acetate (5 g), ammonium citrate (2 g), $KH_2PO_4$ (2 g), $MgSO_4 \cdot 7H_2O$ (0.1 g), $MnSO_4 \cdot 4H_2O$ (0.05 g), Tween-80™ (1 mL), and deionized water to make 1 L; pH was 6.5 at 37°C. The centrifuged cells were placed in 6% sodium alginate and extruded into 0.5 M calcium chloride. The beads (2 mm dia.) were hardened at room temperature for 30 min. Changing the ratio of bacterial cell mass to 6% sodium alginate from 1:2 to 1:5 had only an insignificant effect on lactic acid production. The cell concentration in the immobilized mass, at the steady state, was $10^9$ cells/g (Linko, 1985).

On the other hand, Tipayang and Kozaki (1982) immobilized *L. vaccinostercus* into calcium alginate (3%) gel for lactic acid production from 2% xylose containing 1% peptone and 0.5% yeast extract. Lactic acid levels of 0.9 to 1% (w/v) in the broth were achieved in 5 days. Gestrelius (1982) fermented malolactic into lactic acid using calcium alginate gel-entrapped *L. oenos* cells in a continuous column reactor. He suspended one volume of wet cell (15% dry matter) in 5 to 12 volumes of 5% sodium alginate and dripped the cell suspension into 2% calcium chloride. The beads, thus formed, were hardened for 2 h at room temperature, and placed in a sterile filtered grape juice/glycerol medium (reconstituted grape juice with 30 mM malate, 10% ethanol, and 35% glycerol at pH 5.0). This was incubated overnight at 6°C, and stored at −20°C.

Calcium alginate (8%) gel-immobilized *S. cremoris* and *L. helveticus* cells were used for the pre-fermentation of skim milk. Linko (1985) reported that immobilized *L. helveticus* produced less D-acid than free cells, which produced L- and D-acids at a ratio of about 2:1; the immobilized cells exhibited a ratio of about 4:1.

Ramakrishna et al. (1988) continuously produced ethanol with immobilized yeast cells in a packed bed reactor from cassava starch hydrolysate. The principle of ionatropic gelation of polyelectrolytes was used for entrapment of the yeast. The cells were mixed with 2 to 4% sodium alginate solution and extruded through a nozzle into a $CaCl_2$ solution (1 to 2%). The gelled droplets were cured for 3 h and stored at 4°C. The average cell loading was found to be $3 - 3.5 \times 10^8$ per bead.

### 6.5.1.6 Chitosan

Vorlop and Klein (1981) used two types of chitosan and the following polyanions for the tryptophane synthetase activity of *E. coli*: polyphosphates, poly(aldehydo-carbonic acid) and poly(1-hydroxy-1-sulphonate-propen-2). A few grams of chitosan were dissolved in dilute acetic acid at a final pH of 5. The cells were mixed into the chitosan acetate solution, and added dropwise to 1.5% (w/v) sodium polyphosphate at pH 8.5.

### 6.5.1.7 Polyacrylamides

These are synthetic polymers with a wide range of properties. Gels are prepared by reacting a monomer (acrylamide or methylacrylamide) with a bifunctional cross-linking agent ($N$, $N'$-methylene-bis-acrylamide). The mechanical strength of the gel increased in proportion to the square root of the concentration of acrylamide monomer (Chibata and Tosa, 1976). Many synthetic polymeric gel systems were used for the cell entrapment. Copoly-(styrene-maleic acid) showed good gelling properties and aluminum sulphate was a satisfactory co-electrolyte.

### 6.5.1.8 Celluloses

Substituted celluloses are prepared by replacing the hydroxyl groups with various substituents. According to Tsao and Chen (1977) cellulose can be converted to beads by mixing it in an organic solvent (1:10 to 1:6 w/v). The beads can be hardened by treatment with cross-linking reagents.

Omata et al. (1979) entrapped microbial cells into photo-cross-linkable polymers. The polymers were formed during brief illumination of photo-cross-linkable prepolymers such as poly-butadiene, polypropylene glycol and maleic polybutadiene. Microbial cells were added in a water-acetone and prepolymer mixture. After mixing, it was spread on a transparent polyester sheet and irradiated from an ultraviolet (UV) light for 3 min. Polymerization formed an entrapped cell matrix. The gel was then divided into smaller pieces (5 × 5 mm) for further applications.

Sonomoto et al. (1980) entrapped microbial cells into polyurethane matrices. Polyurethane matrices were synthesized from toluene diisocyanate and polyesterdiol. The urethane was melted at 60°C, cooled to 37°C, cells were mixed, spread on a glass plate, and kept at 4°C for 1 h. The gel was divided into pieces for further applications.

Soluble fibrinogen can be enzymatically converted to fibrin by thrombin, and can be used for the entrapment of anchorage-dependent animal cells. The cells were collected by centrifugation and resuspended in a 5 mL medium. After addition of 5 mL fibrinogen solution (2% w/v) and thrombin (100 $\mu$L, 150 NIH U/mL), the mixture was poured into extracted paraffin oil (30 mL) supplemented with detergents (Triton 100™ and SPAN 85™ both at 0.1% w/v). When the reaction was completed, 25 mL medium was added. After sedimentation of the beads, the oil phase and most of the medium was aspirated. The beads were washed with several changes of medium until they were essentially free from the oil (Nilsson et al., 1987).

## 6.5.1.9  Alginate-Gelatin System

Siva Raman et al. (1982) employed a mixed alginate-gelatin system for yeast immobilization. Cells (*S. cerevisiae* and *S. uvarum*) were mixed to a solution of gelatine (20% w/v) and sodium alginate (2% w/v). The beads were formed by dropping into calcium chloride solution. The alginate was then leached out in phosphate buffer and the porous gelatin beads stabilized by cross-linking with glutaraldehyde.

Horitsu et al. (1988) produced organic acids from immobilized cells of a fungi. Acid production rate of citric acid had a half-life time of 105 days, 28 mg of itaconic acid/h/10 g gel having a half-life time of 30 days, and 52 mg of L-lactic acid/h/10 g gel having a half-life time of 17 days, were observed in the respective continuous methods. For citric acid, beet molasses (sucrose, 10%), pH adjusted to 5.8, was used; and for itaconic acid, a mixture of 10% glucose, 0.37% $NH_4NO_3$, 0.2% $MgSO_4 \cdot 7H_2O$, 0.27% $NH_4H_2PO_4$, and 0.1% corn steep liquor, pH 5.0; and for lactic acid, a mixture of 5% glucose, 0.05% $NH_4NO_3$, 0.03% $KH_2PO_4$, 0.025% $MgSO_4 \cdot 7H_2O$, and 2% $CaCO_3$, pH 7.15, was used. The following methods were used for the immobilization of cells:

(1) Immobilization of cells with polyacrylamide: 10 g of washed cells were mixed with 32 mL physiological saline (0.85 M NaCl); 6 g acrylamide monomer; 0.32 g *N,N*'-methylenebisacrylamide; 4 mL, 5%, 3-dimethyl aminopropionitrite solution; 4 mL, 2.5% $K_2S_2O_8$ solution. The mixture was then gelled at 25°C for 30 min, and cut into 4 mm cubes.

(2) Immobilization of cells with calcium alginate: 10 g washed cells were mixed with 30 mL deionized water and 30 mL, 6% sodium alginate. The mixture was extruded as discrete droplets into stirred 2% $CaCl_2$ solution by syringe; and allowed to stand for 30 min.

(3) Immobilization of cells with photo-cross-linkable resin: 10 g washed cells were mixed with 10 mL M/10 Tris-HCl buffer (pH 7); 50 g, 40% ENT-3400 solution and 0.2 g benzoin ethylether. The mixture was

extruded as discrete droplets into paraffin liquid by syringe; and illuminated with chemical lamp at 300 to 400 nm for 5 min.

(4) Immobilization of cells with $x$-carrageenan: 4 g $x$-carrageenan was heated to solution in 50 mL physiological saline water; and cooled to 40°C. 10 g cells in 20 mL suspension was mixed in the solution, and filled up to 100 mL with physiological saline solution. This was extruded as discrete droplets into stirred 2% KCl solution by a 50 mL syringe, and allowed to stand for 2 h.

## 6.5.2 Cell Encapsulation

Cell encapsulation is the containment of cells behind a barrier, typically a membrane. In microencapsulation, a selectively permeable or impermeable membrane is created around droplets or particles to prevent the free diffusion of predetermined substances either in or out of the inner compartment (Tampion and Tampion, 1987).

Isotropic membranes have same size pores on both sides of the membranes, while anisotropic membranes have larger pores on the outer face leading to smaller pores on the other side. Anisotropic or hollow fiber membranes are used for cell immobilization.

## 6.5.3 Cell Covalent Bonding

A wide range of inorganic support materials are available for simple adsorption or immobilization. Some of them are aluminum, brick, ceramics, coke, diatomaceous earths, glass, magnetite, porous rock, silica, zirconia, etc. With activation, these bind cells covalently without leakage into the process stream. These are activated with glutaraldehyde after surface coating (such as gelatin).

Navarro and Durand (1977) used porous silica beads (100 to 200 $\mu$m in dia.) to immobilize the yeast *S. carlsbergensis*. The beads were first reacted with $\gamma$-amino propyltrimethoxysilan and activated with glutaraldehyde. Cells were bonded to supports by mixing the supports in a cell suspension for one hour at pH 5.0. Completely untreated silica beads were shown to induce metabolic changes similar to those treated with $\gamma$-aminopropyltrimethoxysilan.

Many metals form gelatinous precipitates of their hydrous oxides. Out of these, hydroxides of titanium or zirconium are suitable for cell immobilization. Kennedy et al. (1980) immobilized *Acetobacter* species on hydrous titanium-chelated cellulose and hydrous titanium oxides to produce acetic acid. Cellulose powder (CF11, Whatman) was reacted with titanium chloride in acid solution for 2 h, then dried and washed.

Cells can be immobilized by cross-linking with bi- or multi-functional reagents such as glutaraldehyde, toluene diisocyanate and others. The following polymer matrices have been tried for entrapping cells: collagen, gelatin, agar, alginate, carrageenan, cellulose triacetate, epoxy resin, polyacrylamide, photo-cross-linkable resin, polyester, polystyrene, and polyurethane. A major disadvantage of polyacrylamide is the toxicity of the acrylamide monomer, the cross-linking agent, the initiator, and the accelerator.

## 6.5.4 Cell Adsorption

Adsorption involves surface interactions between cells and support materials, and binds cells by ionic and hydrogen bonding. The support materials for cell adsorption should be nontoxic, reusable, stable to heat and pH values, and resistant to microbial degradation. These materials should have high cell retention and loading capacities. The cells may be attached on the surfaces by electrostatic attraction and extracellular polymeric products of the cells. Carrier surface charge is mainly responsible for cell attachment. For example, it is better to choose a positively charged support for *A. niger* immobilization, since the cells are negatively charged.

Huysman et al. (1983) provided the selection of porous and non-porous support materials for use in upflow methane reactors. The porous materials were natural sponge (50% porosity), reticulated polyurethane foam (97% porosity), unreticulated polyurethane (30% porosity), and a polymethane foam coated with polyvinylchloride. The nonporous supports used were activated carbon, argex (a fire-expanded clay), glass beads, sepiolite (a natural clay) and zeolite (a natural alumino-silicate). Sepiolite and reticulated polyurethane foam were most readily and extensively colonized. A number of microorganisms such as *Bacillus* and *Pseudomonas* species can be absorbed to ceramic, glass, kieselguhr, plastic, and wood. Cell adsorption depends on cell age, pH, electrostatic interactions between the cell and support, and the retention capacity of the carrier.

## 6.5.4.1 Inorganic Supports

Bland et al. (1982) immobilized *Zymomonas mobilis* on vermiculite for the production of ethanol in attached film-expanded bed reactor. Maximum production was obtained at a dilution rate of 3.6 with 64% conversion of the substrate. Arcuri et al. (1980) immobilized *Z. mobilis* on polystyrene beads (1 mm dia.) and into glass fiber pads (2 × 2 mm) to produce ethanol. Polystyrene beads provided negligible attachment but the bacteria formed flocs. Immobilization of microbial cells to inorganic supports was affected by pH, ionic strength, surface charge and area, support composition, and support surface properties.

## 6.5.4.2  Organic Supports

Thin shavings of wood are used as supports in the acetic acid production by *Acetobacter*. Chen and Zall (1982) used cellulose acetate as support for *Saccharomyces fragi* in a continuous attached film expanded-bed reactor for the fermentation of whey to alcohol. Many irregularly shaped, fibrous, or mesh systems are available for supports. Such supports adsorb cells and grow them into the interstices. For cell self-immobilization, fibrous nylon net is a convenient support. Linko et al. (1986) used it for the production of ethanol from D-xylose by immobilizing *Pichia stipitis*. Cells of many microorganisms (*A. wentii*, *A. oryzae*, etc.) were immobilized by adsorption to fibrous carboxymethyl, phosphoric acid, diethylaminoethyl and mixed amines (Johnson and Cielgler, 1969). Cellulose was washed with 0.05 M $K_2H \cdot PO_4$-NaOH buffer at pH 5.8. Cells were added to the cellulose, and the mixture was poured into a column for settlement. Unadsorbed cells were washed out with the buffer solution.

## 6.5.4.3  Ion-Exchange Supports

Ion-exchange materials are used as supports for cell immobilization because ionic charge interactions play an important role in the early stages of adsorption. These materials include resins—polystyrene, polyacrylic, polymethacrylic, macroreticular, etc.; polysaccharides—celluloses, etc.; and Sephadex.

The most important parameters influencing cell immobilization to these materials are initial cell concentration, pH, the presence of specific ions, and support properties. *Nitrobacter agilis* cells were immobilized to three types of sulfonated resins (Dowex 50 W-X) with ion exchange capacities of 0.007, 0.074 and 6.3 meq/g. *Azotobacter agile* was immobilized to the anion exchanger Dowex resin for succinate oxidation, producing a 25% higher glucose oxidase enzyme activity compared to free cell system. *Pseudomonas aeroginosa* cells were also immobilized to anion exchanger Dowex 1 (Hattori et al., 1972).

Cell growth on glass-coated microcarriers is similar to monolayers grown on glass. On dextran beads, cells grow in single layers and attachment appears to involve the entire cell edge. Cells grown on glass-coated beads are attached via long, slender filopodia, allowing them to be gently detached in a short time at low trypsin concentrations with better recovery of viable cells, typically 95 to 99%. This also permits multiple attachments and layers of cells with certain cell types.

Similarly, the membrane-type matrix was prepared as follows: 50 mL of the mixture was spread on a plate (1 × 250 × 200 mm), and soaked in a cold 0.3M potassium chloride solution. To estimate enzyme activity, the membrane was cut to 1 × 10 × 10 mm pieces.

The carrier-binding method is based on direct binding of cells to water-insoluble carriers by physical adsorption, ionic bonds, or covalent bonds. As carriers, water-insoluble polysaccharides (cellulose, dextran, and agarose derivatives), proteins (gelatin and albumin), synthetic polymers (ion-exchange resins and polyvinyl chloride), and inorganic materials (brick, sand, and porous glass) are used. This process of immobilization for cells is not widely used because of the toxicity of the coupling agents involved.

## 6.6 **PLANT CELL IMMOBILIZATION**

Physically restraining the cells on a fixed support is an intermediate state between homogeneous suspension culture and the highly structured tissue matrix of the whole plant. The benefits of plant cell immobilization result from (1) the retention of cells in the reactor, (2) high concentration of cellular activity, (3) microenvironmental control, (4) rapid removal of biocatalyst from the product, and (5) uncoupling of growth and production phases (Rosevear and Lambe, 1985).

Plant cells in culture have been used for many bio-conversions including hydroxylations, methylations, acetylations and glycosylations. However, immobilized plant cells have been used to a limited extent for these purposes.

Plant cells for immobilization can be grown in a culture vessel, and removed by filtration. These are then aseptically immobilized by entrapment or surface binding.

Immobilized cells can be formed in either particles/beads or sheets. Beads are very versatile, as they can be packed into columns or beds of different dimensions. Immobilized plant cells are mechanically strong compared with free cells, and hence, gas supply is not limited by the problems of cell damage nor is gas velocity limited by the problem of cell washout.

Sterility of all equipment, nutrients and gases is required to keep the plant cell reactor free of bacterial or fungal contamination.

### 6.6.1 Entrapment

The entrapment within gel-forming polysaccharides appears to be most suitable, but other polymers were also used such as synthetic (polyacrylamide, epoxy resin, polyurethane), carbohydrates (agar, agarose, alginate, cellulose, ϰ-carrageenan), and proteins (collagen, gelatin, fibrin), etc. Many species of plant cells were entrapped in these matrices (Brodelius,

1985), e.g., *Catharanthus roseus* cells were entrapped into alginate, agarose, carrageenan and polyacrylamide matrices.

The problem of low shear tolerance of plant cells was eliminated by the entrapment of the cells in a protective polymeric matrix (Brodelius, 1988). Most experiments on immobilized plant cells have been performed in shaker flasks, but a few researchers used them in packed bed, bubble column and fluidized bed reactors. The immobilization of plant cells did not change the cell metabolism irreversibly. Since plant cells in suspension culture are relatively large (up to 100 $\mu$m in dia.), they are immobilized by entrapment within various gels or membrane reactors. However, covalent attachment and adsorption have also been tried.

The plant cells were immobilized by entrapment in the following materials:

(1) Agar and agarose: These formed gels upon cooling a heated aqueous solution. A gelling temperature of about 30°C was adequate (Brodelius, 1988).

(2) Alginate: Cells of *C. roseus* entrapped in agarose, agar and carrageenan showed productivity similar to that of freely suspended cells under growth-limiting conditions (Brodelius and Nilsson, 1980). On the other hand, alginate-entrapped cells showed an increased synthesis. Similarly, alginate-entrapped cells of *Morinda citrifolia*, after 21 days of incubation, produced about 10 times as much anthraquinone from sucrose as did freely suspended cells under growth-limiting conditions (Brodelius et al., 1980). Since immobilized cells are closer, metabolically, to cells in the whole plant than are the cells in suspension, they are more productive.

*Digitalis lanata* cells were immobilized in alginate beads to convert $\beta$-methyl digitoxin into $\beta$-methyl digoxin (Brodelius, 1988). In a batch operation, the solution was changed every second day over a period of 59 days, and resulted in a cell productivity higher than that of a free cell system. Thus, the long-term production of extracellular secondary metabolites is feasible using viable, immobilized plant cells.

(3) $\varkappa$-Carrageenan: After heating the $\varkappa$-carrageenan, beads are formed by dripping the cell-carrageenan suspension into a medium containing 0.3 M potassium ions (Brodelius, 1988).

(4) Membrane reactors: Hollow fiber reactors were employed. The plant cells were supplied on the shell side of the reactor, and oxygenated media were supplied at a high flow rate through the fiber lumen. Hollow fiber modules were also used to retain plant cells by putting them on the shell side of the reactor while culture fluid was recirculated through the fiber. Polyphenols were produced in such a reactor using both soybean and carrot (Prenosil and Pedoisen, 1983) over periods of above 700 h.

(5) Polyurethane: Free suspended cells were found to invade such porous particles, and were retained strongly in the foam particles. Polyurethane is biologically inert and thus its use avoids the toxicity problem. Cells can be immobilized in polyurethane by polymerizing the plastic around wet cells. Plant cell suspensions were passively entrapped within the interstices of preformed polyurethane foam (Lindsey and Yeoman, 1987) by using the following procedure: 1 cm cubes of polyurethane foam were cut from foam sheets, washed in hot distilled water and autoclaved. Fresh suspension cell culture was prepared, and 6 cubes were added to 60 mL of medium and cells in a 250 mL Erlenmeyer flask. The culture was maintained under normal conditions on orbital shakers. Cells were entrapped within the foam cubes as the cells divided and expanded within the matrix. After about 14 days, each cube was packed with cells, and the foam was stable in the liquid nutrient medium over several months. The immobilization was unaffected by mechanical degradation due to cell growth, gas formation, chemical degradation, or medium channeling. The rate and extent of entrapment were dependent on both the pore size of the foam and cell aggregate size.

(6) Polyacrylamide: The plant cells were suspended in an aqueous solution of prepolymerized linear polyacrylamide, partially substituted with acylhydrazide functional groups (Brodelius, 1988). Later, the suspension was cross-linked with a controlled amount of dialdehyde under physiological conditions. The gel was then divided into smaller pieces.

Plant cells can be cultured on the surface of fibrous polypropylene matting, and can be immobilized in nylon netting in combination with agar and alginate gels (Lindsey and Yeoman, 1987). First, transfer the cells from callus culture to suspension culture for rapid cell proliferation and more homogeneous cultures. Then prepare agar (Oxoid No. 3) solution (2% w/v). Sterilize by autoclaving and allow it to cool to 35°C. Also sterilize the nylon netting pieces. Cut to cylindrical shape (2−3 × 1 cm). Sieve cells from stationary phase suspension cultures and mix 1:1 (v/v) with the molten agar. Dip the nylon pieces into the agar-cell suspension. The cells are entrapped as the agar is allowed to solidify. The cells can also be immobilized in nylon netting supported by calcium alginate gel. In this, a calcium cross-linked alginate gel is used to entrap the cells. The use of the nylon netting allows the successful formation of larger and mechanically stronger biocatalytic particles than can be produced with gel alone.

Brodelius and Nilsson (1980) compared a number of gels for the immobilization of *Catharanthus* cells. The cell survival was poor in low molecular weight monomers, such as acrylamide and glutaraldehyde, while preformed marine origin polymers (such as alginate and carrageenan) provided viable

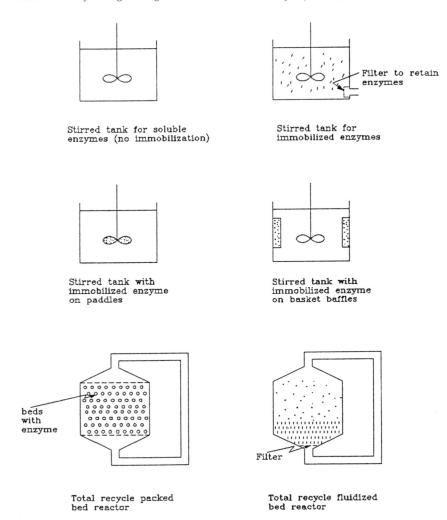

**FIGURE 6.4.** Various types of batch reactors used for immobilized enzymes. Reproduced with permission from "Handbook of Enzyme Biotechnology." 1985. A. Wiseman, ed., Chichester, U.K.: Ellis Harwood, Ltd.

immobilized enzyme can be retained by filtering the product, sedimentation, immobilizing the enzyme onto a magnetically active particle, or immobilizing it to the baffles of the reactor (Figures 6.5 and 6.6). For plug-flow reactors, many designs are available.

Webb et al. (1986) used a spouted bed reactor for cellulase production. A strong basal jet of recycled liquid carried the particles upward in the center of the reactor, and replaced by the downward movement of particles at the periphery of the vessel.

## 6.7.2 Immobilized Cell Reactors

Immobilized cells can be cultivated in conventional fermentation systems, such as stirred tanks or air-lift systems. Oxygenation can be achieved by a

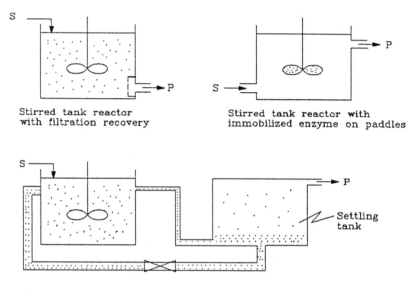

Stirred tank reactor
with filtration recovery

Stirred tank reactor with
immobilized enzyme on paddles

Stirred tank reactor with settling tank recovery

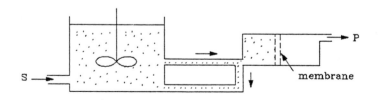

Stirred tank reactor with ultrafiltration recovery

P = Product,    S = Substrate

**FIGURE 6.5.** Different types of continuous flow stirred tank reactors for immobilized enzymes. Reproduced with permission from "Handbook of Enzyme Biotechnology." 1985 (reprinted 1986). A. Wiseman, ed., Chichester, U.K.: Ellis Harwood, Ltd.

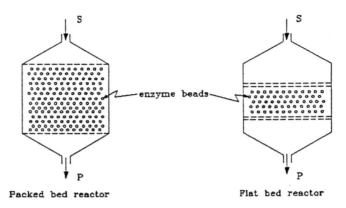

Packed bed reactor          Flat bed reactor

Fluidized bed reactor     Hollow fiber reactor     membrane reactor

P = Product,      S = Substrate

**FIGURE 6.6.** Various types of continuous flow reactors for immobilized enzymes. Reproduced with permission from "Handbook of Enzyme Biotechnology." 1985 (reprinted 1986). A. Wiseman, ed., Chichester, U.K.: Ellis Harwood, Ltd.

perfusion system separate from the growth chamber or by an air-lift system (Werner et al., 1987).

## 6.7.2.1  Air-Lift

Many of the laboratory-scale immobilized cell reactors are air-lift types containing cells immobilized on reticulated foams (Loh et al., 1986). Due to the absence of a stirrer motor, gas-lift reactors have lower capital and operating costs, lower chances of contamination, lower shear, and better aeration. Fluidized-bed reactors have been used widely, where upward flow of liquid creates fluidization. These reactors with self-immobilized mixed microbial populations on dense support materials are economical (Tampion and Tampion, 1987). Collagen beads with entrapped cells can be used in the fluidized bed reactors (Werner et al., 1987).

Margaritis and Wallace (1982) used a fluidized bed reactor to ensure complete mixing for the Z. *mobilis* immobilized in calcium alginate beads. Considerable leakage of cells occurred in the STR system at 3000 rpm, but negligible leakage occurred in the fluidized bed system.

## 6.7.2.2 STR

According to Einsele (1978), for geometrically similar stirred tank reactors (STR), the power input per unit volume decreased, and Reynolds number and mixing time increased with the increase in reactor volume. STR scale-up system for the production of ethanol using Z. *mobilis* had a lower nitrogen demand and better mixing than a gas-lift tower reactor (Fein et al., 1983).

## 6.7.2.3 Tapered

Chen and Zall (1982) employed a tapered expanded-bed reactor to investigate the continuous conversion of whey into alcohol by self-immobilized *S. fragi* on cellulose acetate powder. A productivity of 7 g ethanol/ $(dm^3 \cdot h)$ was achieved.

The tapered column reactor showed a better flow pattern of gas and liquid phases within the reactor, thereby providing a more homogeneous distribution of gas-liquid-solid phases in the reactor without any phase separation, compared to conventional cylindrical column reactor for the ethanol fermentation (Hamamci et al., 1987).

## 6.7.2.4 Rotating Thin Film

In these reactors, the inner cylinder is rotated in a trough of medium and the outer surface is coated with cells (Moser, 1982). In the horizontal rotary reactor, a single cylinder contains the cell suspension and the self-immobilized cells form a film on the inner surface of the cylinder. The liquid film is continuously renewed during rotation of the cylinder. The mass transfer coefficient at the gas − liquid interface is related to the rotational speed.

## 6.7.2.5  Membrane

These have been used less as immobilized cell reactors than as enzyme reactors. Polysulphone membranes are the most stable and can be steam-sterilized.

## 6.7.2.6  Hollow Fiber

In the hollow fiber, cells are either immobilized to the extra capillary space of silicon rubber tubes or in the intracellular space of ceramic cartridges. Hollow-fiber reactor systems provide a useful means for the immobilization of enzymes and microbial or animal cells. In this system an ultrafiltration or microporous filtration membrane can be easily used to separate a catalyst from reactants and products. Hollow-fiber and flat-sheet reactors, and microcapsules all utilize membranes for immobilization. In hollow-fiber and flat-sheet reactors, the membrane is preformed; in microcapsules, the membrane is formed around the cells. These systems protect the delicate outer membranes of cells from the shear forces that result from medium flow or stirring (Heath and Belfort, 1987).

Thus, in hollow-fiber membrane bioreactors, cells can be easily immobilized without much preparation, primary separation of the products can be carried out, and scale-up operation is easier. In these reactors, insoluble substrates cannot be used, substrate pretreatment is generally required to avoid clogging, transportation of gas is difficult, and polymer-based fibers cannot be repeatedly heat-sterilized. Aerobically growing *E. coli*, *Aspergillus niger* and *Norcardia mediterranei* were immobilized in the interstitial space of a dual hollow-fiber bioreactor formed by a parallel arrangement of three microporous polypropylene hollow fibers contained within a silicone tubercle (Chang et al., 1986). The leakage of *E. coli* through the fiber walls was observed. *A. niger* growth expanded the silicone fiber and compressed the inner fibers to reduce the substrate flow rates gradually to zero. *Nocardia mediterranei* immobilization provided 30-fold increase over the productivity of a comparable batch system.

## 6.8  **CHANGES IN CELL AFTER IMMOBILIZATION**

The changes in the cell after immobilization are affected by cell type, immobilization method, matrix type, and cell wall composition. In acrylamide support for immobilization, a ten-fold increase in cell number is generally desired to compensate for the sensitivity to monomer toxicity. The cationic layer of inorganic support can activate cell surface enzymes. On the other hand, metallic ions required for hardening the organic support matrix can activate enzymes of the cell membrane due to oxidation-reduction reactions.

Kolot (1988) summarized changes in cell metabolism after immobilization. There is an increase in enzymatic activity, oxygen consumption, operational stability, cell count, and respiration rate; a shift in pH optimum by 0.5 to 1.0, and in temperature optimum by 10°C; and a decrease in cell growth time.

## 6.8.1 Examining Immobilized Cells

A scanning electron microscope is used to observe porous gels and support materials. Slides are prepared by critical point drying of the sample on a specimen stub and evaporative coating with an electron-conducting layer (Tampion and Tampion, 1987). A transmission electron microscope provides the structure of membrane and organelles. This cannot be successfully used on many inorganic supports.

By determining the number of cells before and after immobilization, one can calculate the approximate number of immobilized cells. This can be done easily with a coulter counter. If cell counting is difficult, then protein content or nucleic acid content will provide indirect measurement.

## 6.8.2 Assessing Viability of Immobilized Cells

Some of the most widely used methods are:

(1) Ability to reduce redox stain: Different derivatives of tetrazolium chloride are used as redox indicators. Functional cells will reduce these indicators by respiratory metabolism. The color change is observed under a light microscope.

(2) Ability to produce ATP: For this, special instruments to measure ATP are required. Many of these instruments are based upon the luciferin which liberates visible light in the presence of ATP.

## 6.9 **MEMBRANE BINDING**

### 6.9.1 Nitrocellulose (NC) Membranes

These are useful media for genetic studies and immunological analyses. These demonstrate high binding capacities (80 $\mu g/cm^2$) of proteins and nucleic acids with low background levels. These membranes contain no cellulose acetate, which will inhibit sample binding. These will support the cell growth essential for colony blotting, plaque lifts and tissue culture applications. Membranes comprise a wide range of pore sizes to immobilize samples of both high and low molecular weight. Their binding characteristics have a unique structure which, when coupled with its lack of growth-inhibiting ingredients, facilitates bacterial growth even when the bacteria are stressed by transformation. These are available in a variety of formats, with about 0.45 μm pore size.

## 6.9.2 Ion Exchange Membranes

These are cellulosic supports modified with functional ion-exchange groups. These are constructed with a membrane support that ensures high wet-strength, a uniform pore size (0.45 μm) and even surface properties for solid phase analyses. These membranes are used to immobilize samples that do not bind well to other media, or very low molecular weight structures. These are available with either of two different exchange moieties:

(1) DEAE (diethylaminoethyl) membranes are ideal for the immobilization of anionic molecules such as nucleic acids and acidic proteins (e.g., glycoproteins and other negatively charged structures). These membranes carry the DEAE groups in their protonated form, and have a binding capacity of up to 35 μg/cm².

(2) CM (carboxymethyl) membranes bind basic molecules, including histones and other positively charged samples. The weak cation exchanger on the surface will bind samples in a pH range of 4 to 7, and can be used for many solid phase techniques involving cationic samples.

## 6.9.3 Charged-Modified Nylon-66 Membranes

These are strong and durable, producing a high density of positively charged binding sites on the nylon matrix. These charged sites augment the matrix's affinity for macromolecules such as nucleic acids and proteins. These membranes can be treated with aggressive solutions such as sodium hydroxide. These allow electrophoretic transfer of nucleic acids, and detection of single-copy genes. These are available in a variety of formats with about 0.45 μm pore size.

(1) *Polysulfone membranes* are made from a tough and durable high-temperature aromatic polymer. These are available in pore sizes of 0.2 to 0.45 μm, with diameters ranging from 25 to 293 mm. These have low protein binding, and can withstand temperatures up to 121°C.

(2) *Modified polysulfone membranes* offer higher flow rates, lower extractables, and greater strength than cellulosic membranes. These have low protein binding and high thermal stability, and are suitable for sterilizing filtration. These are available in 0.2 to 0.8 μm pore sizes, and in 13 to 293 mm diameters. These are also available in the form of capsules and disks.

# CHAPTER 7

# *Biosensor Techniques*

## 7.1 **INTRODUCTION**

*Precise* control of a process is essential if complex and dynamic reaction mixtures and broths are to be used on an industrial scale. Without such control, substrate and products, biocatalysts, ion concentrations, etc., cannot be kept at optimal levels. Rapid and sensitive on-line monitoring and control of such parameters require sensors specific to the substrates, products of fermentation, and the numbers of viable whole cells in the broth (Karube, 1984). Recent developments in sensor and control technology have been greatly influenced by the rapid advance of micro-electronics.

A biosensor uses a biologically derived sensing element as part of a physicochemical transducer. It consists of a biologically selective material, with the substance being measured providing a biochemical signal that is converted by the transducer into an electrical response (Graham and Moo-Young, 1985). Biologically selective materials for biosensors are enzymes, microbial cells, monoclonal antibodies, lectins, triazine dyes, and plant or animal tissues. The immobilization of these materials on the surface of a transducer forms the basis for many biosensors. Classical immobilization techniques can be used; generally it is carried out within a gel such as polyacrylamide, cellulose, or agarose, or a membrane, such as acetyl cellulose. The transducers commonly used are ion selective or gas sensing electrodes.

Immobilized microbial and enzyme sensors are used in various food-biotechnological processes to monitor and control the concentration of many substances such as acetic acid, alcohols, ammonia, cholesterol, glucose, glutamic acid, and lactic acid. These sensors are generally stable for 2 to 4 weeks and have response times of 0.5 to 2 min. Generally, biosensors cannot be sterilized; they are inactivated by strong acids and bases and by other chemicals, and they have a limited life (Schmidt and Saschewag, 1987). These can be used with conventional sensors in flow systems, preferably in off-line operation and on-line in combination with automated sterile sampling.

Many reviews on biosensors (Guilbault, 1982; Suzuki et al., 1982; Lowe et al., 1983; and Mosbach, 1983) detail the numerous configurations of

**171**

biocatalysts. Prototype enzyme probes for glucose and urea, ion-specific probes for the measurement of $NH_4^+$, $Mg^{2+}$, $Ca^{2+}$, $Na^+$, $PO_4^{3-}$, etc., and auto-analyzer-type systems to measure precursor and complex growth regulators are available (Kampen, 1983).

More than 60 enzyme electrode probes have been developed, most of them specific for a single chemical. The following substrates were measured: sugars (glucose, sucrose, lactose, maltose), acids (acetic, formic, lactic, oxalic, uric), various amino acids, acetylcholine, cholesterol, creatinine, penicillin, and urea (Ouellette and Cheremisinoff, 1985).

## 7.2 **TECHNIQUES**

In biosensor development for a particular application, two components are important: (1) a biocomponent, which may be an enzyme, bacterial cell, plant or animal tissue, DNA fragment, or antibody; and (2) a transducer, which may be a thermistor, electrode, piezocrystal, field effect transistor, or an optical or electrical transducer. For example, many enzymatic reactions increase the enthalpy of the substrate by about 15 to 100 kJ/mol. This will increase the temperature in the range of 0.005 to 0.03°C per millimolar substrate (Wagner and Schmid, 1990). This can be measured by an extremely sensitive calorimeter.

Biosensors are generally categorized based on the measured properties. These categories are potentiometric, amperometric, optical or optielectrical, calorimetric or thermal conductimetric, and piezoelectric.

### 7.2.1 Potentiometric

A potentiometric device measures electrical potential (differential charge density). Since many biological reactions affect charge density changes between enzyme and substrate, this property may be used to monitor biotechnological reactions. An ion-selective electrode consists of two electrodes, one

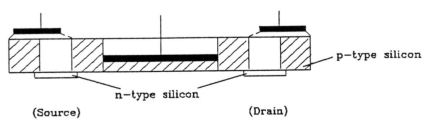

(Source)

n-type silicon

(Drain)

p-type silicon

**FIGURE. 7.1.** Field effect transistor.

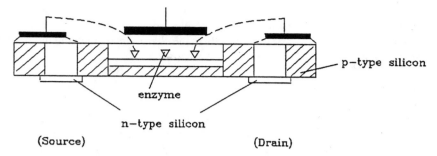

**FIGURE 7.2.** Enzyme field effect transistor.

covered with an ion-selective membrane and the other a field effect transistor (FET). In the FET, the charge on the center lead will generate a field, which will either increase or decrease the effective area for conduction through the p-type material, depending on the charge (Figure 7.1). The center lead of the FET is connected to the electrode covered by the ion-selective membrane. Technological advances have allowed direct integration of the chemical and electronic components of this design, such as the ion selective FET (ISFET), the immunochemical FET (IMFET), and the enzyme FET (ENFET).

IMFET consists of either antibodies or antigens covalently attached to an inert hydrophobic layer at the gate of an FET. The insulated gate FET measures the change in charge resulting from the formation of the antibody/antigen complex. In ENFET, an enzyme is immobilized directly onto the upper plate, while the center lead acts as the reference electrode (Figure 7.2). When the enzyme interacts with the surroundings, its charge varies.

In these sensors, problems have occurred with ion leakage across the semiconductor, and with the response mechanisms.

## 7.2.2 Amperometric

In an amperometric device, at constant potential, the current changes are due to the surface reaction of the electrode. The electrode may be coated with an electron transfer promoter species by adsorption. Electrons travel from the enzyme and substrate, through the promoter, and then to the electrode. The promoter serves both to protect the electrode and to increase the response. In many cases, however, the physical parameters of the analyte prevent measurement in such a manner. The problem may be overcome by using an electron mediator, which operates as a "shuttle," carrying electrons from the analyte to the electrode, at a greatly enhanced rate (Turner and Swaine, 1988). The mediator may be dispersed through the sample—a method that is reliable, but whose response time is relatively slow. Response may be improved, but apparently at the expense of stability, by immobilizing the mediator on the electron surface. The electrons are transferred from the substrate to the enzyme, and then rapidly to the electrode through the mediator (Hall, 1988).

### 7.2.3 Calorimetric

Measurement of the heat generated by an enzymatic reaction, in enzyme thermistors, has been used to determine the concentrations of various biomolecules. These sensors consist of a column containing an immobilized enzyme through which the sample flows either continuously or intermittently. The heat of the reaction between the substance being monitored and a suitable enzyme can be measured by using a thermal probe (Graham and Moo-Young,

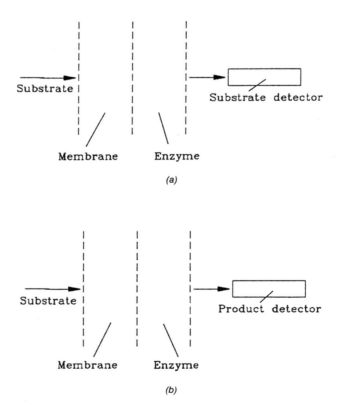

**FIGURE 7.3.** Indirect diffusion—limited enzyme electrode: (a) change in substrate concentration determined with a transducer; (b) product formation measured with a transducer.

TABLE 7.1. *Transducers in enzyme electrodes (modified after Aston and Turner, 1984).*

| Transducer | Principle or Species Sensed |
|---|---|
| Amperometric | $H_2O_2$, $I_2$, NADH, $O_2$ |
| Field effect transistor | $H^+$, $H_2$, $NH_3$ |
| Ion-selective | $CN^-$, $CO_2$, $H^+$, $I^-$, $NH_4$, $NH_3$ |
| Photomultiplier | Light emission |
| Photodiode | Light adsorption |
| Piezoelectric crystal | Mass adsorption |
| Thermistor | Heat of enzymatic and other reactions |

1985). These probes remain stable for several days and have a response time of 1 to 3 min. Enzyme thermistors have been developed to measure ethanol, cephalosporin, galactose, lactose, urea, cholesterol, sucrose, and penicillin.

Figure 7.3 shows two types of indirect diffusion-limited enzyme electrodes. In the former method, a physical or chemical change is followed continuously for a short period. The initial reaction rate is dependent on enzyme concentration, activity, substrate affinity, and the concentration of non-saturating substrates. In the second method, the reaction is allowed to reach equilibrium, making it insensitive to the physical and chemical conditions affecting enzyme activity. This technique is used for substrate determination, while the former is applied to inhibitor and activator assays. Care is required to control other conditions affecting the reaction rate, especially pH and temperature (Aston and Turner, 1984). Table 7.1 lists transducers commonly used in indirect enzyme electrodes.

Oxidoreductases that may be coupled to modified electrodes detect enzymes and monitor inhibiting and activating reactions. Direct electron transfer from enzymes and whole organisms to materials compatible with silicon-based devices (enzyme-effected transistors) are becoming popular in monitoring and control systems (Aston and Turner, 1984).

A modification of the enzyme thermistor is the thermometric enzyme-linked immunosorbent assay (TE-LISA). The thermistor uses a column filled with an immobilized immunosorbent such as monoclonal antibodies (Graham and Moo-Young, 1985). Faster response times are achieved by immobilizing a biologically selective material (e.g., enzyme) directly onto the electrode surface. Enzyme field effect transistor (ENFET) and immunologically sensitized FET have been developed by incorporating enzymes or antibodies, respectively, into the gate layers of FETs.

## 7.2.4 Optical

An enzyme reagent is immobilized on the surface of the optical fiber. When it comes into contact with the analyte, an optical change occurs. These are useful when a biological system induces an optical change, such as fluorescence and adsorption.

The fluorescence capillary fill device is constructed of two thin plates of glass separated by a narrow gap. On one plate, an immobilized layer of antibody is coated, and on the other an antigen. A sample reagent is then drawn into the space by capillary action, where it competes with the fluorescently labelled antigen. Since molecules bonded to the surface will fluoresce at shorter wavelengths, it is possible to determine the quantity of reagent bound to the surface (Opie, 1987). The advantages of this method are that it is less costly, flexible, and easy to use.

## 7.2.5 Conductimetric and Piezoelectric

The conductimetric biosensor measures non-specific ion conductance. Some enzymes produce ionic products, and therefore increase solution conductance. This increase is measured by recording the time dependence of the conductance when an AC signal is switched between a pair of electrodes (Merten et al., 1986).

The piezoelectric biosensor is based on a mass change induced by a specific reaction. This is accomplished by inducing an acoustic resonance in a piezoelectric crystal coated with an immobilized antibody. When an antigen bonds to the surface, the effective mass of the crystal is increased, and therefore the frequency that induces resonance changes (Turner and Swain, 1988).

One type of transducer is composed of a piezoelectric or surface acoustic wave detector operating in an oscillator mode, and the resulting biosensor, containing enzymes or antibodies (Guilbault and Suleiman, 1990). The piezoelectric crystal is a piece of oscillating quartz, onto which is placed an adsorbent that selectively interacts with the analyte compound.

A frequency change qualitatively indicates the presence of the analyte, and the total frequency change, $F$, is proportional to the amount of analyte present, according to the Sauerbrey Equation:

$$\Delta F = (- 2.3 \times 10^6)\, F^2\, \Delta m / A$$

where:

$\Delta F$ = change in fundamental frequency of the coated crystal
$F$ = resonant frequency of the crystal
$A$ = area of the electrode surface of the crystal
$\Delta m$ = the change in mass due to the deposited coating

A vibrating quartz crystal can be an extremely sensitive microbalance, with a detection limit of $10^{-12}$ g.

Two types of oscillator sensors are available—the surface acoustic wave

(SAW) device and the piezoelectric crystal. The SAW devices can operate either on the Raleigh wave propagation principle at solid thin film boundaries or as a bulk wave piezoelectric device. The latter follows the Sauerbrey principle, i.e., small thin films can be treated as equivalent mass changes on the crystal.

Immobilized biological compounds can be used directly for the assay of gaseous substrates or inhibitors, simply by coating a piezoelectric crystal with the soluble or immobilized biological reagent.

*Glucose-biosensor* by enzyme immobilized on the quartz crystal microbalance (QCM) (Lasky and Buttry, 1990): The operating principle of QCM derived from oscillating piezoelectric devices is based on the converse piezoelectric effect, in which application of an electric field across a piezoelectric material induces a deformation. If the electric field is rapidly alternated in polarity, it is possible to excite the piezoelectric material into oscillation at a mechanically resonant frequency.

The resonant frequency is dependent on the total mass of the resonator, including any mass that may be coupled to the surface of the resonator. Because of its excellent mechanical and electrical properties, single-crystal plates of alpha quartz are frequently used in QCM devices. The key to the use of the QCM in biosensor applications is to find a way of producing mass changes at the surface of the QCM. This usually entails the immobilization of some chemical or biochemical species at the QCM surface which will bind to the analyte, thereby producing a mass increase of the QCM resonate and a corresponding decrease in its oscillation frequency.

Following immobilization, the QCM crystals are placed into a flow cell of local design, which allows for reasonably rapid changes in the solution exposed to the sensor and reference crystals. These data indicate that the biosensor can respond selectively to glucose in the presence of other sugars, with some changes in response time, depending on the exact conditions of the experiment. The longest period of time they have successfully operated a biosensor continuously was 4 days (Lasky and Buttry, 1990). The biosensor was relatively stable towards prolonged storage at low temperatures. Thus, the QCM biosensor for glucose is an attractive method for real time measurements at moderate concentration of glucose, based on response time, selectivity, and sensitivity.

## 7.3 **MICROBIAL COUNT SENSOR**

Continuous and accurate measurement of cell concentration is required to develop maximum yield during continuous fermentation. The conventional methods for measuring the microbial count in fermentation include (1) colony count, (2) optical measurement of turbidity, (3) the amount of adenosine triphosphate (TPA), (4) measuring the charged metabolites, and (5) measuring electrical properties using an electrochemical method (Suzuki and Karube, 1981).

Microbial electrode sensors were developed by immobilizing a cell. The substrate concentration was measured by sensing the product of reaction. For example, BOD (biological oxygen demand) can be sensed by using an oxygen sensor and immobilizing *Trichosporon cutaneum*; ethyl alcohol concentration can be detected by immobilizing *Trichosporon brassiere* and an oxygen sensor. Similarly, using a carbon dioxide sensor and immobilized *E. coli*, glutamic acid concentration can be sensed (Ouellette and Cheremisinoff, 1985).

The impedance of the culture will change, up to $10^6 - 10^7$ organisms/mL, as a result of the metabolic activity of the microbial population. The performance of an impedance assay for the microorganisms count in food samples requires the inoculation of special media, and then impedance measurement over a period of 20 h.

Since all known microorganisms contain ATP, an assessment of the microbial content of a sample can be made by measuring the amount of ATP present. Bioluminescence is often used for such an assay.

To determine the populations of *Saccharomyces cerevisiae* and of *Lactobacillus fermentum*, a system with two similar electrodes was employed, each consisting of a platinum anode and a silver peroxide cathode. One electrode was the determination electrode, and the other (reference) was covered with a cellulose dialysis membrane (Matsunaga et al., 1979). The electrolyte was phosphate buffer (0.1 M, pH 7.0). An anion exchange membrane was used as a separator.

Piezoelectric membrane was used for continuous determination of a population of *Saccharomyces cerevisiae* in a fermenter (Ishimori et al., 1981).

## 7.4 ENZYME ELECTRODE PROBES

In principle, an enzyme-based electrode is a device that consists of an electrochemical sensor in contact with a layer of an immobilized enzyme. A conventional electrode such as an oxygen or ion-specific electrode is situated in close proximity to the enzyme, and detects the substrate or product of the enzymatic reaction; for example, it detects glucose oxidase-based electrodes that assay glucose by monitoring the consumption of oxygen, which occurs in the production of hydrogen peroxide.

To construct an enzyme electrode probe, first an enzyme is selected that reacts with the substance to be measured. For example, for a urea electrode, urease can be used; for a glucose electrode, glucose oxidase, etc. The selected enzyme may be commercially available or can be purified in a laboratory.

The enzyme is then immobilized using a suitable method (see Chapter 6). The technique of chemical immobilization is preferred for this purpose. The enzyme is treated with the aldehyde and an inert support (such as glass beads), which results in a rigid layer of bound enzyme. The immobilized enzyme is then attached to the sensor of an electrode that is sensitive to the product of the enzyme-substrate reaction.

The stability of the enzyme electrode depends on the type of entrapment, the optimum conditions of the enzyme, and the content of enzyme in the gel. On the other hand, the response time of the enzyme electrode is affected by the concentrations of substrate and enzyme, optimum conditions (pH, temperature) and the stirring rate of the solution.

Enzyme-based biosensors can be used for the selective and sensitive analysis of a wide variety of compounds (Davidson, 1990). Some of these are discussed below.

## 7.4.1. Electrodes Based on $O_2$

Oxygen electrodes can be applied to a large number of enzymic and microbial processes. Many enzymes are classified as oxidoreductases with $O_2$ as acceptor. Koyama et al. (1980) used a 12.5 $\mu$m Teflon™ membrane over a platinum cathode, followed by the enzyme membrane (cellulose triacetate) and protected by an ultrafiltration membrane preventing the passage of molecules $> 10^4$ daltons. This sensor had a linear sensing range of 0.003 to 2 m·mol/L and a response time of 10 s.

Enfors (1981) used enzymes mixed with albumin and cross-linked with glutaraldehyde on a platinum gauze attached to a galvanic electrode. The reference electrode was another $O_2$ electrode. The current through the cell was controlled by a proportional-integral controller. It had a linear operating range up to 2 g/L of glucose concentration.

Enzyme electrodes for sensing amino acids utilize amino acid oxidases. Karube et al. (1980) used monoamine oxidase electrode to evaluate meat freshness. Monoamine oxidase was immobilized in a collagen membrane in combination with amperometric determination of oxygen consumption to determine the response for various amines known to be produced at putrefaction. The linear operating range of the electrode was 0.05 to 0.2 m·mol/L of tyramine.

## 7.4.2. Electrodes Based on Hydrogen Peroxide

In these enzyme electrodes, the $H_2O_2$ produced by the reaction is measured by a polarographic electrode. Most of the commercial glucose analyzers use such electrodes. Three commercial glucose analyzers were evaluated by

Keyes et al. (1979), all using glucose oxidase and amperometric measurement of $H_2O_2$. One of the instruments (Yellow Springs, Ohio, U.S.A.) used an enzyme probe consisting of a platinum anode polarized at 0.7 V, and a silver cathode. The probe was covered by three membranes. The inner cellulose acetate membrane allowed passage of $H_2O_2$ and $H_2O$ to the probe. The middle layer contained immobilized glucose oxidase, and the outer polycarbonate layer was used to prevent the leakage of glucose oxidase, as well as to prevent the leakage of other proteins, macromolecules and catalase into the enzyme layer.

### 7.4.3 Electrodes Based on Ammonia

Gorton and Ogren (1981) measured urea concentration by converting urea to ammonia in a reactor with immobilized urease. The ammonia was detected by an ammonia gas membrane electrode.

Oxidative determination of amino acids by amino acid oxidases generates ammonia, which can be sensed by ammonia or ammonium electrodes (Danielsson, 1985).

### 7.4.4 Electrodes Based on pH and Other Factors

Many enzymic reactions directly decrease or increase the proton count and thus these reactions can be followed by measuring the pH. Urea and glucose can be measured by this method.

White and Guilbault (1978) cross-linked L-lysine decarboxylase and albumin with glutaraldehyde on the gas-permeable Teflon™ membrane of a $pCO_2$ sensor. This provided an enzyme probe to measure L-lysine concentration in the linear range of 0.05 to 100 m·mol/L with a response time of 5 to 10 min.

### 7.4.5 Enzyme Transistors (Chem FETs)

Danielsson et al. (1979) combined immobilized enzymes with a hydrogen and ammonia sensitive FET named an "enzyme transistor." They studied the application of a hydrogen and ammonia sensitive palladium-coated semiconductor device (Pd-MOSFET) in enzymic analysis. The $p$-channel semiconductor was similar to a field effect transistor of the MOS type (metal-oxide semiconductor). This chip contained a heater and a controller to provide a constant temperature (100 to 150°C) to increase the reaction rate at the Pd surface. Winquist et al. (1982) used Pd-MOS capacitors to measure hydrogen.

Table 7.2 summarizes the characteristics of various biosensors based on immobilized enzymes (Guilbault and Neto, 1985).

TABLE 7.2. *Characteristics of various immobilized enzyme biosensors.*

| Type | Sensor | Immobilization | Stability | Response Time | Enzyme Amount (U) | Range (m) |
|---|---|---|---|---|---|---|
| Acetic acid | Pt(O$_2$) | Chemical | >4 months | 30 s | 10 | 10$^{-1}$ to 10$^{-4}$ |
| Amygdalin | CN$^-$ | Physical | 3 days | 10–20 min | 100 | 10$^{-2}$ to 10$^{-5}$ |
| Alcohols | Pt(H$_2$O$_2$) | Soluble | 1 week | 12 s | 10 | 0.5 to 100 mg% |
| | Pt(O$_2$) | Chemical | >4 months | 30 s | 10 | 0.5 to 100 mg% |
| Cholesterol | Pt(H$_2$O$_2$) | Soluble | – | 2 min | – | 10$^{-2}$ to 10$^{-4}$ |
| D-Amino acid | Cation | Physical | 1 month | 1 min | 50 | 10$^{-2}$ to 5 × 10$^{-5}$ |
| Glucose | pH | Soluble | 1 week | 5–10 min | 100 | 10$^{-1}$ to 10$^{-3}$ |
| | Pt(H$_2$O$_2$) | Physical | 6 months | 12 s | 10 | 2 × 10$^{-2}$ to 10$^{-4}$ |
| | Pt(H$_2$O$_2$) | Chemical | >14 months | 1 min | 10 | 2 × 10$^{-2}$ to 10$^{-4}$ |
| | Pt(O$_2$) | Chemical | >4 months | 1 min | 10 | 10$^{-1}$ to 10$^{-5}$ |
| | I$^-$ | Chemical | >1 month | 2–8 min | 10 | 10$^{-3}$ to 10$^{-4}$ |
| | Gas(O$_2$) | Physical | 3 weeks | 2–5 min | 20 | 10$^{-2}$ to 10$^{-4}$ |
| L-Amino acid | Pt(H$_2$O$_2$) | Chemical | 4–6 months | 12 s | 10 | 10$^{-3}$ to 10$^{-5}$ |
| | Pt(O$_2$) | Chemical | >4 months | 1 min | 10 | 10$^{-2}$ to 10$^{-4}$ |
| | Cation | Physical | 2 weeks | 1–2 min | 10 | 10$^{-2}$ to 10$^{-4}$ |
| | NH$_4^+$ | Chemical | >1 month | 1–3 min | 10 | 10$^{-2}$ to 10$^{-4}$ |
| | I$^-$ | Chemical | >1 month | 1–3 min | 10 | 10$^{-3}$ to 10$^{-4}$ |

*(continued)*

TABLE 7.2. (continued).

| Type | Sensor | Immobilization | Stability | Response Time | Enzyme Amount (U) | Range (m) |
|---|---|---|---|---|---|---|
| L-Asparagine | Cation | Physical | 1 month | 1 min | 50 | $10^{-2}$ to $5 \times 10^{-4}$ |
| L-Glutamic acid | Cation | Soluble | 2 days | 1 min | 50 | $10^{-1}$ to $10^{-4}$ |
| L-Glutamine | Cation | Soluble | 2 days | 1 min | 50 | $10^{-1}$ to $10^{-4}$ |
| L-Tyrosine | Gas($CO_2$) | Physical | 3 weeks | 1–2 min | 25 | $10^{-1}$ to $10^{-4}$ |
| Lactic acid | Pt[Fe-$(CN)_6^{4-}$] | Soluble | <1 week | 3–10 min | 2 | $2 \times 10^{-3}$ to $10^{-4}$ |
| Nitrate | $NH_4^+$ | Soluble | – | 2–3 min | 10 | $10^{-2}$ to $10^{-4}$ |
| Nitrite | $NH_3$ (gas) | Chemical | 3–4 month | 2–3 min | 10 | $5 \times 10^{-2}$ to $10^{-4}$ |
| Succinic acid | Pt($O_2$) | Physical | <1 week | 1 min | 10 | $10^{-2}$ to $10^{-4}$ |
| Sulphate | Pt | Chemical | 1 month | 1 min | 10 | $10^{-1}$ to $10^{-4}$ |
| Urea | Cation | Physical | 3 weeks | 30 s – 1min | 25 | $10^{-2}$ to $5 \times 10^{-5}$ |
| | Cation | Chemical | >4 months | 1–2 min | 10 | $10^{-2}$ to $10^{-4}$ |
| | pH | Physical | 3 weeks | 5–10 min | 100 | $5 \times 10^{-3}$ to $5 \times 10^{-5}$ |
| | Gas($NH_3$) | Chemical | 20 days | 1–4 min | 0.5 | $10^{-2}$ to $10^{-4}$ |
| | Gas($CO_2$) | Physical | 3 weeks | 1–2 min | 25 | $10^{-2}$ to $10^{-4}$ |
| Uric acid | Pt($O_2$) | Chemical | 4 months | 30 s | 10 | $10^{-2}$ to $10^{-4}$ |

Reprinted by permission, Guilbault, G. G. and G. O. Neto. 1985 ''Enzyme, Microbial and Immunochemical Electrode Probes,'' in Enzymes and Immobilized Cells, A. I. Laskin, ed., Menlo Park, CA: The Benjamin/Cumming Pub. Co.

## 7.5 **MICROBE AND ORGANELLE PROBES**

These probes are similar to "enzyme probes," in which, in place of enzymes, the bioselective membranes or reactors contain organelles, cells, or tissue slices (Danielsson, 1985). Karube et al. (1980) employed immobilized nitrifying bacteria on an acetylcellulose membrane, which was held against the $O_2$ probe by a gas-permeable Teflon™ membrane. This was used to measure ammonia gas. Arnold and Rechnitz (1982) developed a tissue (rabbit liver)-based membrane electrode for guanine. On the other hand, Rechnitz (1981) developed a probe to measure glutamine using porcine kidney tissue in combination with an ammonia gas-sensing electrode.

Table 7.3 shows the characteristics of various microbe-based biosensors (Guilbault and Neto, 1985).

## 7.6 **DNA PROBE**

The use of nucleic acid sequences as probes in hybridization assays for the detection of specific microorganisms has tremendous potential. Virtually any organism of interest can be detected in complex samples: viruses, bacteria, protozoans, etc., in foods and fluids (Fitts, 1986). The requirements for a DNA hybridization assay are an appropriate DNA sequence to use as a probe, a method for labeling that probe molecule, and a method for detecting the label.

*DNA probes* are segments of single-stranded DNA that will bind to complementary DNA or RNA. Their length may vary from 10 to several thousand bases. DNA probes can be designed with varied specificity. Random DNA fragments are used to detect certain species, while DNA fragments from specific toxin genes are used to detect pathogens. After selecting the appropriate DNA probe, the target nucleic acid is prepared for hybridization.

The target nucleic acid is immobilized onto a solid support, such as nitrocellulose or nylon filters, for hybridization with a single-stranded probe. The bacterial colonies are transferred to nitrocellulose filter and lysed by soaking with 0.5 M NaOH and 1.5 M NaCl. After neutralization, the filter is air-dried and the DNA is fixed by incubating at 80°C for 2 h under vacuum (Schleifer, 1990). Cell lysates are placed onto nitrocellulose filters for dot blotting. The hybridization of two single-stranded nucleic acids occurs at about 25°C below *Tm* (the temperature at which the strands in 50% DNA are dissociated) (Schleifer, 1990). DNA probes are labeled with a detector group, generally with an isotope such as $^{32}P-$ , $^{35}S-$ , or $^{125}I-$ . Nucleotide analogues are introduced into nucleic acid *in vitro* using DNA polymerases by nick-translation (Rigby et al., 1977). The labeled DNA probe hybridized to the target DNA is identified by autoradiography (for detail see Chapter 2).

TABLE 7.3. *Characteristics of microbial biosensors.*

| Sensor | Species | Electrode | Response Time | | Stability Range (ng/L) |
|---|---|---|---|---|---|
| | | | Min | Days | |
| Acetic acid | T. brassicae | $O_2$ | 15 | 30 | 10 to 100 |
| Ammonia | N. europea | $O_2$ | 8 | 15 | 0.2 to 1.2 |
| Arginine | S. faecium | $NH_3$ | 10 | 21 | 2 to 200 |
| Aspartate | B. cadaveris | $NH_3$ | 20 | 10 | 0 to 20 |
| BOD | T. cutaneum | $O_2$ | 20 | 17 | 0 to 60 |
| Cholesterol | N. erythropolis | $O_2$ | 1 | 30 | 3 to 40 |
| Ethanol | T. brassicae | $O_2$ | 10 | 30 | 3 to 22.5 |
| Formic acid | C. butyricum | Fuel cell | 20 | 20 | 10 to 1000 |
| Glucose | P. fluorescens | $O_2$ | 10 | 14 | 2 to 20 |
| Glutamic acid | E. coli | $CO_2$ | 5 | 15 | 8 to 800 |
| Glutamine | Porcine kidney | $NH_3$ | 5 to 7 | 30 | 3 to 300 |
| Nicotinic acid | L. arabinosus | pH | 60 | 30 | 0.05 to 5 |
| Nitrate | A. vinelandi | $NH_3$ | 10 | 20 | 0.6 to 50 |
| Nystatin | S. cerevisiae | $O_2$ | 60 | 20 | 0.5 to 80 |
| Sugars | B. lactofermentum | $O_2$ | 10 | 20 | 20 to 200 |
| Vitamin $B_1$ | L. fermenti | Fuel cell | 360 | 60 | 1 to 500 |

Reprinted by permission, Guilbault, G. G. and G. O. Neto. 1985. "Enzyme, Microbial and Immunochemical Electrode Probes," in *Enzymes and Immobilized Cells*, A. I. Laskin, ed., Menlo Park, CA: The Benjamin/Cumming Pub. Co.

Many DNA probes have been used for the rapid detection of food-borne pathogens. More than 350 strains of *Salmonella* can be detected in this way.

## 7.7 **MISCELLANEOUS SENSORS**

### 7.7.1 Acetic Acid Sensor

The microbial sensor comprised an $O_2$ electrode consisting of a Teflon™ membrane (50 $\mu$m thick), a platinum cathode, an aluminum anode, and a saturated potassium chloride electrolyte. A porous membrane bearing the immobilized *Trichosporon brasicae* was fastened to the surface of the Teflon™ membrane and covered with a gas-permeable Teflon™ membrane (Karube, 1984). This was used for the continuous determination of acetic acid in fermentation broths (Hikuma et al., 1979).

### 7.7.2 Alcohol Sensor

Based on the principle that some microorganisms utilize alcohols, a sensor has been developed by Hikuma et al. (1979) using immobilized yeast or bacteria, a gas-permeable Teflon membrane, and an $O_2$ electrode to detect respiratory activity (Karube, 1984). Generally, *Trichosporon brassiere* is immobilized to sense the concentration of ethyl alcohol.

Guilbault and Lubrano (1974) employed the alcohol oxidase to determine the ethanol concentration over the range 0 to 10 mg/100 mL.

### 7.7.3 BOD Sensor

The ability of microorganisms, including yeast, to consume $O_2$ in the presence of an organic source provided a basis for the development of a BOD sensor (Kulys and Kadzianskiene, 1980). *Trichosporon cutaneum* is generally immobilized for this purpose.

### 7.7.4 $CO_2$ Sensor

The $CO_2$ sensor design is based on the measurement of pH change in a film of electrolyte solution trapped between a glass electrode surface and a hydrophobic polymer membrane (Shoda and Ishikawa, 1981). The $CO_2$ gas from the sample diffuses across the membrane until its partial pressure equilibrates within the electrolyte film in front of the pH glass and the bulk of the internal electrolyte. The gas reacts with water to form bicarbonate and hydrogen. The resulting potential is observed by a pH circuit (Gary, 1989).

## 7.7.5 Glutamic Acid Sensor

Certain microorganisms contain glutamate decarboxylase, which catalyzes the decarboxylation of glutamic acid to produce $CO_2$ and amine (Karube, 1984). Based on this, Hikuma et al. (1980) developed a microbial sensor for glutamic acid incorporating immobilized *E. coli*, as a source of glutamate decarboxylase activity, in conjunction with a $CO_2$ sensing electrode.

## 7.7.6 Glucose Sensor

Assimilation of glucose by microorganisms, and the related respiratory activity, can be determined with an $O_2$ electrode. Based on this, a microbial sensor consisting of immobilized whole cells of *Pseudomonas fluorescens* and an $O_2$ electrode was used for the monitoring of glucose in molasses (Karube, 1984).

Lubrano and Guilbault (1978) developed the glucose electrode using membranes containing immobilized glucose oxidase. The enzyme was co-cross-linked with bovine serum albumin using glutaraldehyde. On the other hand, Thevenot et al. (1979) used glucose oxidase enzyme immobilized on collagen membrane, employing a modified gas electrode in which the pH sensor was replaced by a platinum anode and the porous selective membrane by the enzyme membrane.

Quinoproteins possess a covalently or tightly bound PQQ (Pyrrolo-quinoline quinone), prosthetic group. Recently, a PQQ-containing NAD (P)-independent glucose dehydrogenase from *Acinetobacter calcoaceticus* has been used to construct an amperometric glucose sensor using a ferrocene-mediated electrode.

## 7.7.7 Methane Gas Sensor

Microorganisms that oxidize methane with the consumption of $O_2$ can be used in a methane sensor system. When a gas containing methane contacts the sensor, methane is assimilated by the microorganism with $O_2$ consumption so that current from the $O_2$ electrode decreases to a steady state (Karube, 1984).

## 7.7.8 Nicotinic Acid Sensor

Nicotinic acid can be determined by using *Lactobacillus arabinosus* immobilized in 2% agar gel and a combined glass electrode to measure lactic acid produced (Matsunaga et al., 1978). As the membrane potential of the glass electrode is proportional to the log of proton activity in the solution, the lactic acid produced by the immobilized whole cells can be determined (Karube, 1984).

### 7.7.9 Nystatin Sensor

A yeast electrode composed of a membrane supporting immobilized yeast (*Saccharomyces cerevisiae*) attached to an $O_2$ electrode, with a collagen membrane to prevent leakage of the yeast cells, was developed to measure nystatin (Karube et al., 1979).

### 7.7.10 Redox Sensor

Oxidation/reduction (redox) is a measurement of the overall oxidation or reduction power of the solution. The sensor used to measure this electrical potential is a noble metal (such as Pt, Au, Ag) which is capable of absorbing and releasing electrons (Gary, 1989). Effective coupling of redox proteins to electronic systems by direct electron transfer is used to develop biosensors (Aston and Turner, 1984). These electrodes require an accompanying reference electrode. Redox electrodes can monitor the point where a fermentation shifts from aerobic to anaerobic conditions.

### 7.7.11 Vitamin B₁ Sensor

This sensor incorporates a silver peroxide cathode, a platinum anode, and a phosphate buffer electrolyte. An anion-exchange membrane is used to separate the electrochemical cell from the culture broth (Karube, 1984).

### 7.7.12 Food Freshness Sensors

A number of biosensors have been developed to measure the freshness of fish and meat products. Luong et al. (1988) used an $O_2$ electrode as a transducer, and nucleotidase/nucleoside phosphoryl/xanthine oxidase as a biocomponent to determine the fish freshness, by measuring IMP/hypoxanthine/xanthine. Similarly, to monitor the meat freshness a knife-shaped sensor was developed, which measures glucose concentrations at different depths of the meat product by a series of four glucose oxidase-coated amperometric electrodes (Turner et al., 1986).

### 7.7.13 Urea Sensors

Nilsson et al. (1973) developed an urea electrode employing entrapped urease and a glass electrode. The response time was 7 to 10 min and the linear range was from $5 \times 10^{-5}$ to $1 \times 10^{-2}$ M, with a change of about 0.8 pH per decade. The life of the electrode at the room temperature was about 2 to 3 weeks. On the other hand, Guilbault and Tarp (1974) used a thin layer of urease chemically bound to polyacrylic acid (pH 8.5) and an air-gap ammonia electrode. A response time of 2 to 4 min, and a linear range of

$3 \times 10^{-2}$ to $5 \times 10^{-5}$ M was obtained with a slope of 0.75 pH per decade. Alexander and Joseph (1981) measured urea by immobilizing urease on an antimony electrode, providing a linear range of $5 \times 10^{-4}$ to $1 \times 10^{-2}$ M urea with a sensitivity of 44 mV per decade change in urea concentration, and a response time of 1 to 2 min. Tran-Minh and Broun (1975) developed an enzyme electrode for urea, using glutaraldehyde-bound urease and either $CO_2$ or glass electrodes.

### 7.7.14 Glutamine Sensor

Guilbault and Shu (1971) developed an electrode for glutamine by immobilizing glutaminase on a nylon net between a layer of cellophane and a cation electrode. The sensor provided a response time of 1 to 2 min and a concentration range of $10^{-1}$ to $10^{-4}$ M.

### 7.7.15 L-Amino Acid Sensor

Nanjo and Guilbault (1974) developed an enzyme electrode for L-amino acid by immobilizing the enzyme and using a platinum-based $O_2$ electrode. The electrode responded to L-crysteine, L-isoleucine, L-leucine, L-lysine, L-methionine and L-phenylalanine. Havas and Guilbault (1982) developed an L-tyrosine selective probe using immobilized ap-L-tyrosine decarboxylase on a $CO_2$ sensitive membrane. A linear range of $2.6 \times 10^{-3}$ to $4 \times 10^{-5}$ M was reported.

### 7.7.16 Cholesterol Sensor

Bertrand et al. (1981) developed an enzyme electrode to measure cholesterol concentration using collagen-immobilized cholesterol oxidase and by detecting $H_2O_2$ produced in the enzymatic reaction.

# CHAPTER 8

# *Down-Stream Processing Techniques*

## 8.1 **INTRODUCTION**

*The* commercialization of a new biotechnology product will depend on sequential developments in basic biological sciences (i.e., genetic engineering), in biological systems (i.e., fermentation), and in down-stream processing. Down-stream processing refers to the unit operations for the recovery, purification, separation, and concentration of the essential products from a fermentation process at the lowest cost, and highest recovery factor and quality. It represents a major part of the overall costs of a process, as much as 40% of the total costs (Smith, 1985). Thus, its cost efficiency is a key factor in the production of biotechnological compounds. The cost of the selected down-stream method will usually determine the economic feasibility of the process. Consequently, improvements in these techniques will significantly enhance the viability of a bioprocess. This requires chemical synthesis and characterization of new separation techniques, as well as the application of engineering fundamentals.

Down-stream processing utilizes many techniques, and the selection will depend largely on the chemical and physical properties of the product. The biomass component of the process may be a required product (e.g., a single cell protein), the repository of the product (e.g., intracellular enzymes), or a byproduct or waste product that is no longer required (Smith, 1985). These considerations will determine the method of choice for biomass separation. Industrial-scale separation processes require a high degree of purity in the desired products. Care is required to avoid contamination during separation processes, since most spent broths still contain a rich source of nutrients suitable for a wide range of common contaminants.

Several methods are employed to perform the steps necessary to purify biological materials from complex solutions:

- product recovery, i.e., removal of cell, cell debris, and insoluble matter, by means such as centrifugation
- product concentration, i.e., removal of water, by means such as reverse osmosis

**189**

- preliminary purification, i.e., separation of precipitated impurities, by means such as precipitation
- secondary purification, i.e., separation of precipitated desired product by means such as chromatography
- product conditioning, e.g., drying

The following is a list of down-stream processing techniques:

(1) Solid-liquid separations

    a) Mechanical methods

- filtration
- ultracentrifugation
- crystallization
- sedimentation

    b) Membrane methods

- liquid membranes
- microfiltration
- ultrafiltration
- reverse osmosis
- pervaporation
- electrodialysis

    c) Electrical methods

- electrocoagulation
- electrofiltration
- electromagnetic separation
- electrophoresis
- electrostatic separation

(2) Multi-stage equilibrium separations

    a) Extraction methods

- absorption
- adsorption
- chromatography
- solvent extraction (liquid-liquid)
- precipitation
- supercritical fluid extraction
- ion exchange
- aqueous two-phase systems
- foam fractionation
- solid-liquid extraction

    b) Thermal methods

— distillation
— drying
— evaporation
— freeze drying

In the following sections, commonly used methods are discussed along with cell disruption techniques.

## 8.2  **CELL SEPARATION AND DISRUPTION OR RUPTURE**

### 8.2.1  Cell Separation

The initial recovery phase of the processing includes the gross separation of components such as cells and broth. The recovery of intracellular products involves concentration of cells to a 50% v/v solution, cell rupture to break the cells and release the product, and partial purification yielding an intermediate product. Extracellular products are small molecules or macromolecules that are transported through the cell membrane. Products are concentrated, usually from 0.1 to 100 g/L of broth, because the products are often present in low concentrations (Grabner, 1986).

Cells can be separated or concentrated by (Chmiel, 1987):

- sedimentation or flocculation: useful when the density difference between cells and suspension medium is large
- centrifugation-decanter, solid-injecting, and nozzle separators, used when the density difference is not too small
- filtration for diminishing density differences
- flotation
- coagulation

The decanter consists of a rotating bowl supported at both ends and an internally arranged screw conveyor rotating at a slightly lower speed in the same direction as the bowl (Larson, 1974). Solid-ejecting separators are disk-bowl machines that eject the solids intermittently during operation. In the nozzle separators, the concentrate is continuously discharged through nozzles under pressure at the top of the machine.

In some cases flotation is a feasible alternative for cell separation. This process has been used in wastewater treatment and for the recovery of single cell proteins.

In coagulation, the effluent stream with microbial cells is withdrawn from the broth and the pH is lowered (Frankenfeld et al., 1969). The acidified effluent stream is heated to 45 to 90°C before passing through a separator.

## 8.2.2 Cell Disruption

Cell disruption is required to release the intracellular enzyme or other biologically active compounds, and so facilitates their recovery. This involves the mechanical, physical, chemical, or microbial breakage of cell wall or membrane. Neither plant nor animal cells are difficult to rupture. However, disruption of microbial cells requires a more violent method (Patel, 1985). Typical cell disruption techniques are shown in Table 8.1. Generally, the following methods are employed.

### 8.2.2.1. Shearing by Pressure

Shear presses to disrupt cells are available in either small or in larger sizes. The cell solutions are subjected to high pressures, hence disrupting the cell structures.

### 8.2.2.2 Induction of Lysis

This involves non-mechanical means for the hydrolysis of the cell wall. Cell membranes can be ruptured by physical stresses, such as a sudden depression, osmotic shock, or rupture with ice crystals. Certain lytic agents such as cationic and anionic detergents are used to damage the lipoprotein of cell membranes. Such chemicals may destroy the intracellular materials thereby limiting their use.

### 8.2.2.3 Ultrasonic

Liquid processors of about 500 W power, working on ultrasonics, are commonly used for research, analytical, and light industrial applications. The sonicator models W380 and W385 (Mandel Scientific Co., Canada) consist of a 12.7 mm tapped titanium disruptor horn and a cool-running 20 kHz ultrasonic convertor. Alternating on and off cycles prevent sample heating at high intensities. This provides both continuous or discrete (1 to 5 s cycles) operation.

TABLE 8.1. *Cell disruption techniques.*

| Mechanical | Physical | Chemical |
|---|---|---|
| – Liquid shear/high pressure homogenizer<br>– Solid shear/ bead or ball mill | – Ultrasonic/sonication<br>– Osmotic shocks<br>– Freezing/thawing | – Alkalis<br>– Enzymes: lysozyme or EDTA<br>– Detergents: anionic, cationic or nonionic |

## 8.2.2.4 Mini-Bomb Cell Disruptor

Cells are disrupted through the sudden depressurization of cell suspensions. First, the mini-bomb is pressurized with air, nitrogen, or carbon dioxide. After the cells have equilibrated, the pressure is released. The cells rupture, which results in a uniform homogenate of whole cells and sub-cellular components. A short treatment at moderate pressure will rupture the outer cell membranes leaving the nuclei intact. After separating these fractions, a second treatment can be used to rupture the nuclei.

Mechanical methods are simple, effective, and readily scalable, and are preferred for the large-scale disruption. The homogenizer is used in well over half of all large-scale installations; the bead mill in about 10%, and the remaining use a variety of chemical and physical methods (Lund, 1987). Disruption by ultrasonics produces heat, and overheating can decrease product activity (Chmiel, 1987). Other physical methods require relatively more time and energy. Microbial enzymes can digest away the outer wall of microbial and plant cells, thus allowing easier disruption of the cell membrane. However, these enzymes are relatively expensive. The main problems associated with chemical treatment are those of cost, possible toxicity, and recovery of chemicals.

The choice of method for large-scale cell disruption depends on the:

- microbial characteristics, i.e., the susceptibility of cells to breakage, and ease of extraction from cell
- cost
- speed
- sensitivity of the product to heat and shear
- location of the desired product within the cell

Operating parameters such as capacity, pressure, temperature, and residence time should be optimized.

Bond et al. (1987) investigated the performance of six methods of cell disruption:

(1) Freeze-thaw: cells were frozen at $-27°C$ and thawed after 1 week

(2) Freeze-dry: cells were lyophilized after freezing in liquid nitrogen and pH 7 buffer was used to rehydrate the dried cells

(3) Pestle and mortar: cells were ground with acid-washed sand for 5 min

(4) Polytron blender used for 5 min

(5) A sonicator

(6) A bead mill: using two bead sizes for 5 min

The total protein releases (mg/mL) for the six methods were 14.7, 39.6,

21.2, 36.1, 48.9, 19.5 (large beads) and 15.4 (small beads), respectively. Thus, sonication appeared to be the most efficient method, although it is too expensive at industrial scale.

## 8.2.2.5 Tissue Grinder

This combines both conical and cylindrical surfaces to effectively reduce tissue and produce a uniform homogenate. As the pestle is bottomed within the tube, tissue is ground to a smaller particle size in the tapered section.

## 8.3 **MECHANICAL METHODS**

### 8.3.1 **Filtration**

Filtration separates solids from liquids by passing the mixture through fine pores, which are smaller than the diameter of solid particles but large enough to allow the liquid to pass through. The solid liquid mixture is known as the feed slurry, the liquid component that passes through the filter medium is called the filtrate, and the separated solids are known as the filter cake. The purpose of filtration can be sterilization, clarification, bacterial reduction, particulate removal, and so on.

The equipment must provide a support for the filter medium; a space for the accumulation of the solids; and piping for the feed slurry, wash liquid, steam, and for air to remove the filtrate, washings, and exhaust air or steam. Tanks are required for the storage of feed slurry, filtrate, and washings.

The flow of filtrate can be accomplished by gravity, by the application of a pressure upstream of the medium (pressure filtration), or by applying a vacuum down-stream from the medium (vacuum filtration).

#### 8.3.1.1 Filter Media

Proper filter medium is required for the satisfactory operation of a filtration process. Filter media are made from cotton, polymers, cellulose, and other fiber-forming materials. These may be fixed or movable, rigid or flexible, metallic or non-metallic. The primary filter material is a heavy cloth of cellulose material (Brown et al., 1987).

#### 8.3.1.2 Filter Aid

Filter aids are required to prevent slow filtration rate, rapid medium blinding, and unsatisfactory filtrate clarity. Heating the slurry and decreasing the viscosity of the solution can aid filtration. Other filter aids are diatomaceous silica—a pure silica prepared from deposits of diatom skeletons; and expanded perlite—particles of "puffed" lava that are principally aluminum alkali silicate. These may be used as precoats for protecting the filter medium or may be mixed with the slurry to produce more open cake building or both.

## 8.3.1.3 Design

The rate of filtration is calculated by dividing the driving force by the filter resistance. Thus, the equation for flow through the filter, under the driving force of the pressure drop is:

$$\frac{dV}{dt} = \frac{A \cdot \Delta P}{\mu \cdot r\,(W \cdot V/A + L)}$$

where:

$dV/dt$ = the rate of flow through the filter
$\Delta P$ = pressure drop across the filter
$A$ = area of the filter surface on which the cake forms
$\mu$ = viscosity of the slurry
$r$ = specific resistance of the filter cake
$L$ = equivalent thickness of the filter
$W$ = solid content per unit volume of liquid
$V$ = volume of fluid passed through the filter
$t$ = filtration time

Once the initial cake has been built up, flow occurs under a constant pressure differential. For this, the equation can be written as:

$$t \cdot A/V = V \cdot \mu \cdot r \cdot W/(A \cdot \Delta P) + \mu \cdot r \cdot L/\Delta P$$

These equations assist in selecting various process parameters. Sometimes, constant rate filtration is also used, which can be achieved by adjusting various parameters.

Testing the filtration character of the slurry is required in selecting the filter and determining the filter size. This can be done with a test leaf, which is performed with a shallow pan covered with filter media and attached to a drainage pipe. The pipe is connected to a suction flask for collecting the filtrate, and the flask is connected to a vacuum source. The slurry to be tested is contained in a vessel with an agitator. The filter leaf is immersed in the slurry with the filter media on the underside, and the filtration is started. At the end of the test, the leaf is removed from the slurry and inverted to allow the residual filtrate to drain into the suction flask (Boss, 1983a). The data collected from this test will include the mass of dry cake collected per m²·h, the volume of filtrate collected per m²·h, the moisture content of the cake, the wash quality, and the filtrate clarity.

## 8.3.1.4 Influence of Process Variables

The factors affecting the filtration process are (Charm, 1978; Perry and Chilton, 1984; Anon, 1988b):

(1) Cake thickness: The average flow rate during a filtration is inversely proportional to the square of the filtering area. Thus, the average filtration rate for a given quantity of filtrate or cake is inversely proportional to the square of the thickness of the cake at the end of the filtration. The economic choice is a cake of appreciable thickness.

(2) Viscosity or apparent viscosity: The filtrate flow rate is inversely proportional to the filtrate viscosity.

(3) Temperature: Higher temperatures permit higher filtration rates.

(4) Particle size: Even a small change in particle size affects the specific cake resistance, and larger changes affect the compressibility of the cake. Decreased particle size results in a lower filtration rate and higher moisture content of the cake.

(5) Filter medium: In the selection of the filter medium, a balance is required for optimum pore size to prevent excessive "bleeding" of fine particles and to reduce filter resistance.

(6) Slurry concentration: The change in slurry concentration may affect the rate of medium plugging.

(7) Pressure: Flow rate is directly proportional to the increase in pressure in the filtration of slurry with granular solids. In flocculent or slimy precipitates, filtration rate increased marginally with an increase in pressure.

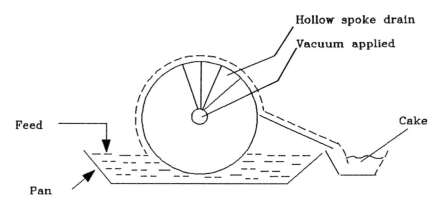

**FIGURE 8.1.** A rotary vacuum filter.

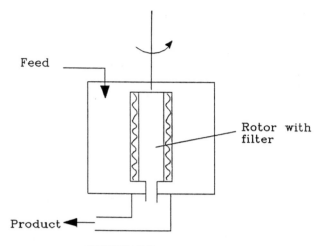

**FIGURE 8.2.** An axial filter.

## 8.3.1.5 Equipment

The equipment selection will depend upon the degree of filtrate clarity required, particle size, slurry concentration, capacity, and the physical characteristics of liquid and solid. Important factors in filter selection are the slurry character, process conditions, desired results, required capacity, and cake formation time (Boss, 1983a).

Plate-frame filters have more robust construction, and are easier and cheaper to clean and sterilize than other types. Rotary vacuum filters are used in fermentation, for separation of yeast from broths, and for many waste treatment processes (Smith, 1985). These are used to separate solids, which can be maintained in suspension with relatively gentle agitation. The simplest system is a drum with a hollow spindle connected to a vacuum pump, and with hollow spokes evacuating through the perforated metal circumference of the drum (Brown et al., 1987) (Figure 8.1).

In the axial filter, a membrane is wrapped around a rotor spinning in a chamber, into which feed is introduced under pressure. The rotor is perforated, and passages are provided for filtrate to exit through the axis (Figure 8.2) (Griffith et al., 1981). Biopolymer separation using microscreens and axial filters is less costly and more energy efficient than separation using diatomaceous earth filters or centrifugation.

Cross-flow filtration systems are used to separate whole cells and cell debris from fermentation broths. This consists of a porous membrane across which the feed slurry passes at high velocity (2 to 6 m/s) and through which a fraction of the broth passes (Brocklebank, 1987). This offers lower operating costs than rotary vacuum filtration, and is preferred to centrifuges where protein adherence to cell debris occurs.

A microstrainer is a rotating cylinder housed in a horizontal position in an open tank (Green and Kramer, 1979). The cylinder is covered with a fine mesh nylon or metal screen (20 to 65 $\mu$m pore size). The solution is forced through the screen by a pressure gradient. Potential problems are slime formation due to the growth of microorganisms and poor performance due to shock loading.

## 8.3.1.6  Other Considerations

Define the conditions that exist before filtration and the specifications you need after filtration. Components and characteristics of the solution should be considered before filter selection. Some examples are chemical composition, viscosity or rheological parameters, surface tension, and fluid temperature. The operating cost is related to the degree of filtration, because the smaller the filter pore size, the shorter the filter life, and higher pressure required to maintain the same flow rate. The degree of filtration will be based on (1) required flow rates, (2) available pressure, (3) filter change frequency—the filters are changed when the differential pressure exceeds the filter capabilities, and (4) equipment requirements.

TABLE 8.2. *Centrifuge characteristics (modified after Brocklebank, 1987, reprinted by permission of Elsevier Applied Science Pub. Ltd., Barking, U.K.).*

| Centrifuge Type | Advantages | Disadvantages |
|---|---|---|
| Tubular | High force (16,000 to 60,000 g's), lower denaturation, constant temperature operation | Low capacity, batch operation, low solids capacity, manual solids removal |
| Chamber | Lower denaturation, constant temperature operation | Low force (up to 3500 g's) batch operational manual solids removal |
| Decanter | High capacity, high solids loading, continuous solids discharge | Low force (up to 4000 g's), wet solids needed, foaming of solids at discharge |
| Nozzle bowl | High liquid and solids capacity, moderate force (8000 g's) | Feed solid content >2%, poor dewatering |
| Self-cleaning bowl | High liquid capacity, high force (up to 15,000 g's), good dewatering | Low solids capacity, expensive |

## 8.3.1.7 Prefiltration

Prefiltration is used to delay filter blockage, because filter life may be decreased by loading with large particles.

## 8.3.2 Centrifugation/Ultracentrifugation

A large centrifugal force is used to separate subcellular fractions such as yeast, proteins, and cell debris from fermentation broths. These are used to separate a heavy phase from a light phase or suspended solids from a liquid. For biotechnological applications, most centrifuges develop 14,000 to 15,000 g's, while ultracentrifugation utilizes up to 400,000 g's. These are generated by the centrifugal forces of a rapidly rotating rotor (Ouellette and Cheremisinoff, 1985). Centrifuges generally consist of (1) a rotor or bowl in which centrifugal force is exerted upon the contents, (2) a system to rotate the bowl about its longitudinal axis, (3) a frame to support the system, and (4) an enclosure to contain the rotor and keep the separated products (Boss, 1983b).

Figure 8.3 illustrates a typical centrifugal separator. Table 8.2 provides characteristics of different types of centrifuges. Figure 8.4 can be used for choosing an appropriate type of centrifuge. The selection process will probably require centrifugation tests and consultation with equipment vendors. The test can be conducted in a laboratory-scale centrifuge using test tubes to determine whether sedimentation will achieve a satisfactory separation (Boss, 1983b).

## 8.3.3 Crystallization

In crystallization, an orderly and dense aggregation of particles is formed from a super-saturated solution (Chmiel, 1987). Crystallization consists of (1) the formation of a super-saturated solution, (2) the appearance of crystalline nuclei, and (3) the growth of nuclei to size (Glasgow, 1983). For some systems, crystallization has some advantages over distillation. The heat of fusion is much lower than the heat of vaporization, for instance, and almost-pure products can be achieved in a single step. The disadvantages are that refrigeration is frequently required, and that handling solids is more costly than handling liquids (Busche, 1985).

The selection of crystallization equipment depends upon the solubility characteristics of the solute. Typical equipment includes an evaporative crystallizer, a cooling crystallizer, a vacuum-cooling crystallizer, and a batch crystallizer.

Data required for a proper crystallizer design are solubility data, the

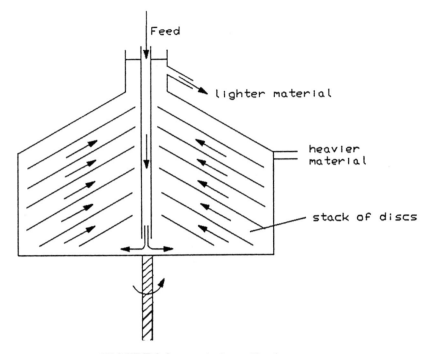

**FIGURE 8.3.** A typical centrifugal separator.

physical properties of the solution, the heat of crystallization and of concentration, the capacity required, and the pilot scale test data.

A basic approach to sizing a crystal growth container is given elsewhere (Glasgow, 1983).

## 8.4 **MEMBRANE SEPARATION METHODS**

Membrane filtration can concentrate soluble substances and remove water without evaporation at ambient temperatures, thus avoiding protein degradation. Membrane separation is based on the molecular size, and is used for its relatively high capacity and low resolution. A variety of membranes are available with a wide selection of porosity, configuration, and composition to maximize purification and product recovery (Grabner, 1986). Membranes can be operated in various modes as listed in Table 8.3.

These membranes are plastic films, about 0.1 to 0.2 mm thick, with pores of controlled size. About 80% of the membrane volume is air, due to the large number of pores. These are used to retain molecules for analysis or to produce clean filtrates.

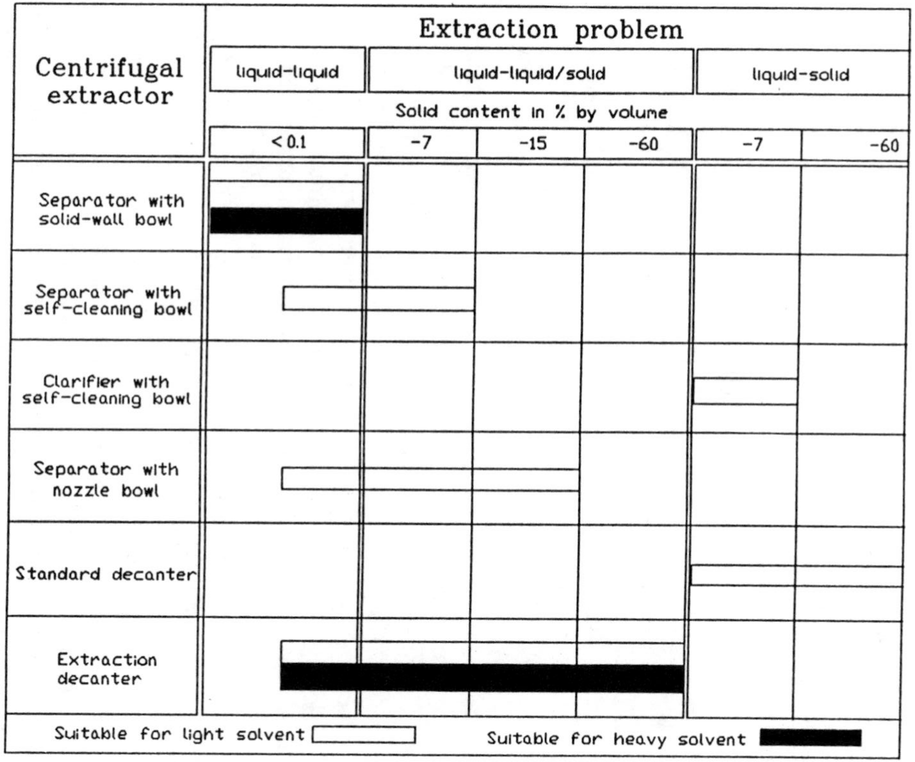

**FIGURE 8.4.** Characteristics of centrifugal separators (Brunner, 1987). Reprinted by permission of Gustav Fischer Verlag, Germany.

TABLE 8.3. *Membrane processes used in bioproduct recovery (Strathmann, 1987, reprinted by permission of Gustav Fischer Verlag, Germany).*

| Process | Membrane | Driving Force | Application |
|---|---|---|---|
| Microfiltration | Symmetric microporous $(0.05 - 10 \mu m)$ | Pressure, 1 to 5 bar | Sterilization, clarification, cell harvesting |
| Ultrafiltration | Asymmetric microporous $(1 - 50 nm)$ | Pressure, 2 to 10 bar | Separation of macromolecules |
| Reverse osmosis | Asymmetric microporous skin type | Pressure, 10 to 100 bar | Concentration of microsolutes |
| Liquid-liquid membrane | Microporous | Solubility, active transport | Organic solvents removal |
| Pervaporation | Asymmetric microporous skin type | Vapor pressure gradient | Organic solvents removal |
| Electrodialysis | Ion-exchange | Voltage | Separation of ions and proteins |

Membrane modules are required to install a membrane for an industrial separation process. Typical membrane modules used in the down-stream processing of bioproducts are plate and frame, tubular, capillary, hollow fiber, and spiral wound (Strathmann, 1987). Table 8.4 summarizes the characteristics of these modules. Fouling problems have been partially solved by a periodic back-flushing of membranes. Aggregation of protein molecules, influenced by pH, ionic strength, etc., affects membrane transport behavior (Sirkar and Prasad, 1986).

The manufacturer is an important variable in membrane selection (McGregor, 1986). The membrane's physicochemical characteristics and its interaction with the processing environment are important factors in the separation process.

Hammes (1987) reviewed the applications of membrane processes in food processing, including biotechnology. Reverse osmosis, ultrafiltration, electrodialysis, pervaporation, and microfiltration have been used in the dairy, meat, fish, brewing, starch, and vegetable and fruit processing industries.

## 8.4.1 Liquid-Liquid Membranes (LLM)

Liquid-liquid membranes are of two types: the support type and the emulsion type. The former utilizes a microporous membrane where pores are filled with liquid, while the latter is operational when the membrane forms tiny droplets suspended in a liquid phase. These can be used to separate hydrocarbon mixtures as well as aqueous solutions.

The liquid emulsion membrane (LEM) consists of an emulsion of two immiscible phases which is dispersed in a third phase (Figure 8.5). One of the two phases comprising the emulsion is completely encapsulated by the other phase. The non-dispersed emulsion phase acts as a membrane between the interior and exterior phases. The feed is emulsified in the membrane-forming solution and added to the continuous phase later on. Surfactants such as saponin, and additives such as glycerol, are used as film-strengthening agents. By controlling the amount and types of surfactants and additives used, stability, permeability, and selectivity of the membranes can be controlled, thus LEM can be produced for specific applications (Cahn and Li, 1974).

Due to dispersion, small emulsion droplets are formed when continuous phase of the emulsion is agitated. With an increase in agitation, more droplets of smaller sizes will be formed, thus increasing the surface-to-volume ratio of the emulsion and resulting in a larger product separation rate. No separation will occur if agitation disintegrates the LEM around the droplets. The emulsion particles usually range from 0.5 to 5 mm in diameter, with each emulsion containing smaller encapsulated droplets with a typical diameter of 1 $\mu$m.

LEM or any separation is based on a separation factor that indicates the degree to which the products of separation will differ. The greater the degree of separation the larger the separation factor and the more acceptable the process is likely to be. The separation factor ($\alpha_{AB}$) is

TABLE 8.4. *Typical membrane module characteristics.*

| Module | Membrane Installation | Use | Advantage/Limitation |
|--------|----------------------|-----|----------------------|
| Plate and frame | Flat sheets | MF, UF | Easily disassembled for cleaning |
| Tubular | Porous support tubes | MF, UF | Easy control of polarization and fouling |
| Capillary | Capillary membranes in a shell tube | UF, PV | — |
| Hollow fiber | Capillary membranes in a shell tube | RO | Sensitive to fouling; fast and economical recovery of microbes from broths |
| Spiral wound | Sandwich on a porous support, then wound in spiral configuration | RO | High surface area per unit volume, low cost |

MF = micro-filtration; PV = pervaporation; RO = reversed osmosis; and UF = ultrafiltration.

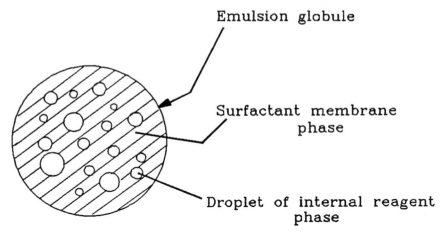

**FIGURE 8.5.** Liquid emulsion membrane structure.

defined as $\alpha_{AB} = \Delta Y_A \cdot X_B / (\Delta Y_B \cdot X_A)$, where $Y$ is the concentration of components after passing through the membrane, $X$ is the initial concentration of the mixture, and subscripts $A$ and $B$ denote materials $A$ and $B$. Liquid membrane separation results from chemical reactivity, ion exchange, and permeability, which in turn is a function of molecular weight and molecular volume. The enrichment, which can be achieved between the two components, is a function of the relative permeability of the components being separated.

LEM has been used for the separation and recovery of different compounds, and has also been found to be an economical way of separating commodity-type biochemicals such as propionic acid and acetic acid (Thien et al., 1986). Successful immobilization of enzymes by liquid membrane encapsulation has been accomplished with many enzymes.

### 8.4.2 Ultrafiltration (UF)

Ultrafiltration, in contrast to reverse osmosis (RO), does not retain many of the smaller molecules, and is more of a separation process like ion-exchange and electrodialysis, than a concentration process. The tubular UF systems usually remove molecules in the molecular weight ($Mw$) range from $10^3$ to $10^5$ daltons, but allow passage of medium-sized organic molecules and ions (Parkinson, 1983). An operating pressure of 0.1 to 1.0 MPa is generally used. UF membranes can be made from a much wider range of polymers since these membranes do not have to be water-absorbent. UF is also a more expensive process than RO.

In ultrafiltration, a solution is introduced into and pumped through a membrane. Molecules smaller than the membrane pores are forced through the membrane under hydrostatic pressure and are retained as filtrate. Molecules larger than the membrane pores are swept across the membrane surface in a cross-flow pattern and are returned to the feed reservoir as retentate. There is no build-up of materials on the membrane filter. Ultrafiltration utilizes membranes with small pore sizes ranging from 0.015 to 8 microns. The retention efficiency of membranes is dependent on particles and concentration; membrane pore size, length, and porosity; and overall flow rate (Ouellette and Cheremisinoff, 1985). Flux rate increases with transmembrane pressure drop and cross-flow velocity, but falls with increasing solute concentration and fouling (Brocklebank, 1987).

Hollow-fiber UF modules consist of hollow fibers (such as polysulfone) in polycarbonate and styrene-acrylonitrile housings. The molecular weight cutoff for the hollow fibers is 50,000 daltons. These are used for the concentration of proteins, viruses, and bacteria. These modules can be used in either of two modes: (1) dead-end filtration, and (2) cross-flow filtration. A 100-fold increase in the concentration of yeast cells is achieved in about 5 hours.

Ultrafiltration is used to remove water, inorganic and small organic molecules from protein and enzyme solutions. Membranes with up to 2 to 10 m² area, hollow-fiber and flat-sheet plate-stack membranes (10 mm to 15 m diameter), as well as plate-stack systems are used (Brocklebank, 1987).

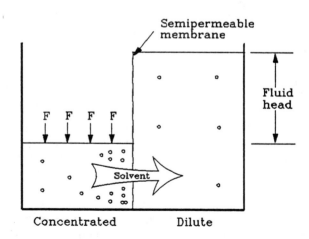

°: Solute    F = osmotic pressure + fluid head

**FIGURE 8.6.** Principle of separation by reverse osmosis.

Tangential-flow filtration is a technique that utilizes either microporous or ultrafiltration membranes to manipulate microbes or subcellular components. In typical tangential-flow filtration, a single pump is utilized to move fluid tangentially across the membrane (Ludwig and O'Shaughnessey, 1989).

## 8.4.3 Reverse Osmosis (RO)

Movement of the solvent through the membrane due to a difference in solute concentrations is osmotic flow (Figure 8.6). When a force is applied that is greater than the difference in the osmotic pressure, the solvent is transferred from a solution with high solute concentration to a solution with low solute concentration (Kaup, 1973) (Figure 8.7).

Reverse osmosis is used to concentrate many protein solutions. Salt, low molecular weight organic molecules, and ionic species of 0.01 $\mu$m diameter and less than 300 molecular weight are retained with reverse osmosis. It operates at relatively high pressures (0.70 to 14 MPa) to overcome osmotic pressure of the retentate. RO membranes are composed of special polymers that allow water molecules to pass through while holding back most other types of molecules. Commercially available membrane types are cellulose acetate, aromatic polyamide, and composite membranes. Cellulose-acetate membranes are prone to hydrolysis at extreme pH, and are subject to compaction at operating pressures. Aromatic polyamide membranes are prone to compaction, while composite membranes are sensitive to free chlorine. RO mostly uses either spiral-wound modules or hollow fibers. A hollow fine-fiber module can pack 16,600 m$^2$ of surface area into one m$^3$, while a spiral-wound module can only contain about 1000 m$^2$ in the same volume (Ouellette and Cheremisinoff, 1985).

RO is an efficient and energy-saving separation process (up to 20% w/w) compared to evaporation. Problems with RO include concentration polarization of the solute at the membrane surface, membrane fouling due to precipitation, and loss of water flux due to compaction, hydrolysis, or pH (Lee, 1979).

Generally, operating pressures are higher in RO than UF. UF and RO membranes are asymmetric in structure with skin at the surface being the selective barrier. UF membranes provide high fluxes, and all rejected components are accumulated at the surface, where they are removed by appropriate agitation. RO membranes are also asymmetric in structure, but the substructure is "sponge-like," with the core diameter increasing from the membrane surface to the membrane bottom (Brunner, 1987).

The water flux through a particular membrane is a function of its physical characteristics (porosity, membrane thickness, chemical composition) and the system conditions (temperature, pressure, fluid viscosity, concentration of dissolved solids, etc.).

Using RO, a brewery can make a series of beers of various alcohol levels from one master brew.

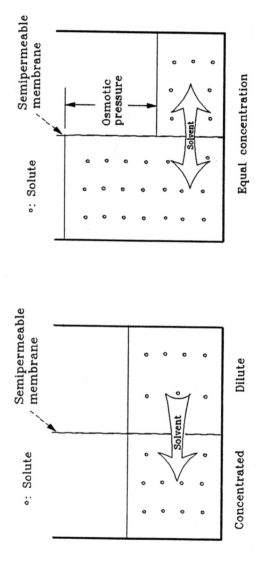

**FIGURE 8.7.** Characteristics of the reverse osmosis separation.

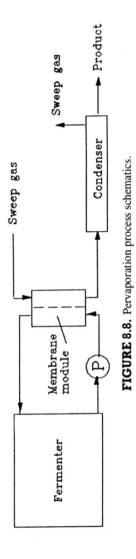

**FIGURE 8.8.** Pervaporation process schematics.

### 8.4.4 **Pervaporation**

In pervaporation, a liquid mixture directly contacts one side of a thin, non-porous polymer membrane and the permeate is removed in vapor phase from the opposite side (Figure 8.8). The transmembrane flux is caused by a partial pressure gradient across the membrane. This can be achieved by creating a vacuum in or sweeping a carrier gas through the down-stream compartment (Gudernatsch et al., 1987).

Pervaporation can be used to remove volatile components such as organic solvents from a fermentation broth. All constituents of the broth except some $CO_2$ and other low molecular weight organic solvents are more or less completely retained by the membrane (Strathmann, 1987).

### 8.4.5 **Electrodialysis**

Electrodialysis is a membrane process like UF and RO, but it removes only the ionic species. It uses two types of membranes: (1) the anion-permeable membrane, which permits the passage of anions only, and (2) the cation-permeable membrane, which allows only cations to pass through it. The feed is supplied to alternate membrane compartments, and due to the electrical driving force created by an applied D.C. voltage, the anions are attracted to the anode and the cations are attracted to the cathode. The ions will not reach their respective electrodes since they cannot pass through both types of membranes. This causes the formation of alternate concentrate and dilution compartments. The process is continuous and regeneration is not required. However, it is costly to achieve a high degree of demineralization (Mulligan and Fox, 1976).

Electrodialysis uses semipermeable ion-exchange membranes and electric force to transfer ions across the membrane barrier. Electrical potential gradients allow charged molecules to diffuse in a medium at faster rates than are attainable by chemical potentials. Thus, as shown in Figure 8.9, an electric current transfers ionic components into or out of protein solutions flowing between parallel alternate stacks of anionic and cationic membranes. Commercial electrodialyzers incorporate turbulence promoters and limit current densities to avoid polarization.

Electrodialysis is used for both salting out proteins by the addition of neutral salts, and desalting protein solutions. The advantages are a fast and controlled removal of salts, no product dilution, low membrane area, negligible product adsorption, easy salt recovery, independence of diffusion and protein concentration, and the use of lower concentrations of salting-out agents (Brocklebank, 1987). Electrodialysis can also be used to remove inhibitory end products from fermentations and to increase yield (Hongo et al., 1986).

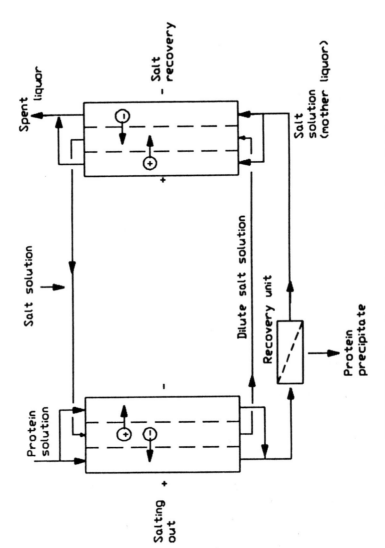

**FIGURE 8.9.** Scheme for protein salting-out using electrodialysis.

210

## 8.4.6 Microdialysis

Microdialysis is used to examine the chemical composition of a wide variety of samples without altering the fluid balance and with minimal damage to the sampling site. A small probe containing a tiny dialysis membrane (500 $\mu$m diameter, and lengths from 1 to 10 mm) is introduced into the sampling site. A suitable perfusion fluid is pumped at a low flow rate (e.g., 2 $\mu$L/min) through the probe and the resulting perfusate is collected and analyzed. Chemicals with molecular weights below the cutoff point of the probe membrane pass into the probe and are quantified in the perfusate. In addition, other compounds (e.g., pharmaceuticals, nutrients, toxins) can be introduced to the sampling site via the perfusion fluid to study effects on the surrounding tissue. The process is driven by a concentration gradient across the dialysis membrane.

The basic requirements for microdialysis include a precise syringe pump capable of delivering $\mu$L/min flow rates, microdialysis probes and the accessories that facilitate handling, precision syringes, and the appropriate analytical instrument (most frequently a liquid chromatograph).

## 8.5 **ELECTRIC METHODS**

### 8.5.1 Electrophoresis

With electrophoresis, electrically charged particles suspended in an electrolyte will move towards the oppositely charged electrode. In general, free boundary electrophoresis; zone electrophoresis on supports such as paper, starch, and cellulose powder; and electrophoresis in agarose and acrylamide gels, are used for lab-scale enzyme purification and characterization (Patel, 1985).

The advantages of electrophoresis are easy process control, high resolution, and no change of state. The disadvantages are electric double layers, thermal gradients and heat generations, hydrodynamic drag, and convention currents (Rudge and Ladisch, 1986; Chmiel, 1987).

Recycling effluent through a thin-film continuous-flow electrophoresis (CFE) chamber allows separation of a binary feed with negligible dilution of products, and permits throughput to be increased over thin-film CFE. Using the entire transverse thickness of the chamber and assuming square inlet ports increase throughput by 10 times. Reducing the electric field strength and reorienting the device into a more stable configuration allows a further 10 to 100 times increase in capacity. A further 10 times increase in capacity is made possible by using multiple feed streams. Thus, an overall increase in the capacity of thin film type CFE separations of 100 to 10,000 times can be achieved by recycling (Gobie and Ivory, 1986).

## 8.5.2 Gel Electrophoresis (GE)

For the separation of proteins and nucleic acids, high-resolution dynamic imaging (HRDI) is a new technology that permits visual monitoring of a gel during the separation process without modification of the samples of interest (Pickett et al., 1990). Sodium dodecyl sulfate-polyacrylamide GE (SDS-PAGE) separates proteins. HRDI produces images in which the brightness of protein bands is directly related to protein concentration. HRDI is based on the phase contrast technique. Phase differences in light passing through a gel (caused by local refractive index variations) are converted to intensity differences on an image screen.

Using HRDI, one can monitor separation as electrophoresis proceeds, then stop the run as soon as optimal resolution is obtained for the proteins of interest. The instrument also functions as a camera, allowing the user to record this image on a film. The film image may then be used to locate specific bands on the gel without staining or other modification. The ability to identify protein bands without staining is particularly important when one wishes to retain the biological activity of the protein of interest.

## 8.5.3 Capillary Electrophoresis (CE)

CE brings speed, quantification, reproducibility, and automation to the electrophoresis process. CE accommodates very high voltages (30,000 V) and current densities (5 $A/cm^2$) because of its efficient dissipation of heat resulting from the passage of electric current through the buffer contained in the capillary. CE is well-suited to automation because it allows on-line detection and automation of sample loading. Other features of the technique include the availability of new selectivities, very high separation efficiencies, simplicity, and very small sample requirements.

This technique is even more varied in potential applications than the more traditional techniques of LC or GC. Analytes may be separated by charge, mobility, size, hydrophobicity, ion exchange, etc. Analytes can be large or small, organic or inorganic, water-soluble or not, of biological origin or from the test tube. Capillary electrophoresis is an efficient separation technique in a number of applications including amino acids, peptides, proteins, nucleotides, oligonucleotides, classic pharmaceuticals, and neurotransmitters.

A CE system consists of a thin, open fused-silica capillary tube (10 to 200 $\mu$m I.D.) that runs between two aqueous buffer solutions, a high-voltage power supply, a means of applying the sample onto the column, a detector with CE flow cell, and a means of acquiring data. A sample is introduced into a reservoir containing an electrode. High voltage is applied between the two reservoirs, which drives the different components in a sample through the capillary at different speeds. The separation is based on the movement in the field based on charge and size. Each compound passes as a narrow zone or band through a detector and is measured by sensors of absorbance, fluorescence, conductivity, etc.

Most capillaries are made of silica that has been coated with an external layer of polyimide to make it more rugged and less susceptible to fracture. As the silica surface becomes ionized, an annulus of cation is attracted to the surface of the silica to balance the charge. When the current over the capillary is turned on, the cations of the silica tend to migrate toward the cathode. This migration of cations causes a concomitant migration of fluid through the capillary. This flow of liquid through the capillary is called the electro-osmotic flow (Olechno et al., 1990).

In free solution, the electrophoretic mobility ($u$) of a solute is defined as the electrophoretic velocity of a solute ($Vep$) per unit of electric field strength ($E$), or $u = Vep/E$. The internal diameter of the capillaries is either 50 or 100 $\mu$m, and the external diameter is approximately 360 $\mu$m. The external polymer coating has a thickness of approximately 15 $\mu$m (Figure 8.10).

Because electrophoretic mobility is a very strong function of temperature (approximately 2%/°C), precise temperature control is necessary to ensure the reproducibility of electrophoretic mobility measurements. The high ionic strength of the running buffer allows greater flexibility in the composition of

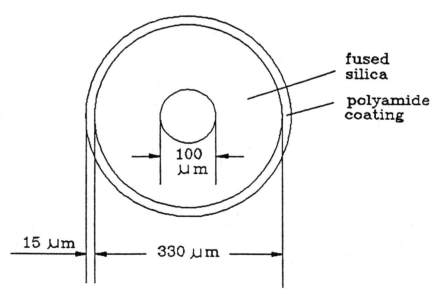

**FIGURE 8.10.** Dimensions of capillaries used in electrophoresis.

the sample, and this system can be run at high field strengths due to efficient liquid cooling of the capillary.

Strickland and Strickland (1990) optimized conditions for the separation of peptides by free-solution capillary zone electrophoresis, based on charge density. Fourteen bioactive peptides were selected on the basis of their diversity in composition and size.

High-performance capillary electrophoresis (HPCE) combines both liquid chromatography and gel electrophoresis. HPCE is much faster and less intensive than gel electrophoresis, and can resolve complex samples not easily separable by LC. This is suitable for (1) the separation of closely related protein variants, (2) high-resolution analysis of nucleic acid fragments, (3) fat amino acid identification in peptide sequencing, and (4) rapid quality assurance of biosynthetic products.

The superior separation power of micellar electrokinetic capillary chromatography (MECC) over ion-pairing liquid chromatography (IPIC) in the separation of fourteen nucleosides and nucleotides was discussed by Lahey and Claire (1990). All fourteen compounds including six neutrals, two monophosphates, two diphosphates, and three triphosphates were separated by MECC in less than 40 min. An optimized isocratic IPLC separation failed to adequately resolve half of the compounds examined in this study. Because temperature affected the mobilities of the phosphates differently from the mobilities of the neutral compounds, temperature optimization provided by the equipment was essential to the success of the separation.

## 8.5.4 Magnetic Separation

The number of biological applications of magnetic separation have increased at a rapid rate. This method separates hybrids from single-stranded DNA in nucleic acid assays, separates cellular sub-populations, and modifies/purifies biological substances (Vaccaro, 1990).

A magnetic separation system has (1) a magnetizable particle or bead that acts as a solid phase for sorting of the cell (material of interest), and (2) a magnet providing movement in the particle. Rapid magnetic particle movement requires a strong, high-gradient magnetic field.

The magnetic separation of cells involves cell incubation with the specific monoclonal antibody for 30 min on ice. Then the particles, with the cells attached, are magnetically pelleted on the side of the vessel, and the remaining unbound cells are decanted.

Any cell surface characteristic can be used for selection, as long as the magnetic particle is conjugated to an appropriate molecule. For small volumes of cells, particular care must be taken to minimize non-specific binding and trapping.

Dunlop et al. (1984) reviewed the basic magnetic principles involved in the separation and labelling techniques. New developments in magnetic labelling techniques for cells and microspheres have extended the useful range of magnetic separation, particularly high-gradient magnetic separation, into biotechnical areas.

## 8.6 **EXTRACTION METHODS**

### 8.6.1 **Adsorption**

Adsorption is a process of interaction between molecules of a fluid mixture and the surface of a solid, e.g., batch adsorption in stirred tanks for industrial production (Chmiel, 1987). Molecules and particulates in a fluid distribute themselves between a solid and the fluid. Strong binding facilitates adsorption even from crude mixtures, minimizing the amount of adsorbent required without too great a loss of resolution. Adsorption can be used for purification as well as some concentration at the early stages of a process.

Adsorption is generally less costly and easier to monitor and control than chromatography, but usually provides lower resolution (Table 8.5). Other advantages include reduced binding site occupancy by unwanted materials, reduced mass action effects, and an uptake that is much less dependent on input concentration (Thompson, 1987). Disadvantages are the inactivation of ligand, and a lower ligand density than in ion exchange.

Activated carbon filters may be used to recover solvents such as acetone, ethanol, freons, etc. Inorganic salts such as ammonium sulphate are used in large-scale enzyme purification processes.

In affinity/specific adsorption, resolution is achieved almost entirely by the specificity of the interaction between ligand and adsorbate. Water-soluble

TABLE 8.5. *Bulk adsorption and chromatography (Thompson, 1987, reprinted by permission of Gustav Fischer Verlag, Germany).*

| Characteristics | Adsorption | Chromatography |
|---|---|---|
| Capacity | High | Low |
| Resolution | Low (unless affinity) | High (if affinity) |
| Elution | Step | Gradient |
| Suspended solids | Good | Poor |
| Cost | Low | High |
| Complexity | Low | High |
| Operational problems | Low | High |
| Scale in use | Large | Low/medium |

polymers bearing appropriate ligand groups are added to a broth for stepwise removal of desired products. The polymers contain precipitation groups that permit quantitative precipitation of the polymer bearing the products by pH changes on the salt addition (Ouellette and Cheremisinoff, 1985).

## 8.6.2 Chromatography

Chromatography is a very common separation technique for analytical and preparative work. The mechanism of the separation will depend in different cases on adsorption, ion-exchange, affinity to immobilized ligands, size exclusion, and molecular sieving effects (Patel, 1985; Grabner, 1986). The systems are based on operating a packed column in a batch mode with the appropriate selection of packing and operating conditions for optimum resolution. The efficiency of liquid chromatography is determined by column length, particle size and packing quality, and mobile phase velocity and viscosity.

Various chromatographic techniques are summarized in Table 8.6 and discussed below.

TABLE 8.6. *Chromatographic techniques for protein recovery (modified from Brocklebank, 1987, reprinted by permission of Elsevier Applied Science Publishers Ltd., Barking, U.K.).*

| Technique | Separation Basis | Sorption Media | Usage Examples |
|---|---|---|---|
| Adsorption | van der Waal's forces, dipole moments | Calcium phosphate, hydroxyapatite, titania | Sorption from crude feedstocks, fractionation |
| Affinity chromatography | Specific surface structure | Agarose matrices with affinity ligands | Fractionation, sorption from crude feedstocks |
| Chromatofocussing | Isoelectric point | Agarose plus buffers | Fractionation |
| Gel filtration | Molecular size | Dextran polymers | Removal of solvents, fractionation |
| Ion Exchange | Molecular charge | Cellulose, dextran agarose matrices; inorganic sorbent matrices with reactive coatings | Concentration by sorption from crude feed stocks, fractionation |
| Reverse phase liquid chromatography (RRLC) | Hydrophilic and hydrophobic interactions | Silica with hydrocarbon coatings | Fractionation |

## 8.6.2.1 Adsorption Chromatography

A variety of compounds such as alumina, bentonite, charcoal, calcium phosphate, celite, and porous glass absorb and fractionate proteins and substances of low polar content (Smith, 1985; Brocklebank, 1987) are used.

## 8.6.2.2 Affinity Chromatography

This method is based on immobilizing an enzyme on a column. When a protein with impurities is passed on this column, the enzyme is covalently bound to the immobilized protein. The desired protein is then recovered by washing the column with an appropriate solvent (Patel, 1985; Birch et al., 1985). The affinity of the interaction is such that only the substance to be purified is adsorbed. A suitable support matrix and a ligand specific for the molecule of interest are required. A number of matrices are commercially available which are supplied in activated form for coupling to specific chemical groups on the ligand of interest. Matrix choice depends on rigidity, insolubility, stability, porosity, ability to couple ligand, and cost (Birch et al., 1985). The ligand must chemically attach to the matrix and have an interaction of appropriate affinity for the molecule to be purified. The interaction must be reversible to permit product recovery. Large-scale users have developed dye-ligand affinity chromatography. One of the major advantages of the affinity mode is the extraordinary range of ligands available. The replacement of very expensive natural ligands by synthetics is clearly a preferred route to anyone involved in the difficult process of developing suitable separation strategies.

Affinity techniques are able to concentrate dilute and stabilize the protein when absorbed onto the process by a single affinity stage (Jones, 1990).

## 8.6.2.3 Gel-Filtration Chromatography

This method allows the selection of an eluate fraction containing molecules of a specific molecular size. This relies on the ability to produce chromatographic gels of precisely controlled pore size using materials such as cross-linked dextrans, agarose, porous glass, or polyacrylamide (Smith, 1985). The fractionation of the proteins depends on the different abilities of various sample molecules to enter pores. The small molecules are partially retained in the matrix pores, whereas large molecules pass more quickly through the bed (Brocklebank, 1987). Scale-up is achieved by increasing the bed width but not the bed height.

## 8.6.2.4 HPLC

High-performance liquid chromatography (HPLC) utilizes small rigid beads (10 to 100 $\mu$m dia.) to provide fractionation of multicomponent feed streams (Brocklebank, 1987). Applications involve separation of molecules of similar size and composition. It offers advantages in resolution, speed, and ease of quantitative sample recovery.

## 8.6.2.5 Centrifugal Partition Chromatography (CPC)

CPC utilizes liquid-liquid partition, countercurrent distribution of solute mixtures between two immiscible liquid phases, in the absence of a solid support, to perform separation of complex mixtures of chemical substances. The stationary phase is solvent that is retained in the column by centrifugal force. The liquid mobile phase flows through the stationary phase, as in conventional chromatography. Solvents can be reclaimed for re-use by either fractional or azeotropic distillation (Cazes, 1988; Cazes and Nunogaki, 1987).

Separation columns are connected in series within column cartridges, and arranged around the rotor of a centrifuge with their longitudinal axes parallel to the direction of the applied centrifugal force. Separation of low-molecular weight substances is accomplished with organic solvent partition systems. Organic solvents widely used are butanol, chloroform and ethyl acetate. Water-miscible organics are generally used as modifiers to adjust the partition coefficients of the sample components. Combination of hydrocarbons with methanol or acetonitrile was used to separate fat-soluble vitamins, fatty acids, steroles, and hydrophobic substances (Cazes and Nunogaki, 1987). CPC can be easily adapted for large-scale continuous separations.

Industrial-scale systems contain circular disk-type partition cells, and the rotor is constructed by stacking several partition cell disks. This has been applied to separate, isolate, and purify a broad variety of substances — food and food additives, natural products, and fermentation products. High-performance centrifugal partition chromatography (HPCPC) is a process-scale separation technology (Cazes, 1989). High loadability and the throughput minimize processing cost.

## 8.6.2.6 Ion-Exchange Chromatography

This method is based on net electrostatic charge, charge density, the molecular size of the proteins, and the pH and ionic strength of the solution. It has been used for protein fractionation and enzyme purification. Resins and celluloses are the two most commonly used ion-exchange materials (Groginsky and Houghten, 1985). The selection of the matrix will depend upon the protein molecular weight, flow rate, ionic strength, and resolution. The scale-up is achieved by increasing column cross-sectional area. In ion exchange experiments, the first stage is sample application and adsorption. Unbound substances can be washed out from the exchanger. In the second stage, enzymes of interest are eluted from the column and collected in separate fractions. The fractionation is achieved since different proteins have different affinities for the ion exchanger due to differences in their charge (Patel, 1985).

### 8.6.2.7 Reversed Phase Chromatography

This method is based on a nonpolar stationary phase, a porous microparticulate chemically bonded alkyl silica, and a polar mobile phase. Separation is based on hydrophobic contacts between the molecule of interest and the stationary phase (Ouellette and Cheremisinoff, 1985).

### 8.6.3 **Solid-Liquid Extraction**

Solid-liquid extraction occurs because of the solubility of the solute in the solvent. The parameters that form the basis of the design are the solvent, the process temperature, the desired overflow and hold-up compositions, the type of extractor, the method of contact, and the size of the system.

### 8.6.3.1 Solvent

In most cases, the fact that "like dissolves like" provides guidance as to the general types of solvents that can be used. Other important solvent properties are selectivity for the solute, physical properties such as surface tension, viscosity, density, and boiling point, as well as thermal stability, toxicity, and cost (Mark et al., 1980).

### 8.6.3.2 Process Temperature

The process temperature affects the solubility, vapor pressure, and selectivity of the solvent, as well as solute diffusivity and product sensitivity. Thus, the temperature will depend on the types of solvent, solute, and solid involved, and is generally determined experimentally.

### 8.6.3.3 Overflow and Underflow Compositions

The terminal compositions of the overflow and underflow are usually related to some arbitrary physical limitation of the processing plant, such as

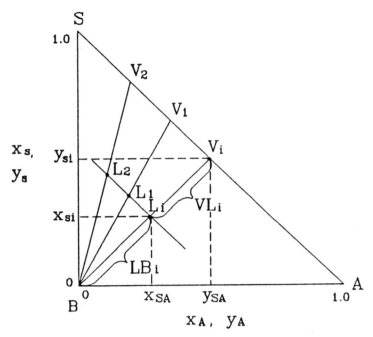

**FIGURE 8.11.** Design of a solid-liquid extraction process.

the rate of extract production needed or the rate of raw material throughput (Perry and Chilton, 1984).

### 8.6.3.4 Type of Extractor and Method of Contact

Percolation and extraction rate data are collected experimentally to determine whether the extraction should be done by percolation or by immersion, and how much extraction time is needed.

### 8.6.3.5 Extractor Size

For a given throughput rate, the size of the reactor depends on the number of stages that are required to achieve the desired underflow and overflow compositions (Perry and Chilton, 1984).

### 8.6.3.6 Design

The design is usually based on the concept of a multiple contact countercurrent extraction unit (Blackadder and Nedderman, 1971).

In countercurrent systems the overflow, $V$, is taken to flow in a direction opposite to that of the underflow, $L$. The right triangle is used as a convenient method of analyzing the equilibrium conditions of solid-liquid extractions (Figure 8.11) (Blackadder and Nedderman, 1971). The rectilinear sides of the triangle represent the mass fractions of $S$ (solvent) and $A$ (solute); $B$ represents inerts. For an overflow $V$ stream, the $Y$ (mass fractions) subscripted with $S$ or $A$ is used to represent the mass fraction in that stream, while $X$ (mass fractions) is used for the underflow or $L$ stream. All $V$ streams fall on the hypotenuse of the triangle, since this represents solution of the solute in the extracting solvent.

The location of the $L$ stream on the diagram is determined from experimental data. The $V$ is located on the hypotenuse of the triangle for each solution

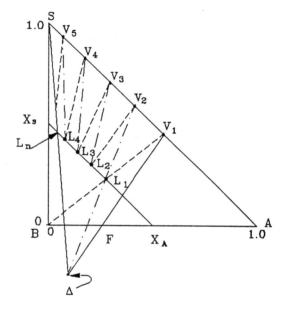

In this case 5 stages
(ideal) are needed.

- - - - -   Tie line (joins $V_1$ to B)
— · —   Construction line

**FIGURE 8.12.** Determination of the number of stages of the solid-liquid extraction process.

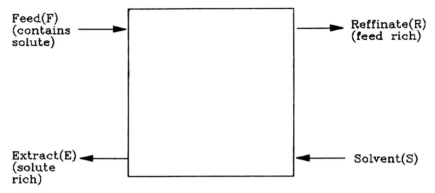

**FIGURE 8.13.** Liquid-liquid extraction system.

composition. $V$ is joined with $B$ and the point $L$ lies on the line $VB$ because it is a mixture of inerts and solution. The exact position of $L$ for a given retention is determined by the Lever rule: mass of solution retained/mass of inerts $= LB/VL$ (Blackadder and Nedderman, 1971). This is carried out for many compositions and a locus of $L$ is determined (Figure 8.12).

After locating $L$, the number of theoretical stages of the extraction process is determined. Generally, the initial compositions of the feed and the solvent, as well as desired final overflow and underflow compositions are known. A point "$\triangle$" is then located by $\Delta = F - V_1 = L_1 - V_2 = \cdots = L_{n-1} = V_n = L_n - S$. A tie line is drawn from $V_1$ to $B$, and its intersection point is $L_1$. *A construction line is drawn from* $\triangle$ through $L_1$ to yield $V_2$. This is repeated until steps fill the diagram, and the number of steps is equal to the number of theoretical stages.

A modified McCabe-Thiele plot is also used for this purpose (McCabe et al., 1985).

### 8.6.4  Liquid-Liquid Extraction

For the recovery of large-molecule products, and purification and concentration of products, liquid-liquid extraction is commonly used. This consists of the removal of a soluble constituent (solute) from one liquid (feed) into another (solvent) (Figure 8.13). It depends on the transfer of a solute from one liquid to another, formed by two immiscible solvents in contact with each other. In two aqueous phases, the principal factors influencing the distribution of proteins between the phases are molecular size, shape, and charge of the dissolved polymers and proteins (Chmiel, 1987). A further separation is required to recover the solute from the solvent, so that solvent can be reused in the extractor. For this, other separation methods such as distillation, evaporation, crystallization, and filtration can be used.

In the liquid-liquid extraction process, basically two immiscible liquids are brought into contact for mass transfer of components from one phase (aqueous) to the other (solvent). The following example indicates the characteristics of the extraction process. If a solution of acetic acid is agitated with an ethyl acetate, some of the acid — but little water — will enter the ester phase. Since at equilibrium the densities of the aqueous and ester layers are different, they will settle after agitation has stopped and may be decanted from each other. A certain degree of separation will have occurred since the acid-to-water ratio in the ester layer is different from that in the original solution, and also different from that in the residual water solution. This is an example of a stepwise contact, and can be carried out in a batch or in a continuous manner.

Liquid-liquid extraction is used primarily when distillation is impractical or too costly to use. If the relative volatility for the two components falls between 1.0 and 1.2, then liquid extraction is more practical than distillation. Extraction may also be attractive as an alternative when very low temperatures would be required to avoid thermal decomposition. Transfer of the dissolved component (solute) may be enhanced by the addition of "salting out" agents to the feed mixture or by adding "complexing" agents to the extraction solvent.

In considering a liquid-liquid extraction process, an understanding of the phase equilibrium relationship is vital. The liquid-liquid equilibrium is governed by the phase rule: $F = C - P + 2$, where $F$ is the degree of freedom, $C$ is the number of components, and $P$ is the number of phases. Thus, for a binary system, the composition of each phase is fixed if pressure and temperature are held constant. For a ternary system with fixed temperature and pressure, setting one composition fixes the other two.

The phase diagram (Figure 8.14) represents an isothermal system and indicates that $B$ (product or solute) is completely miscible with both $A$ (water) and $S$ (solvent), but the two solvents are only partially miscible with each other. Only those points lying within TMPNU will produce a two-phase system. The concentration of $B$ in the raffinate lies on the curve TP, while the concentration of $B$ in the extract (after solvent stripping) lies on the curve PNU. Feeds in the range DB cannot be processed because the addition of solvent $S$ will not produce a two-phase system. For a particular tie line, the distribution coefficient is determined by drawing a straight line through that point (representing the particular tie line) and the origin. The slope of that straight line is then the distribution coefficient ($m$) for $B$, which is defined by (Figure 8.15):

$m$ = concentration of $B$ in $S$ ($X_{BS}$)/concentration of $B$ in $A$ ($X_{BA}$), at equilibrium

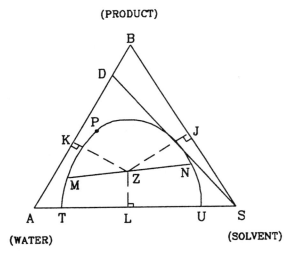

MN -- tie-line
Composition at Z -- A:B:S = ZJ:ZL:ZK
P -- plait point where the two phases merge into one
AT -- solubility of S in A
US -- solubility of A in S

**FIGURE 8.14.** Phase diagram for an isothermal liquid-liquid extraction process.

Thus, at equilibrium the ratio of solute in two liquid phases is called the distribution coefficient ($m$). The easiest method for determining $m$ is to mix solvent and feed containing varying quantities of solute in a separating funnel, and to analyze each phase for solute after settling. Where feed and solvent are essentially immiscible, the binary plot is useful. If there is extensive miscibility, a ternary plot would be preferable (Todd, 1983). If more than one solute is present, the preference or selectivity of the solvent for one ($A$) over the other ($B$) is called separation factor ($\alpha_{AB}$), i.e., $\alpha_{AB} = m_A/m_B$. This value must be greater than unity to separate $A$ from $B$ by solvent extraction.

Solvent-feed ($S/F$) ratio is used to estimate the goodness of solvent:

$$(S_1/F)/(S_2/F) = K_2/K_1$$

For example, $K_2/K_1 = 10$ means that 10 times more solvent-1 will be required than solvent-2. The selectivity coefficient ($\beta$) is defined as: $\beta = K_B/K_A$, where $K_B$ is the equilibrium coefficient for $B$ and $K_A$ is equilibrium coefficient for $A$.

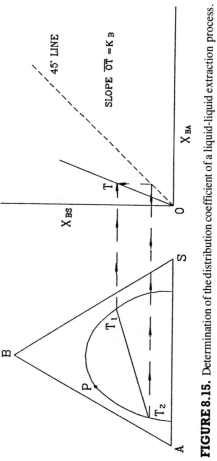

**FIGURE 8.15.** Determination of the distribution coefficient of a liquid-liquid extraction process.

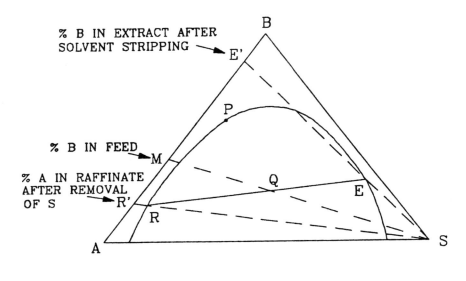

Solvent/Feed (S/F) = MQ/QS

**FIGURE 8.16.** Triangle method to analyze a liquid-liquid extraction process.

After the feed has contacted with the selected solvent $S$ at point $Q$, the mixture $Q$ breaks down to give extract $E$ and raffinate $R$ at opposite ends of the tie line RQE (Figure 8.16).

When the components are immiscible, the triangular phase diagram is replaced by a simple plot of the mass or mole fractions of the solute ($C$) in the nonsolute components ($A$ and $B$) (Figure 8.17). At large dilutions the compositions are proportional, and are considered to follow the distribution law.

## 8.6.4.1 Solvent Selection

In selecting a solvent for the liquid extraction processes, the following characteristics of the solvent should be considered (Todd, 1983):

(1) Selectivity: This is a numerical value describing how well the solvent will preferentially extract the solute from the aqueous phase. No extraction is possible when $\beta = 1.0$. A solvent with a poor selectivity (i.e., near unity) would require a large solvent-per-feed ratio and a large number of extraction stages to get a good separation.

(2) Solvent recoverability: Where the solvent is partially miscible with the aqueous phase, recovery of solvent from the raffinate may be necessary as well. Solvent recovery is usually accomplished by distillation. Hence, the relative volatility of the extraction-solvent to nonsolvent components should be significantly greater than 1.

(3) Density: The difference in density between the two liquid phases in equilibrium affects the countercurrent flow rates that can be achieved in extraction equipment, as well as the coalescence rates.

(4) Distribution coefficient ($m$): This is a numerical value showing the ability of the solvent to dissolve the solute. A high $m$ value indicates that a lower amount of solvent is needed for a specific requirement of the solute recovery. Furthermore, it directly affects the selectivity.

(5) Solvent solubility: This refers to the miscibility of the solvent and water. A high immiscibility of solvent and the aqueous phase is desirable since this indicates a small $K_D$ for water and thus leads to a high selectivity value for the solute. Furthermore, separation costs are inversely proportional to the selectivity. High immiscibility of solvent and water also implies two important factors to be considered—density difference, and interfacial tension between the two phases. Density difference between the organic and aqueous layers directly influences the flow.

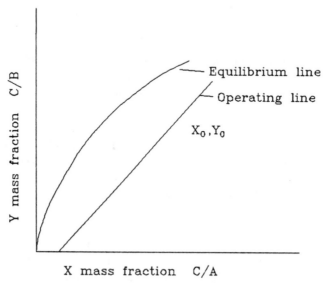

**FIGURE 8.17.** $X$ versus $Y$ when components are immiscible in a liquid-liquid extraction process.

Large interfacial tension is preferable since this implies a high coalescence rate of dispersed droplets. However, large interfacial tension also indicates a necessity for mechanical agitation.

(6) Other: Solvent properties such as low corrosiveness, non-flammability, low cost, etc., should also be considered in selecting a particular solvent. Solvent mixtures can be employed in liquid-liquid extraction. Solvent mixture can give $m$ and $\beta$ values substantially different from those measured for individual solvent constituents (Munson and King, 1984). However, before picking a particular solvent mixture, the foregone solvent criteria should be considered.

## 8.6.4.2 Agitation

Agitation depends on the fluid properties and extraction column geometry and size. With agitation, the mass transfer rate of the solute can be promoted. Different extraction processes require different types of agitators. For example, in a plate extraction column, a variable-speed motor with a cam is required for the up-and-down motion of the perforated and baffle plates. Mechanical agitation is required for a high interfacial tension between the two phases. Furthermore, such factors as the impeller diameter and Reynolds number also have an influence on the requirement and efficiency of the degree of agitation. If a non-Newtonian liquid is encountered, pilot-plant (batch, fed-batch, or continuous) experiments are conducted before scaling up.

## 8.6.4.3 Extraction Systems

In liquid-liquid extraction, the following systems are generally used.

(1) Simple contact: This is the most economical method. The extraction is fixed by equilibrium relations and the quantity of solvent used (Figure 8.18).

(2) Crosscurrent multi-stage contact: The best results are obtained when equal volumes of solvent and feed are used at each stage. A large amount of solvent is required to obtain a high degree of extraction (Figure 8.19).

(3) Countercurrent multi-stage contact: The feed and solvent enter at opposite ends of the series of extraction stages (usually 3 to 6). For a given amount of solvent and a fixed number of stages, this method is more efficient than the crosscurrent (Figure 8.20).

(4) Countercurrent extraction with reflux: Due to feed dilution in countercurrent systems, a limited amount of extract-product composition is obtained. The extract composition can be increased by using an extract reflux after solvent separation. This requires that the extract reflux be returned at an intermediate point in the extraction

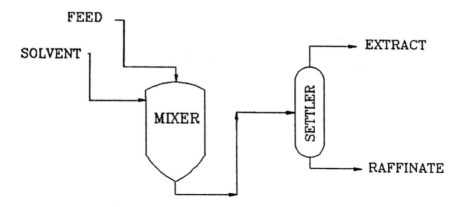

**FIGURE 8.18.** Schematic diagram of a simple contact liquid-liquid extraction.

column, i.e., not mixed with the feed stream. The reflux return must not set the system to the plait point (Figure 8.21).

(5) Double solvent (fractional) extraction: If feed is a multi-component mixture with $n$-components, then $(n - 1)$ countercurrent cascades will be required. The reflux of separated solutes may be used at either or both ends of the extraction columns to reduce the number of stages required (Figure 8.22).

(6) Controlled cycle operation (Breuer et al., 1977): This operation allows only one phase flow at a time, and only for a controlled interval. The flow frequency of this operation is much less than that in pulse operation. The operating algorithm is shown in Figure 8.23.

By installing extraction plates along the inside extraction column, the "separation period" in controlled cycle operation can be eliminated. The function of extraction plates is to enhance the mass transfer rate by repeated dispersion and coalescence of the two phases (Steiner and Hartland, 1980; Perry and Chilton, 1984). Baffle blades are only applied on the top ends of the extraction plates, to reduce the effects of dead zones arising from non-ideal mixing.

### 8.6.4.4  Design Criteria for Countercurrent Multi-Stage Extraction

Most countercurrent multi-stage extraction columns employ combinations of mixing and settling zone arrangement to give the desired mass transfer rate of solute. In the mixing zones, one phase is dispersed in the other by impeller action. There are several kinds of continuous countercurrent liquid-liquid

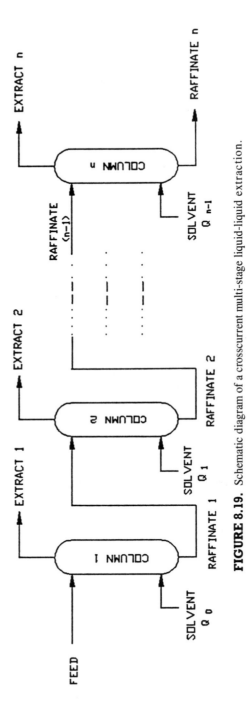

**FIGURE 8.19.** Schematic diagram of a crosscurrent multi-stage liquid-liquid extraction.

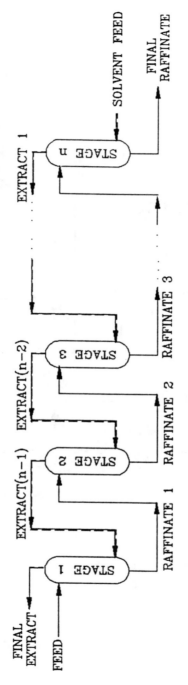

**FIGURE 8.20.** Schematic diagram of a countercurrent multi-stage liquid-liquid extraction.

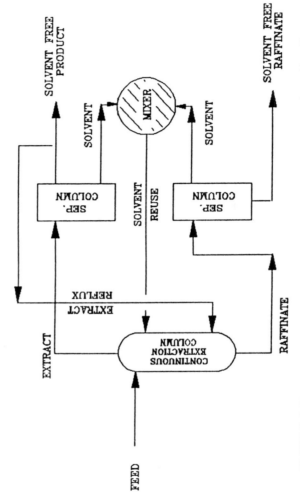

**FIGURE 8.21.** Schematic diagram of a countercurrent liquid-liquid extraction with reflux.

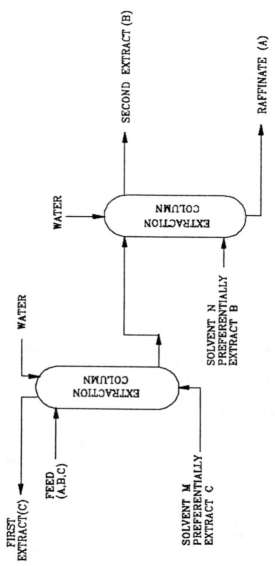

**FIGURE 8.22.** Schematic diagram of a double solvent liquid-liquid extraction.

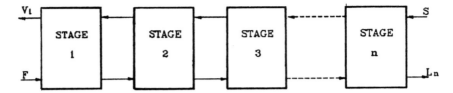

**FIGURE 8.23.** Schematic diagram of a controlled cycle liquid-liquid extraction.

columns available, such as reciprocating plate extraction column, rotating disc contractor, Schelbel-York extractor, Treybal extractor, etc. The design of these columns in commercial scale is not an easy operation.

The transfer of the desired product(s) from one phase to another with the least cost and with the highest efficiency is the utmost objective in liquid-liquid extraction. In designing a countercurrent liquid-liquid extraction column, design variables include column diameter and height, agitator speed, impeller-to-column diameter ratio, the number of compartments, geometry and orientation of baffles, power requirement, and liquid flow rates.

Most of the commercial extractors are cylindrical in shape with an impeller, driven by a motor, sitting in the centerline of the vessel, to provide dispersion of drops of one phase in another by impeller action. The turbulence enhances the mass transfer rate, i.e., it lowers the mass transfer resistance of the solute between phases. The mass transfer rate, in turn, depends on the concentration gradient of solute between the phases, the size of droplets, the rate of repeated coalescence and dispersion, etc. The greater the concentration difference between the two phases, the larger the mass transfer rate will be.

The degree of agitation controls drop sizes. At high rotor speeds, very fine droplets are formed. The revolving motion leads to a lower column capacity, but a higher mass transfer rate. Furthermore, axial mixing, which consists of eddy diffusion or backmixing may arise due to intense agitation. A decrease in concentration difference will increase axial mixing but decrease mass transfer rate. Intense agitation hinders the coalescence of droplets, and thereby lowers stage efficiency.

Emulsion may also occur at the settling zone, usually at the top and bottom of the column, because of high rotor speed. Small bubble size, reduced density difference between phases, and/or low interfacial tension also cause a slower rate of sedimentation of the coalesced emulsion. Emulsion is generally not desirable in extraction processes because it lengthens the residence time. The rate of repeated dispersion and coalescence of droplets promotes the mass transfer rate. Generally, the stage efficiency increases up to a maximum and decreases as the rotor speed goes up (Soares et al., 1982).

Baffle geometry and its orientation have a definite effect on the performance of the column. Typically, vertical baffles are installed inside the vessel to reduce fluid rotation. The presence of these baffles improves mixing effects. An advantage to installing horizontal baffles is that the approximate mixing height, and hence the size of the eddy, can be reduced during scale-up (Scheibel, 1956).

Rotating seals are especially important in preventing leaks as well as contaminations.

## 8.6.4.5 Design

Liquid extraction process design and analysis requires liquid-liquid equilibrium data. A typical solvent extraction stripping tower is shown in Figure

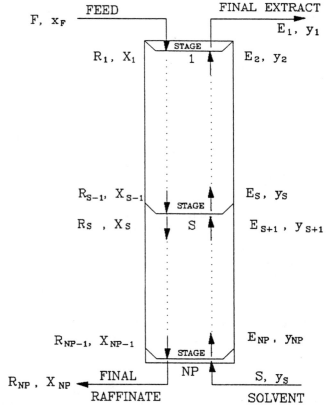

**FIGURE 8.24.** A typical solvent extraction stripping tower.

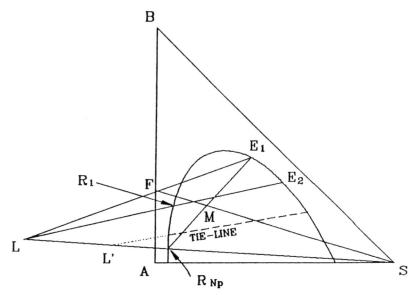

**FIGURE 8.25.** Phase equilibrium diagram to locate "$M$" for a liquid-liquid extraction process.

8.24. A material balance (Brunner, 1987) around the whole system and for solute provides the location of the point $M$ (Figure 8.25) within the two-phase region in the phase equilibrium diagram. After determining the locations of $F$, $S$, $E_1$, $L$, and $R_{NP}$, $R_1$ can be located by the tie-line that extends from $E_1$. A line from point $L$ through $R_1$ provides $E_2$ and so on.

If point $L$ is located in such a way that one of the lines connecting it coincides with a tie-line, it will require an infinite number of stages to reach the condition, while it also indicates the minimum of solvent rate requirement. However, tie-line data are not always available at the specific $E$ points. An alternate method, which requires the construction of an operating line, is used to evaluate the number of theoretical stages. An operating curve can be located on the corresponding distribution diagram. A few random lines are drawn from point $L$ (between lines $LE_1$ and $LS$) as shown in Figure 8.26(a), to intersect the equilibrium curve. The corresponding solute concentrations in raffinate and extract phases are located in the distribution diagram. The number of theoretical stages is then counted by stepping off graphically between the distribution curve and the operating curve as indicated in Figure 8.26(b).

If the two liquids are essentially immiscible, e.g., toluene and water in a toluene-propionic acid-water system, then the operating curve becomes a

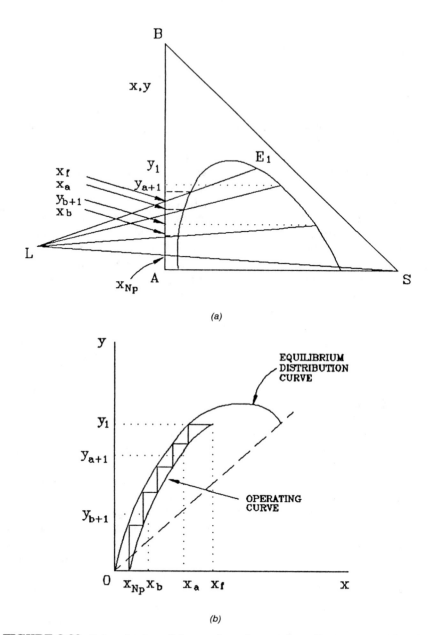

FIGURE 8.26. Determination of the number of stages for a liquid-liquid extraction process.

straight line with slope equal to the solute-free phase ratio in the distribution diagram.

$$\text{Solute free phase ratio } (w) = F(1 - X_F)/S$$

This operating line must pass through the point $(X_F, 0)$ in the distribution diagram since this is the final raffinate exit point, where fresh solvent is fed in. Again, the number of theoretical stages are counted by the stepping-off method illustrated above.

Height equivalent to a theoretical stage (HETS) is calculated by dividing the column height by the number of theoretical stages obtained. The stage efficiency is a measure of actual stage performance and is defined as the ratio between the number of theoretical and actual stages.

Maximum stage efficiency and minimum HETS are the most desirable features of an extraction column. As the efficiency per stage goes up with increasing column height, the value of HETS also goes up. Therefore, a compromise between stage efficiency and HETS value should be realized. Factors such as column diameter, compartment height, phase ratio, and rotor speed also influence stage efficiency and HETS. Since these variables are interrelated and interdependent, experimental data on a small extraction column should be obtained prior to scale-up.

The commercial extractors include spray, packed, and perforated columns; and centrifugal extractors accelerate phase separation by the use of centrifugal force.

### 8.6.5 Precipitation

Precipitation as a means of product purification and separation is accomplished by the aggregation of molecules caused by a change in the environmental factors including pH, temperature, ionic strength, and dielectric constant (Chmiel, 1987). This is achieved by using inorganic salts (e.g., ammonium sulphate), organic solvents (ethanol, acetone) or high molecular weight polymers. By differential precipitation, the product can be retained in solution or collected as a precipitate (Smith, 1985). This is a low-cost and high-yield process usually used to separate a crude protein isolate from solution. It requires little energy but is slower than mechanical separation.

Precipitation is affected by altering the solubility of proteins using various precipitating agents. Precipitation by non-ionic polymers such as polyethylene glycol (PEG) and dextran is attractive in the recovery of enzymes because of its ability to preserve the proteins' structure.

The precipitation curve of the bovine serum albumin using PEG was measured (Anon, 1988a). The amount of polymer reached a minimum at the isoelectric point of the protein and increased with a decrease in its molecular weight and a decrease in the initial protein concentration. At a pH greater than the pI of the protein, increasing the ionic strength shifted the precipitation curve to a higher PEG concentration whereas the effect of ionic strength was the opposite at a pH below the pI.

To achieve any degree of fractionation by precipitation, values of pH, temperature, and ionic strength must be selected which minimize protein solubility for each of the components of the protein mixture (Brocklebank, 1987).

Batch, continuous-stirred tank reactor (CSTR), and tubular reactors may be used for protein precipitation. Factors to be considered for their design are summarized by Brocklebank (1987).

## 8.6.6 Super-Critical Extraction

For the extraction of apolar substances, super-critical $CO_2$ extraction can be used. In this process, the removal of the extractant from the product is simple, and high diffusivities and low viscosities of super-critical solvents result in higher rates of extraction in fixed beds and columns in comparison to normal liquid solvents (Chmiel, 1987). It is an energy-efficient process, and can replace some distillation processes.

$CO_2$ is a very poor solvent. However, with an increase in pressure, its density increases and its extractive power improves substantially. The addition of solute to the $CO_2$ changes the super-critical conditions. These changes are required to obtain super-critical solutions for substances of different molecular weights and in the presence of polar groups that can interact with the $CO_2$.

## 8.6.7 Ion Exchange

Ion exchange involves the reversible interchange of ions between a functionalized insoluble resin and an ionizable substance in a solution. Ion-exchange resins are a special class of poly electrolytes, and consist of cross-linked polymer matrices (Dechow, 1983), such as zeolite (a group of hydrated aluminum complex silicates with cation-exchange properties). The resin porosity determines the size of the ions that may enter, and their rate of diffusion and exchange. The resin selectivity is affected by the starting ionic form of the resin and the degree of cross-linking of the resin. When liquid passes through the filter, the ion-exchange material replaces mineral ions in the fluid with ions of the material. Eventually the ion-exchange capacity is exhausted and it is necessary to regenerate the resins by chemical addition. The process is costly, due to the large consumption of regenerative chemicals (Morris, 1982).

The salts in the fermentation feed stream can be removed by passing the stream through a cation exchange column in the hydrogen ion form, followed by an ion-exchange column in the hydroxide form (Dechow, 1983). Many fermentation products can be purified by adsorbing them on the ion-exchange resins to separate them from the fermentation broth. It can also be used to concentrate desired products of fermentation in a manner similar to purification.

Process operating capacity and column efficiency can be evaluated on a laboratory-scale experiment. The resin column should be >2.5 cm in diameter to avoid wall effects (Dechow, 1983). Subsequent process optimization in the laboratory calls for setting up a packed resin column of approximately the bed depth to be used in the final equipment (1 to 2 m).

### 8.6.8 Aqueous Two-Phase Systems

Aqueous two-phase systems are generated by mixing aqueous solutions of two water-soluble polymers, or a polymer and a salt. These systems offer mild conditions for the separation of cells, organelles, proteins, and other biomolecules, in biochemical processes. The biocompatibility, fast separation process, and the ease to scale-up are some of the characteristics that make aqueous two-phase systems attractive alternatives to other techniques (Mattiasson and Kaul, 1986).

### 8.6.9 Foam Fractionation

Protein concentration using foam fractionation is based on the selective adsorption of proteins at the gas – liquid interface generated by bubbles rising through the solution. The foam carries proteins with it. Foam fractionation provides high separation efficiency from very dilute solutions, reasonable capital investments, and low maintenance and operational costs.

There exists an optimum bubble size for maximum enrichment, based on the inlet protein concentration and superficial gas velocity. The dependence of protein enrichment on pH and ionic strength is reported to be insignificant because of stronger dipole interactions than are produced with an electrical double layer.

Protein enrichment and recovery were measured for a model system of bovine serum albumin (BSA) (Anon, 1988a). Protein enrichment was found to increase with decreasing feed concentration and gas velocities, and increasing bubble size and pool height.

## 8.7 **THERMAL METHODS**

### 8.7.1 Distillation

Distillation is a separation process based on the difference in boiling points of liquids. It can recover a large amount of ethanol from waste brewery effluents (Ouellette and Cheremisinoff, 1985). The three main methods for recovering energy from distillation plants include the multiple effect method, indirect vapor recompression, and direct vapor recompression. Direct vapor recompression is the most energy-efficient choice (Danziger, 1979).

### 8.7.2 Drying

Dryers such as spray, flash, and fluidized bed supply the heat of vaporization by passing hot gases through or across the wet solids. In direct drying systems (tray and rotary), vaporization heat is supplied through a heated wall to the wet solids. Indirect heat dryers (rotary drum, vacuum, and vacuum freeze) transfer heat to the wet material by conduction.

Advantages associated with flash dryers are their low capital cost, short residence time, simultaneous drying, conveying and product recycle, and low space requirements. Disadvantages are their unsuitability for highly abrasive materials and for high moisture (80%) materials. In flash drying, a material to be dried is introduced to a hot gas stream in a duct, and the product is separated from the spent gas at the discharge duct end.

Advantages associated with fluid-bed dryers are their medium retention time, simultaneous drying and size classifying, and small space requirements. Disadvantages are a high distributor plate temperature, and particle-size-dependent drying (Ouellette and Cheremisinoff, 1985). To scale up flash dryers, the cross-sectional area of the air duct is increased proportionally, while for a fluidized bed dryer, the bed area is increased proportionately (Quinn, 1983).

Advantages associated with spray dryers are the one-step process and minimum heat degradation due to short residence time. Disadvantages are their large space requirements and high initial capital cost. The proper gas flow for evaporation in spray drying can be determined by the use of the special psychometric charts.

Advantages associated with tray dryers are their ability to handle anything from thick slurries to fine powders, small space requirements, and a closed circuit system to recover vapors. Disadvantages are their higher capital cost, and fouling of heating surfaces.

Advantages of a rotary dryer are a reasonable capital cost and close temperature control. Disadvantages are the large space requirements and the fact that it requires a dust collection unit (Ouellette and Cheremisinoff, 1985).

Drum dryers cannot process lumpy materials. Advantages of vacuum dryers are low-temperature processing and solvent recovery. Disadvantages are the high capital, maintenance, and operating costs. Advantages of a vacuum freeze dryer are little chemical change, minimum loss of volatile components, and elimination of case hardening and oxidation. Disadvantages are high capital, operating and maintenance costs (Ouellette and Cheremisinoff, 1985).

Dryer selection will depend upon many factors, including feed condition, cost, and required retention time.

### 8.7.3 Evaporation

Evaporation is the removal of solvent as a vapor from a solution by means of an energy input to the process (Figure 8.27). Important properties of liquids affecting evaporator design and performance are concentration, foaming, salting, scaling, fouling, corrosion, product quality, and temperature sensitivity (Freese, 1983). Other solid and liquid properties required to be considered are specific heat, toxicity, radioactivity, explosion hazards, freezing point, and the ease of cleaning.

Evaporators are particularly useful for heat-sensitive materials. These can operate batchwise, continuously, or in a semi-batch. Commonly used types are jacketed vessels, horizontal tube, long and short vertical tube, falling film, plate, and multiple effect evaporators (Freese, 1983).

Advantages associated with thin-film evaporators are the large heating surface, small space requirements, and suitability for heat-sensitive materials. Disadvantages are their unsuitability for salting or severely scaling liquids, and liquids with suspended solids (Ouellette and Cheremisinoff, 1985).

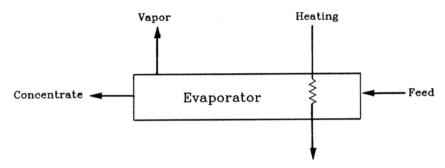

**FIGURE 8.27.** A typical evaporation process.

Advantages associated with multiple effect evaporation are the low maintenance cost, reduced steam requirements, and short residence time. The disadvantage is high capital cost.

Mechanical recompression is most practical for low temperature differences and low boiling point elevations. Vapor recompression evaporators are suitable when evaporating large quantities of water from solutions that do not result in particularly viscous products or give rise to high boiling point elevations.

## 8.8 **TECHNIQUE SELECTION**

Table 8.7 summarizes the size ranges for which various separation techniques can be used. Molecular size or configuration determines a suitable separation technique.

Separation technique selection depends on initial broth characteristics, such as particle size, concentration, consistency, and impurities, as well as on the final product properties such as concentration and form. If you do the right things upstream, you can solve 90% of the problems of down-stream processing. Changing molecular biology or fermentation parameters can reduce or even eliminate problems (Knight, 1989).

Non-thermal physical separation methods are relatively better in biological processes than are chemical methods. Final choice, generally, depends on product quality, time, economics, and availability of suitable equipment. Centrifuges can be used to separate two components if they have a considerable density difference. Similarly, solvent extraction can be used to separate them if they have large differences in solubility. The selection of solid-liquid

TABLE 8.7. *Size ranges of various separation techniques (modified after Smith, 1985).*

| Technique | Particle or Solute Size Range ($\mu$m) |
|---|---|
| Centrifuge | 0.6 to 1000 |
| Chromatography | 0.005 to 1 |
| Distillation | 0.001 to 0.9 |
| Electrodialysis | 0.0008 to 0.08 |
| Filter (fiber) | 10 to 1000 |
| Foam fractionation | 0.03 to 1000 |
| Freeze concentration | 0.001 to 0.9 |
| Ion exchange | 0.0008 to 7 |
| Microfiltration | 0.01 to 10 |
| Reverse osmosis | 0.0008 to 0.08 |
| Sedimentation | 100 to 1000 |
| Solvent extraction | 0.001 to 2 |
| Ultracentrifuge | 0.005 to 1 |
| Ultrafiltration | 0.005 to 10 |

separation technique is also influenced by (1) capacity, (2) solid concentration in feed, (3) process requirements, (4) particle size and density difference from mother liquor, and (5) precipitate handling characteristics.

Many times a combination of different techniques is used to separate the desired product. For example, the separation of a highly purified product may require the following techniques: (1) filtration or centrifugation to remove insoluble particles, (2) liquid-liquid extraction to separate and concentrate substances of different polarities, (3) chromatography for product purification, and (4) centrifugation or crystallization to achieve final product separation.

# CHAPTER 9

# *Fermentation Techniques*

## 9.1 **INTRODUCTION**

*Fermentation* performs complex transformations upon organic materials by using the metabolic activity of microorganisms. It is a reaction or series of reactions in which a biocatalyst (a microbial cell or an enzyme) is used to convert a substance into desirable products, e.g., alcoholic beverages, cheeses, and breads. Fermentation technology originated with the use of microorganisms for the production of foods such as yogurt, cheeses, alcoholic beverages, vinegar, fermented pickles and sausages, fructose, soy sauce, etc. Pasteur's work on the "diseases of beer" and the recognition of the involvement of yeasts in the fermentation were provided scientific basis, and by 1896 the Carlsberg brewery in Copenhagen was using pure yeast cultures (Brown et al., 1987).

A fermenter is a bioreactor used to provide optimum conditions for the controlled growth of microorganisms by regulating agitation, temperature, and aeration of the fermentation broth. To monitor and control fermentation process parameters such as pH and dissolved oxygen concentration should be measured. The maintenance of aseptic operating conditions is important for a successful fermentation. The fermenter vessels are constructed of stainless steel and are jacketed and pressure-coded so that the equipment can be safely steam-sterilized *in situ* (Wernerspach, 1986).

Fermenters can be operated on a batch, semi-continuous, or continuous basis. Fermentation can occur in static or agitated cultures, in the presence or absence of oxygen, and in aqueous or low moisture conditions. The biocatalysts (microorganisms, enzymes, cells) can be free or can be attached to surfaces by immobilization (see Chapter 6). In general, fermentation industries require a bioreactor that can meet a number of different operating conditions including varying consistency, aeration rate, agitation intensity, and fermentation volume (Smith, 1985).

## 9.2 **FACTORS AFFECTING THE FERMENTATION**

A fermenter provides optimum conditions for the microorganism growth. Hence, to design such fermenters a thorough understanding of the require-

**245**

ments and limitations of the organism's growth is required. Requirements for the optimum growth of microorganisms are as follows.

## 9.2.1 Temperature

Fermentation occurs over a wide range of temperatures as long as the temperature is relatively constant. Once an optimum temperature range is selected for a microorganism, fluctuations can result in an inefficient process (Price, 1985).

The overall growth rate is the difference between the growth and death rates of the organism given by the Arrhenius equation (Russell, 1987):

$$dm/dt = M[A \exp(-E/RT) - A' \exp(-E'/RT)]$$

where $M$ is the cell mass; $A$, $A'$ are constants for growth and death, respectively; $E$ and $E'$ are activation energies for growth and death, respectively; and $T$ is the absolute temperature (Figure 9.1). Data on the effect of temperature on a range of organisms has been given by Atkinson and Mavituna (1983). The quantity of heat generated depends on the balance between cell growth and maintenance requirements. This heat must be removed from the fermenter to maintain optimum conditions. The generated heat can be estimated from the oxygen uptake rate (Cooney et al., 1969): $Q = 0.12\ (OU)$, where $Q$ is the heat generation rate; and $OU$ is the oxygen uptake rate.

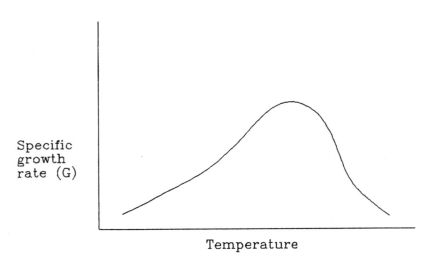

**FIGURE 9.1.** Specific growth rate of microorganisms with regard to broth temperature (Russell, 1987). Reprinted by permission of Elsevier Applied Science Publishers, Ltd., Barking, U.K.

## 9.2.2 pH

Most organisms will retain viability at pH 7, but cells will remain viable over a range of several pH units. Several organisms produce different metabolites at different pH conditions, and in such cases better pH control is needed for optimum yield (Russell, 1987). Optimum pH range of 6.4 to 7.6 has been reported for the methane bacteria (Price, 1985). Optimum pH for fermentation and ethanol production has been reported as pH 4.0, and maximum denitrification is reported to occur at pH 7.0.

## 9.2.3 Moisture

Moisture is required by all bacteria, and they can tolerate conditions ranging from slight amounts of moisture to dilute solutions of nutrients.

## 9.2.4 Nutrients

A microbial cell contains a C:N:P:S ratio of approximately $100:10:1:1$ (Price, 1985). Many organic compounds affect fermentation. These include organic solvents, pesticides, alcohols, and high concentrations of long-chain fatty acids. The Monod equation provides a relationship between growth rate and the concentration of each substrate:

$$G = Gm \, [S/(K_s + S)]$$

where $G$ is the specific growth rate; $Gm$ is the maximum $G$; $S$ is the substrate concentration; and $K_s$ is a reaction constant, equal to the concentration of substrate when $G$ is half the maximum rate. Figure 9.2(a) shows this equation graphically. When $S > 10 \, K_s$, the $G$ becomes constant. At higher nutrient concentration, $G$ is reduced due to the organism inhibition [Figure 9.2(b)]. This equation does not always predict the organism behavior at low $S$, since low $S$ may also start a different reaction. Most organisms will utilize one growth nutrient when using commercial substrates, and will consume the second nutrient when the first has been depleted. This would increase the cycle time for a batch fermenter, but in a continuous fermenter the loss of substrate will be high (Russell, 1987). Thus, the optimization of the nutrient will increase $G$ and the yield. In batch fermenters, nutrient should be continuously added to maintain optimum level; while in continuous fermenters, additional care is required in the feed distribution.

## 9.2.5 Cations

All cations are capable of producing toxic effects in any organism if the concentrations are high enough, but the relative toxicity of the cation varies

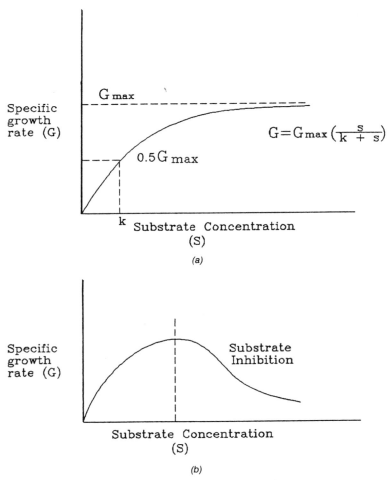

Specific growth rate (G)

$G = G_{max} \left( \dfrac{s}{k+s} \right)$

G max

0.5 G max

k Substrate Concentration (S)

(a)

Specific growth rate (G)

Substrate Inhibition

Substrate Concentration (S)

(b)

**FIGURE 9.2.** Specific growth rate of microorganisms in a fermenter: (a) predicted by Monod equation, (b) affected by substrate inhibition (Russell, 1987). Reprinted by permission of Elsevier Applied Science Publishers, Ltd., Barking, U.K.

(Price, 1985). Low concentrations of cations have a stimulating effect on microbial metabolism. Cations play a nutritional role in the metabolism of all organisms, serving as metabolic activators for a wide variety of enzymes.

## 9.2.6 Oxygen

In aerobic fermentation, since microorganisms can only use dissolved oxygen and oxygen is less soluble in water, mass transfer of oxygen into the broth is generally the rate-limiting factor. The oxygen required for cell growth can be estimated by (Mateles, 1971):

$$\text{kg oxygen/kg cells} = \frac{32 \cdot C + 8 \cdot H + 16 \cdot O}{Y \cdot M} + 1 \cdot O'$$

$$- 2.67 \cdot C' + 1.7 \cdot N' - 8 \cdot H'$$

where $Y$ is the cellular yield/substrate; $M$ is the molecular weight of the carbon source; O, C, H, N are numbers of atoms/substrate molecule; and $O'$, $C'$, $H'$, $N'$ are fractions present in the cell.

Oxygen availability to an organism is measured by the dissolved oxygen tension (DOT), expressed as a percentage of oxygen saturation. Above a critical DOT the growth rate and formation of products are independent of the DOT (Figure 9.3). When the rate of oxygen transfer into the liquor exceeds the cell's demand, then the DOT will rise significantly (Russell, 1987).

## 9.2.7 Rheology

The effect of cell concentration on apparent viscosity or consistency varies depending on the form of the organism. At low concentrations, apparent viscosity is approximately proportional to the concentration, but at higher cell

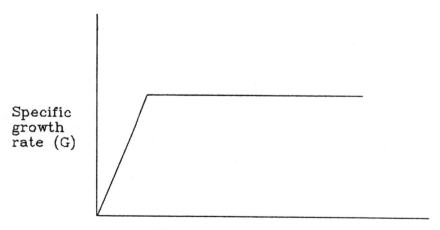

FIGURE 9.3. Specific growth rate of microorganisms in a fermenter with regard to dissolved oxygen tension (Russell, 1987). Reprinted by permission of Elsevier Applied Science Publishers, Ltd., Barking, U.K.

concentrations the apparent viscosity rapidly increases (Russell, 1987). Most fermentation broths are pseudo-plastic, represented by:

$$\tau = b \cdot \gamma^n \text{ and } \mu_a = b \cdot \gamma^{n-1}$$

where $\tau$ is the shear stress; $\gamma$ is the shear rate; $\mu_a$ is the apparent viscosity; $b$ is consistency coefficient and $n$ is the flow behavior index.

As fermentation progresses, $b$ will increase and $n$ will decrease [Figure 9.4(a)]. This increases the apparent viscosity, which in turn will reduce the

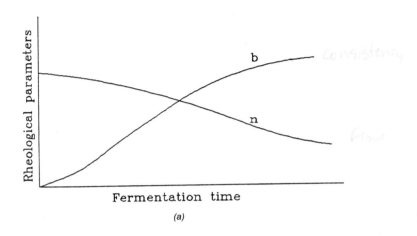

(a)

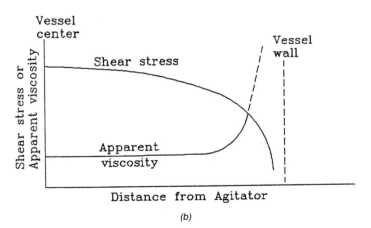

(b)

**FIGURE 9.4.** Fermenter rheology. (a) Typical viscosity change during fermentation; (b) variation of viscosity across a fermenter (Russell, 1987). Reprinted by permission of Elsevier Applied Science Publishers, Ltd., Barking, U.K.

oxygen, heat, and nutrient transfer rate to the cell. In a highly pseudoplastic broth, the degree of agitation decreases rapidly away from the agitator [Figure 9.4(b)]. Thus, aeration may only occur in the zone local to the impeller, and the rate of heat transfer through the vessel wall will be drastically reduced (Russell, 1987).

### 9.2.8 Other Constraints

In deep fermenters or those that are pressurized to enhance oxygen transfer, the pressure may cause damage to cells or increase the $CO_2$ solubility up to an inhibitory level. Increasing shear will damage the cells and their viability will fall depending on the cell type. Shear damage can be caused both by high-speed agitators and by pumps.

## 9.3 **MICROORGANISMS**

### 9.3.1 Lactobacilli (L.)

Many microorganisms for food fermentation fall into the lactic acid bacteria group: the *Lactobacilli*, the *Leuconostocs*, the *Streptococci*, and the *Pedococci*. *L. bulgaricus* and *S. thermophilus* are used to produce yogurt; *L. plantarum* is used to prepare sausages and pickles (cucumbers); and *L. bulgaricus* or *helveticus* is used in cheeses (Chassy, 1986). *Lactobacilli* alter flavor, texture, and appearance of foods; enhance nutritional value; and retard spoilage and reduce contamination.

To improve *Lactobacilli*, the classical techniques of bacterial genetics, i.e., mutagenesis followed by mutants selection with desired traits, have been applied (see Chapters 2, 5 and 11). Phage-mediated transduction is used to move genes between strains. Bacteriophages are common among *Lactobacilli*, and the majority of strains are isolated from natural sources. Conjugal exchange of genetic information between bacteria is also applied to strain development. An erythromycin resistance-encoding conjugal plasmid has been transferred into and between several *Lactobacillus* species (Chassy, 1986).

### 9.3.2 Fungi or Molds

Molds are used for food fermentations, because they (1) are generally capable of rapid growth, (2) can be used with immobilized support systems, (3) offer greater flexibility and a wider range of biochemical capabilities, and (4) can be readily used to integrate recombinant and classic genetics

(Buchanan, 1986). Researchers have used recipient strains of *A. nidulans*. Transformation was detected by transferring the missing genetic element from a wild-type strain of *A. nidulans*, and then selecting the vigorously growing colonies in the absence of the required nutrient. Thus, the transfer of genetic information among fungal genera is feasible. To produce transformants in *A. nidulans* involved cloning the desired gene into a plasmid from *E. coli*. The recipient strain's protoplasts were then formed and mixed with the plasmid. After incubating for a short time and regenerating, the cells were screened for the presence of the specific nutritional marker (Buchanan, 1986). These techniques are discussed in Chapter 2.

The selection of culture medium depends on the specific nature of the cell. Monolayer and suspension cultures are used to produce animal cells in mass culture. Monolayer cultivation systems require a solid support on which the cells may grow. Suspension culture systems grow freely throughout the bulk of the culture.

## 9.4 **EQUIPMENT**

Fermentation equipment can be classified (Sfat, 1986) as aerobic or anaerobic, aseptic or sanitary, liquid or solid. The basic configurations for food fermenters are (1) aerobic-aseptic-liquid, used to produce monosodium glutamate, (2) aerobic-sanitary-liquid, used to produce baker's yeast, (3) aerobic-sanitary-solid, used for koji, (4) anaerobic-sanitary-liquid, used for beer and pickles, and (5) anaerobic-sanitary-solid, used for cheese and bread. A typical fermenter includes (Soderberg, 1983) nutrient feed tanks, sterile air filters, an air compressor, pumps, cooling equipment, an agitator, valves to maintain sterility, and environment controls.

Anaerobic fermenters are used in methane production from waste fermentation. Novel fermenters are being designed to improve oxygen transfer, cell concentration, agitation, heat transfer, and sterility. A significant part of the fermentation cycle is used in sterilization, filling, emptying, and inoculation. The majority of commercial fermenters are batch-stirred tanks containing free aerobic organisms.

The fermenter types can be based on the types of flows and vessels such as batch, semibatch, continuous, stirred-tank reactors, tubular, recycle, etc. The fermenters are also classified based on phases present in the fermenter. Some typical fermenters are discussed below.

### 9.4.1 Air-Lift Fermenters

The air-lift fermenter uses air to agitate the fluid by means of draft tubes or external recycle, eliminating the need for expensive mechanical mixtures

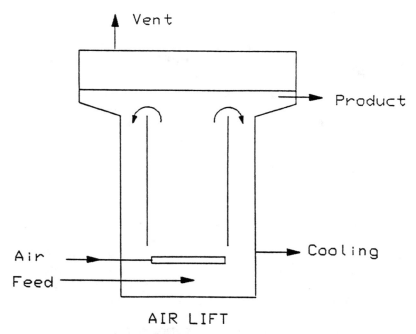

**FIGURE 9.5.** Air-lift fermenter.

(Bungay, 1984) (Figure 9.5). It creates a rapid, low-shear movement by aeration. The liquid movement is initiated by injection of air at the foot of the riser column (Brown et al., 1987). Pilot-scale application of this is to culture plant and animal cells, by avoiding the shear effects of the stirring gear.

This also creates internal recirculation and utilizes sparged air efficiently. Air is sparged into the base of the riser section, and the decrease in density caused by the gas-liquid mixture results in circulation of the liquid. Cooling is by internal coils, and internal baffles are fitted to improve mass transfer (Russell, 1987). These fermenters can achieve, without problems, the high energy input required for aeration and mass transfer in large systems.

Tower fermenters with aeration from below take advantage of the higher hydrostatic pressure, resulting in a higher oxygen solubility. The residence time of the gas in the reactor and thus the oxygen utilization rate is also enhanced, making it possible to lower the energy consumption per kg of the oxygen transferred (Wandrey, 1987). Anchorage-independent mammalian cells can be cultivated in suspension culture reactors, e.g., in an air-lift reactor, in a vibrio-mixed suspension culture reactor, or in a perfusion reactor with a vertical rotating filter in the center of the reactor to achieve a steady state environment and medium exchange (Schmidt-Kastner, 1987).

## 9.4.1.1 Gas-Lift Cascade

Inside a tower, baffles are arranged rising alternately from opposite sides, thus forming chambers one upon the other. The gas flow is supplied to the lowest chamber at that side of the bottom where the first baffle is fixed. The gas rises by its buoyancy up to the baffle, which directs it to the next chamber above. In each chamber, the gas flow rises at one side, causing downflow of the liquid-phase on the other side. Thereby small bubbles can be drawn, forming a whole circulation flow (Blenke, 1987).

The excess air is fed into the fermenter through an air sparger at the base area. It is less efficient than a stirred tank; foaming is a problem and a large amount of sterile air is required.

## 9.4.2 All-Glass Fermenters

These are cheap to install and free of metal contamination, resistant to corrosion and easily adaptable. However, these cannot be operated much above atmospheric pressure, which eliminates steam sterilization; and higher pressure cannot be used to increase oxygen solution rate, reduce foaming, or minimize contamination (Baker, 1978). These are fitted with stirrer assembly, pH and temperature controllers, heat exchanger, sample port, and various valves.

## 9.4.3 Batch Fermenters

Figure 9.6 shows a typical batch fermenter vessel. The vessel is constructed of a suitable grade of stainless steel, with butt-welded joints polished flat on the inner surface (Brown et al., 1987). Two disadvantages are the high energy requirement for agitation, and the damaging effect on cells due to shear.

Thus, a batch, stirred-tank fermenter is used for a multi-product plant since the design is adaptable for different conditions.

The product and cell concentrations increase as the substrate is depleted during the fermentation cycle. The fermenter and substrate can be sterilized between batches, scale-up from laboratory and pilot-plant scale is relatively straightforward, and production may be easily adjusted to suit demand, and can often be used for different products.

The stirred tank is a commonly used fermenter, where air is introduced under the agitator via a sparger. It can be readily adapted for multi-purpose use, and is suitable for low-viscosity broths. To improve the mixing efficiency, hydrodynamically shaped agitators are provided.

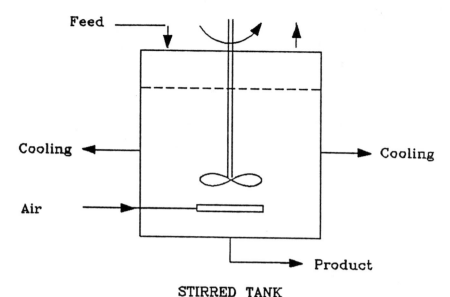

**STIRRED TANK**

**FIGURE 9.6.** A typical batch fermentation vessel.

### 9.4.4 Continuous-Stirred Fermenters

In a continuous fermentation, the conditions in the fermenter remain constant during operation so that fermentation is controlled easily. Continuous operation is generally selected for the production of low-value, high-volume products because of the cost benefits. Sterile medium is supplied at constant flow rate to the homogeneously stirred culture vessel, and culture overflows at the same rate for harvesting. This has been applied to the production of microbial single-cell protein.

### 9.4.4.1 Continuous-Stirred Tank Reactors (CSTR)

Many configurations of CSTRs are used for enzyme-catalyzed reactions (Figure 9.7). For inexpensive enzymes, these can be continuously added to the feed solution. To retain costly enzymes, an ultrafiltration membrane, with smaller pores to prevent escape of the large enzyme molecules, is placed in the effluent stream. If enzymes are immobilized on insoluble particles, then a screen in the effluent line will be sufficient. In another approach, the enzymes can be attached to the agitator shaft using screen baskets. The same effect can be achieved by circulating reaction mixture from a well-mixed reservoir through a short-packed column of immobilized enzyme.

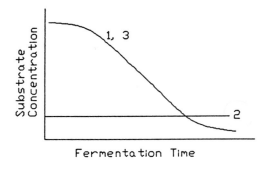

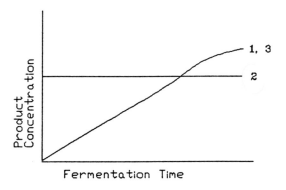

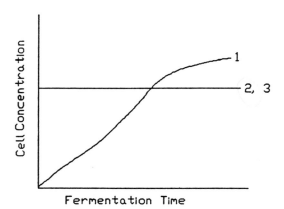

**FIGURE 9.7.** Changes in the concentrations of substrate, product, and cell with time in (1) batch, (2) continuous-stirred tank, and (3) immobilized packed bed fermenters.

## 9.4.4.2 Fluidized Bed Reactors

In this type, liquid flows upward through a long cylinder, and biocatalyst particles (enzymes or cells) are suspended by drag forces. Biocatalyst particles are released at the top of the tower by the reduced liquid drag at the larger cross section. Thus, biocatalyst is retained in the reactor while the medium flows through it continuously.

## 9.4.4.3 Trickle-Bed Reactors

These reactors are three-phase systems consisting of flowing gas and liquid, and a packed bed of biocatalyst. The performance of the reactor is affected by the packing surface area, gas-liquid flow pattern, the degree of wetting of the biocatalyst by the liquid phase, mass transfer of reactants from the gas to the liquid phase, mass transfer of reactants to the biocatalyst, etc. (Bailey and Ollis, 1986).

## 9.4.5 Deep-Shaft Fermenter

The deep-shaft fermenter is based on air-lift principles, but provides large aeration by using hydrostatic pressure. A long thin vessel loops far below the sparger (Figure 9.8). Air is compressed as the medium flows downward to regions of great hydrostatic pressure, and the vent is kept higher than the feed inlet. The net effect on air is an expansion, which causes the circulation in the broth (Bungay, 1984).

## 9.4.6 Roto-Fermenter

It consists of a fermentation vessel, a rotating microporous membrane, and a filter chamber. Fresh medium and air are injected at the fermenter bottom. The cells grow in the annular space around the rotating membrane, and cells can be retained inside the fermenter, while the cell-free filtrate is continuously removed through the rotating microporous membrane. It forms the dual function of cell growth and concentration with the simultaneous removal of metabolic products. Thus, the roto-fermenter can serve as both a fermenter and a cell separator (Margaritis and Wilke, 1978).

## 9.4.7 Solid-State Fermenter

Solid-state fermentation is used to culture on solid nutrients such as bran through which air can diffuse. Composting of organic wastes such as garbage to produce fertilizer is a solid-state fermentation.

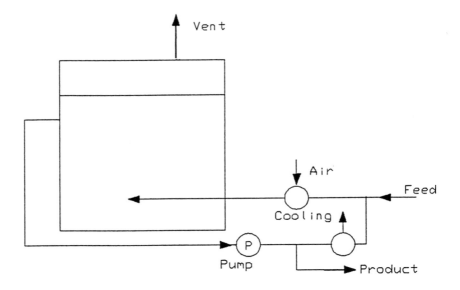

Pumped Recycle

**FIGURE 9.8.** Deep-shaft fermenter.

Commonly used substrates in solid-state fermenters are cereal, legume, bran, wood, straw, and other animal and plant materials. The design is controlled by the heat and mass transfer characteristics. Inter-particle and intra-particle diffusions are limiting factors. Satisfactory inter-particle oxygen transfer is achieved by proper mixing and aeration. $CO_2$ build-up in the void spaces should be avoided. Intra-particle mass transfer is the transfer of nutrients and enzymes within the fermentation substrates, and concerned with the role of enzymes in hydrolyzing the water-insoluble polymers into soluble substrate (Smith, 1985). Microbial heat generation is much greater per unit volume than for liquid fermentations. Heat can be removed by increasing the aeration rate or the frequency of aeration for systems that are not continuously aerated.

## 9.4.8 Scraped Tubular Fermenter

The mechanically scraped plug flow fermenter consists of a horizontal hollow tube through which fermentation medium flows with the assistance of an internal wall scraper. This scraper partially or wholly separates the fluid contained in the tube into moving compartments along the tube. Aeration is provided by orifices located along the tube bottom (Moo-Young et al., 1979).

## 9.4.9 **Membrane Bioreactors**

Cheryan and Mehaia (1986) reviewed membrane bioreactors. The biocatalyst is physically trapped on one side of a suitable membrane. The homogeneously dissolved enzyme is circulated in a loop reactor. The substrate is brought into contact with the biocatalyst, and the resulting product permeates through the membrane. The biocatalyst is retained for further reaction (Figure 9.9). The dense skin layer at the lumen wall should be impermeable to the biocatalyst molecules, which diffuse through the inner wall of the fiber to the spongy part, where the reaction takes place. Applied transmembrane pressure and axial flow rate control the reactor performance. The availability of improved immobilization techniques permits design of continuous flow reactors where reaction could be achieved without biocatalyst loss in the effluent stream. Biocatalysts are more protected and less exposed to denaturation (Drioli, 1986).

The homogeneously soluble enzyme can be introduced via a sterile filter into the reactor, which is easily kept sterile. If the enzyme is deactivated, a constant activity can be maintained in the reactor by supplying additional biocatalyst (Wandrey, 1987). The characteristics of this reactor are (1) continuous operation, (2) no back-mixing, (3) usefulness as a multi-enzyme reactor, (4) usefulness for very diluted solutions, e.g., for the treatment of wastewater, and (5) low residence time of substrates and products (Schmidt-Kastner, 1987).

For successful operation, good control of concentration polarization and fouling effects minimization is needed. The concentration polarization is controlled by operating parameters such as recirculation rate, temperature, and transmembrane pressure, while the fouling is controlled by membrane feed interactions (Cheryan and Mehaia, 1986).

## 9.4.10 **Hollow Fiber**

The basis of this is the hollow-fiber membrane. Thousands of these hollow fibers are formed into a bundle and contained in a housing. In hollow-fiber bioreactors, cells are cultured on the exterior surface or extra-capillary space (ECS) of the fibers. Medium is circulated through the inner lumen or intra-capillary space (ICS). Diffusion of medium across the porous membranes from the ICS to the ECS supplies the cell with fresh nutrients and oxygen, while waste products may exchange in the opposite direction, away from the cells. In the conventional systems, the flow of medium through the hollow fibers is unidirectional. On a small scale, as the size or length of the reactor is increased, the axial pressure drop can result in unequal distribution (Wang et al., 1989). Low molecular weight species such as glucose, vitamins, and amino acids are readily transported across the membrane and nourish the cells (Donofrio, 1989).

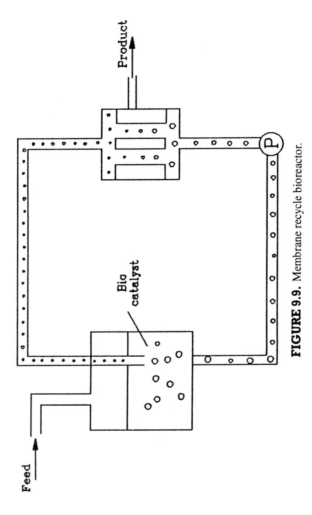

**FIGURE 9.9.** Membrane recycle bioreactor.

## 9.4.11 Other Types

Many variations of the above-mentioned reactors have been developed for fermentation. For example, 25,000 L continuous fermenter employing high-cell-density technology was used by the Provesta Corp. to produce yeast-based flavor enhancers (Mermelstein, 1989). This allowed high oxygen and heat transfer rates, resulting in high cell density and increased yeast production. This design utilized the advantages of foaming, high gas velocity, and high operating pressure to produce very high oxygen transfer rate. An agitator at the bottom of the fermenter provided a high shear rate. Mechanical foam breakers at the top of the vessel controlled the foam level in the fermenter. An efficient heat exchanger system was used. The combination of high oxygen and heat transfer resulted in a rate of productivity three times higher than is possible with other conventional processes.

## 9.4.12 Selection

The goals in fermenter selection are flexibility, minimum cost, thermal control, and product selectivity (Kabel, 1985) (Figure 9.10). Fermenter selection depends on factors such as the biological constraints of the organism, production scale, the level of technology available, economics, and the range of products to be fermented.

Methods suitable for anchorage-dependent cells include various forms of roller bottle systems, plate and packed-bed bioreactors, and several forms of hollow-fiber surfaces. The surface-to-volume ratios of these forms, generally used, are plastic bags (5), spiral films (4), glass beads (10), artificial capillaries (31), and microcarrier suspensions, 20/L (122). Other forms include treated polystyrene and polypropylene or diethylaminoethyl-substituted cellulose fibers (Smith, 1985).

## 9.5 **AGITATION AND MIXING**

The complete and uniform mixing of the medium components within a fermenter can be very complex for a large system. The mechanism of dispersion throughout the fermenter will involve the flow and mixing via the eddies generated by mechanical agitation or the turbulent flow of the liquid. Mixing mechanisms should optimize both axial and radial dispersion.

The agitation distributes the cells and nutrients evenly throughout the medium. In an aerobic fermenter, the agitation also breaks large air bubbles into smaller ones, increasing surface area and assisting the transfer of oxygen from air into the culture, and distributes the dissolved oxygen to the organisms (Rehm and Reed, 1973).

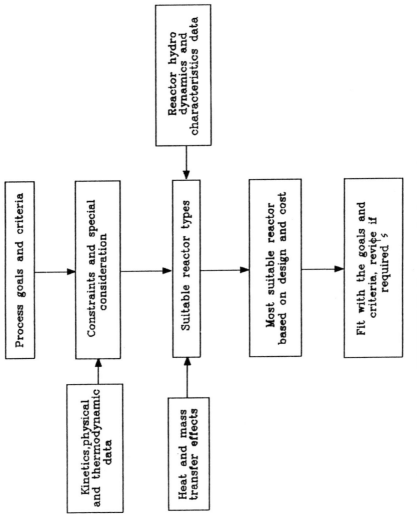

**FIGURE 9.10.** Fermenter selection procedure.

The shape of the stirrers in the fermenter affects fluid flow within the vessel and microbial processes. Peddle-type stirrers are used for low-viscosity liquids and for low lift. Propeller types provide large and longitudinal flow. Flat-bladed turbines are very efficient at gas-liquid dispersion.

Various designs of the agitator have been used, but the most common is a centrally rotating shaft supporting several adjustable impellers. The tendency for vortex formation is prevented by vertical baffles fitted to the vessel walls. The motor-driven shaft can be magnetically or mechanically coupled and either top or bottom mounted (Wernerspach, 1986).

Two types of impellers are generally used. The first is an impeller with no baffles, in which a vortex is created from the liquid surface at high impeller speeds. This will draw air into the fermenter, providing a high degree of aeration at lower power. However, it is difficult to scale up because change in fermenter size will change flow patterns. The second type is an impeller with baffles. These impart power to the liquid in the form of turbulence and flow. Axial flow impellers provide flow parallel to the rotating shaft axis, while radial flow impellers generate flow in radial and tangential directions.

A single impeller can only adequately agitate a fermenter with a height/diameter ratio ($h/D$) of 1. For higher ratios, additional impellers are required. Additional impellers are spaced at intervals equivalent to one tank diameter along the shaft to avoid a swirling type of liquid movement (Rehm and Reed, 1973).

The following ratios of geometrical dimensions are generally used for design purposes (Oldshue, 1985):

$s/D = 1/10$ for turbine and paddle stirrers
$h/D = 1/3$ for all types of stirrers
$d/D = 1/3$ for turbine and paddle stirrers
$d/D = 1/2$ to $1/3$ for propeller stirrers

Where $s$ = radial length of the baffle, $D$ = vessel diameter, $h$ = distance from the vessel bottom, and $d$ = stirrer diameter.

## 9.6 OXYGEN TRANSFER

In aerobic fermentation, introduction of oxygen into the fermentation broth is required to maintain the microorganisms in an active growth condition. In addition to providing oxygen, aeration is also required to purge the fermentation broth of unwanted volatile products of metabolism. In many liquid fermentation processes, a gas, in the form of fine bubbles, is required to reduce the level of dissolved carbon dioxide.

The oxygen transfer is one of the bottlenecks in conventional fermentation techniques, and little work is reported with regards to high cell densities and immobilized cells. The oxygen requirements can be experimentally measured by materials balance techniques before designing the fermenter aeration and agitation.

Sterile air or oxygen is introduced into a fermenter either through an orifice producing bubbles in a hollow agitator shaft, a nozzle, or a sintered-type sparger in the form of a ring or grid. The addition of baffles to the fermenter vessel increases the turbulence uniformity, and the efficiency of dissolved oxygen transfer. High-pressure oxygenated water is also used to oxygenate the broth (Wilson, 1977). The jet of water injected in the fermenter is vigorous enough to cause uniform mixing. The volume of the air or oxygen is generally metered by a variable area flowmeter.

Agitation rate, air flow rate, and partial pressure of oxygen are the dissolved oxygen control variables. Kobayashi et al. (1980) achieved the optimal control of dissolved oxygen by first varying the agitation rate and then the aeration rate, only after the agitation rate was at its limit. Economic optimization was made possible by using an intermediate setting of the two. The optimal arrangement of the stirring and aeration equipment should be determined for each different type of fermentation by pilot plant studies.

The rate of mass transfer between gas and liquid phases is influenced by the solubility of the gaseous components in the liquid. The supply of oxygen to the broth is affected by (Brown et al., 1987) agitation type, i.e., the shape, number, and arrangement of impellers and baffles; agitation speed, i.e., tip speed rather than rpm; liquid depth in the fermenter (a common height-to-diameter ratio is $3:1$ or $4:1$); sparger type; air flow; and physical properties of the broth, i.e., temperature, viscosity, surface tension, and the organism type. These will affect oxygen solubility directly or by bubble size and turbulence.

Oxygen transfer rates can be enhanced in fermentation using oxygen vectors. The oxygen vectors generally used are liquids that are insoluble in the fermentation media. Their utilization in an emulsified form can significantly increase the oxygen transfer coefficient between the gas and aqueous phases. It seems that the vector acts as an active intermediate in the oxygen transport from gas bubbles to aqueous phase (Rols and Goma, 1989). These compounds generally used in biotechnology are hemoglobin (Adlercreutz and Mattiasson, 1982), hydrocarbon (Mimura et al., 1969), and perfluorocarbon (Mattiasson and Adlercreutz, 1987). Rols et al. (1990) used hydrocarbon (*n*-dodecane) and perfluorocarbon (Forane F 66 E) in which oxygen is highly soluble (55 mg/L, and 118 mg/L, respectively). They found that the use of *n*-dodecane emulsion in a culture of *Aerobacter aerogenes* increased the $K_L a$ 3.5-fold.

The oxygen transfer rates can be improved by increasing the bubble surface area $(a)$, maximizing the liquid film mass transfer coefficient $(K_L)$, and maximizing the driving force, which is the difference between the equilibrium concentration of oxygen. The rate of oxygen transfer is limited by the liquid film around the air bubble.

## 9.6.1 Measurement of $K_La$ (Volumetric Oxygen Transfer Coefficient)

The rate of oxygen transfer ($OT$) from the bulk gas phase to the bulk liquid phase is calculated by $OT = K_La (C^* - C)$, where $K_L =$ mass transfer coefficient based on the liquid phase, $C^* =$ dissolved oxygen concentration in equilibrium, $C =$ actual dissolved oxygen concentration in the bulk liquid phase, $a =$ interfacial area between liquid and gas per unit volume of liquid, and $K_La =$ volumetric oxygen transfer coefficient. At atmospheric pressure, $C^*$ is only about 0.25 mM, and $K_La$ varies from 0.02 $s^{-1}$ to 0.25 $s^{-1}$, giving the maximum transfer rates of 18 to 225 m·moles/(L·h). A power input of 1.3 kW/m³ is required to obtain oxygen transfer rates of about 20 to 40 m·moles/(L·h) (Charles, 1985). The $K_La$ depends on the agitation intensity, the gas velocity through the fermenter, solution components, and temperature. Measurement of $K_La$ or dissolved oxygen concentration is required to determine the oxygen availability to the organisms in the fermenter. The size of the bubbles and their velocities in the broth will determine the oxygen transfer rate. The dissolved oxygen concentration can be measured with a dissolved oxygen probe, either a polarographic type or galvanic type. The oxygen transfer rate then can be calculated from either changes in dissolved oxygen readings or inlet gas flow and fermenter off-gas analysis.

A galvanic cell consists of an electrolyte (potassium hydroxide) and two electrodes (silver or lead). The oxygen content of the electrolyte is brought into equilibrium with that of the sample. When voltage is applied to the electrodes, electrochemical reactions take place in the presence of oxygen. Electrons are released for each oxygen molecule at the cathode where oxygen is reduced to hydroxide, causing a current flow through the electrolyte, the magnitude of which is proportional to the oxygen concentration in the electrolyte (Liptak and Venczel, 1982).

A polarographic cell is also composed of two electrodes. A polarizing voltage is applied to reduce the oxygen. The dissolved oxygen in the sample diffuses through the membrane into the electrolyte (potassium chloride). The resulting current is proportional to the oxygen content of the electrolyte (Liptak and Venczel, 1982).

Matsumura et al. (1972) determined $K_La$ for transport between the gas and water phases, and between the water and hydrocarbon phases in an air-water-hydrocarbon (mainly $n$-pentadecane) system using the following model:

$$(K_La)_{Gw} = [1.75(10^{-2}) \exp(0.115\alpha) - 0.8(10^{-3}) \exp(-46.9\alpha)] \cdot N \cdot V_s^{1/3}$$
$$(K_La)_{ow} = 1/[1 + H_{ow}(1/\alpha - 1)] \cdot (K_La)_{Gw}$$

where

$(K_La)_{Gw} =$ volumetric oxygen transfer coefficient at gas-water phase

$(K_La)_{ow} =$ volumetric oxygen transfer coefficient at oil-water phase

$\alpha$ = volume fraction of hydrocarbon
$N$ = rotational speed of impeller/agitator
$V_s$ = linear velocity of aeration based on the cross-sectional area of the stirred tank
$H_{ow}$ = partition coefficient of oxygen in oil-water system

Johnson et al.(1990) measured $K_La$ in 1.5 and 5.0 L New Brunswick Scientific Gelligen bioreactors equipped with a cell lift impeller, by dynamic adsorption/desorption technique and using the following relationship:

$$K_La \cdot t = \ln \left[ (CL\infty - CL_0)/(CL\infty - CL_t) \right]$$

where

$CL\infty$ = dissolved oxygen concentration at equilibrium
$CL_O$ = zero oxygen desorption achieved by sparging nitrogen
$CL_t$ = dissolved oxygen concentration at time $t$

The value of $CL_t$ was continuously recorded following the step change in the inlet gas composition (switching from nitrogen to air). The $K_La$ value was calculated by a linear regression of the recorded data transformed as per the above-mentioned equation. The mass transfer rate increased with temperature, mixing speed and aeration rate. Foam breakers decreased the value of $K_La$. More than 60% of $K_La$ was attributed to surface aeration.

The $K_La$ was measured by a gas-phase balance using the following equations (Rols et al., 1990):

$$N_w = K_La \, (C_{w*} - C_w)$$
$$N_w = Q_o \cdot \left[ (P_T - P_{H_2OT})/P_T \right] \cdot (273/T) \cdot (M_{O_2}/V_m) \cdot (1/V_w) \cdot$$
$$\left[ (P_{O_2i} - P_{N_2i} \cdot P_{O_2o}/P_{N_2o})/P_T \right]$$
$$(1/K_L)_{total} = 1/(K_L)_{Gv} + 1/(K_L)]_{Vw}$$

where $C_{w*}$ and $C_w$ are saturation and actual dissolved oxygen concentrations, respectively.

$(K_L)_{Gv}$ = oxygen transfer coefficient for gas to vector
$(K_L)_{Vw}$ = oxygen transfer coefficient for vector to water
$Q_o$ = volumetric flow rate of outlet air before seeding
$M_{O_2}$ = molar mass of oxygen
$N_w$ = oxygen transfer rate in water
$V_w$ = volume of water
$P_{O_2}, P_{N_2}$ = partial pressure of oxygen and nitrogen in the gas
$o,i$ = outlet, inlet
$P_{H_2OT}$ = partial pressure of water in air at temperature $T$
$P_T$ = total pressure
$V_w$ = molar volume at 273 K and 1 atm. pressure

The dissolved oxygen concentration was measured with an Ingold polarographic probe and a $P_{O_2}$ Num Tacussel oxygen analyzer. The effluent gases were analyzed by gaseous chromatography. $K_L$ for gas-water, gas-vector, and vector-water systems were measured in a magnetically stirred device where interfacial area is fixed by the presence of a membrane separating the phases. The $(K_L)_{Gw}$ was obtained by experiments without vector and $(K_L)_{vw}$ by the use of an oxygen-saturated vector volume (by vector recirculating in a bubble column) on the membrane. From the experimental $(K_L)_{total}$, $(K_L)_{Gv}$ was obtained from the above-mentioned equations.

Linek et al. (1988) proposed a dynamic pressure method for $K_L a$ measurement in large-scale aerated and agitated bioreactors. A change in the total pressure in the reactor by approximately 20% led to a simultaneous change in the oxygen concentration in all the bubbles in the dispersion. This procedure suppressed the influence of non-ideal mixing of the gas phase on $K_L a$.

Ho et al. (1987) measured the $K_L$ and correlated with oxygen diffusion coefficients $(D_L)$ for various single and mixed electrolyte solutions. The resulting correlation was $K_L \propto D_L^{0.67}$. The interfacial area $(a)$ increased with ionic strength and then reached a plateau at a value of 0.42 mol/L.

## 9.6.2 Oxygen Solubility

Slininger et al. (1989) developed a calorimetric method to measure oxygen solubility in fermentation media. The method was based on the consumption of oxygen by glucose oxidase activity and production of the pink quinone of syringaldazine by coupled peroxidase activity:

$$\text{glucose} + H_2O + O_2 \xrightarrow{\text{glucase}} \text{gluconic acid} + H_2O_2$$

$$H_2O_2 + \text{syringaldazine} \xrightarrow{\text{peroxidase}} \text{syringaldazine quinone}$$

Syringaldazine quinone has a molar absorption coefficient of $6.5 \times 10^4$ at 526 nm in 50% methanol. Being highly colored, its sensitivity to chromogen allows the detection of low oxygen concentration. Progress of color formation at 526 nm is a linear function of oxygen added to the assay. Maximum sensitivity was recorded at pH 7.0. This procedure provided the solubility data for accurately calibrating in-line electronic probes for monitoring the dissolved oxygen concentration during the fermentation process.

Lu-Kwang et al. (1988) measured the oxygen diffusion coefficient and solubilities in aqueous glucose solutions and various fermentation media using a membrane-covered oxygen electrode. It worked on the principle that when an electrode of noble metal such as platinum or gold is made 0.6 to

0.8 V negative with respect to a suitable reference electrode (e.g., the calomel or Ag/AgCl), in neutral KCl electrode solution, dissolved oxygen is reduced at the cathode surface. By separating the cathode, anode, and KCl electrode solution from a test medium with a plastic membrane permeable to oxygen but not to ions, the polarographic oxygen electrode can be successfully employed to measure oxygen pressure in the gas, liquid, or semi-solid.

By applying the coupled steady state and unsteady state analysis for oxygen diffusion through the glucose solution (Lu-Kwang, 1988), the following relationships were developed:

$$(D_L/D_o) = (\mu_L/\mu_o)^{-0.725}$$

$$\log (K_o/K_L) = 5.93 \ (10^{-4}) \ C_n$$

Where $D$ = oxygen diffusion coefficient, $\mu$ = viscosity, $C_n$ = sugar concentration, $K$ = oxygen solubility (g/L), and $L,o$ = liquid and water, respectively.

### 9.6.3 Air Sterilization

The air supply is usually sterilized by filtration, to remove particulates $>0.2 \ \mu$m. Many types of filters are available for this purpose, e.g., carbon, mineral wool, or polystyrene resin. Small fermenters are fitted with individual air pumps, flow meters, and filters. An exhaust filter prevents back-infection and the escape of organisms into the environment. Air can be sterilized by heating in electrical heaters or by adiabatic compression. It can be sterilized by exposure to UV light or other radiations.

## 9.7 MEDIA STERILIZATION

Continuous sterilization is conducted by heating the medium at the required temperature in a short time interval; then holding the medium at the heated temperature for a certain period; and finally cooling the medium to the culture temperature. The medium can be heated by steam directly using injectors or indirectly using heat exchangers. The cooling also requires a heat exchanger.

At small scale the media are heated at 120°C for 15 min by using moist heat, or at 160°C dry heat for at least 1 h. For large-scale operations, vigorous sterilization is required by increasing time or temperature or both. The acceptable probability of contamination is 0.001, i.e., one survival in 1000 cells (Brown et al., 1987).

The media may be sterilized in a batch vessel, or by a continuous high-temperature short-time process through a heat exchanger. The media is passed through a heat exchanger for sufficient residence time to sterilize, by controlling its flow rate.

Medium can be sterilized by filtration using membrane, sintered-glass, or candle filters, having pore diameter less than 0.22 $\mu$m. The solution viscosity and turbidity will influence the filter holder choice. Similarly, medium sterilization by chemical agents depends on the pH, temperature, surface tension, contact time and concentration of the chemical. Suitable chemicals used are ethylene oxide and 2-oxetanone.

Radiation sterilization of medium is conducted using X-rays, gamma rays, or particulate cathode rays. However, the radiation resistance varies based on the species and the strain of microorganisms.

## 9.8 **ASEPSIS IN FERMENTER DESIGN**

Maintenance of aseptic conditions are important to keep the fermenter contaminant-free. The following design criteria are generally practiced to achieve this goal.

### 9.8.1 **Aseptic Seals**

In the mechanical agitator fermenters, aseptic conditions are achieved with a rotating seal (bronze bearing) and a stationary seal. The surface of the bronze bearing is provided with a radial slot, into which either silicone or petroleum grease is applied for lubrication.

Another type of seal is a stuffing box in which asbestos impregnated with lubricating grease is tightly packed in the space between the wall of the stuffing box. The stuffing box seal is equipped with a lantern ring to allow steam to penetrate to the shaft for sterilization.

### 9.8.2 **Aseptic Operation**

The pipe lines transporting sterile air, seed, and other materials required for aseptic use should be sterilized with steam and should be free from condensate after sterilization. Therefore, lines should avoid bends and elbows. For a high order of sterility, diaphragm types of valve or ball valves are recommended. The diaphragm-type valve has no packing gland and is less liable to contamination. Ball valves can be double-sealed to minimize the chances of contamination (Stanbury and Whitaker, 1984).

### 9.8.3 **Aseptic Inoculation**

The aseptic inoculation can be performed by providing an additional bypass steam line with a valve connection to carry out adequate steam sterilization followed by cooling the line with sterile air, and finally allowing the passage of inoculum to the seed tank and from the seed tank to the fermenter (Aiba et al., 1973).

### 9.8.4 Aseptic Sampling Point

The end of the sampling point should be kept immersed in the 40% formaldehyde solution. An extra valve (C) with a T-joint can be provided to sterilize the pipe line before sampling. The tank valve (A) is then opened to let some broth pass to waste and to cool the line, and then the sample is collected in a sterile bottle. After sampling, the outlet is re-sterilized with steam by closing valve (A) and opening valve (C) (Stanbury and Whitaker, 1984).

## 9.9 INSTRUMENTATION AND CONTROL

In fermentation, instruments are required for measuring, recording, and controlling process parameters. This data can also be used to optimize or improve the process. For process parameter control, a parameter is measured by a transducer, which is then compared with a set value for feedback criteria. The fermentation process requires controls for airflow, agitation, pressure, temperature, pH, foam, oxygen, and carbon dioxide levels. These should maximize dissolved oxygen transfer, ensure proper agitation rate due to broth consistency changes, and maintain optimum temperature.

Measurement and control of gases at the inlets and outlets of a fermenter are important to control microbial growth and to increase the productivity.

### 9.9.1 Instrumentation

Fermentation sensors are divided into physical, chemical, biochemical and biological categories. Important physical sensors are those designed to measure temperature, agitation, power input, heat generation, gas flow rate, liquid flow rate, foaming, heat transfer, viscosity, and broth volume. Chemical measurements include pH, redox potential, oxygen concentration, carbon dioxide concentration, ionic strength, broth composition, and nutrient composition. Biochemical measurements are cell mass composition, enzyme activity, DNA/RNA level, etc. A few of these sensors are described below.

#### 9.9.1.1 $CO_2$

On-line $CO_2$ is measured with gas chromatography, mass spectrometry, or spectrophotography. Membrane electrodes for dissolved $CO_2$ are also available.

## 9.9.1.2 Foam

During fermentation, foaming can develop due to agitation and aeration. Foaming is largely a result of the stabilization of the liquid film by proteins or proteins adsorbing to the gas/liquid interface. Protein concentration above 1 mg/L can induce foaming. Foaming can be a serious problem in fermentation, particularly in a large-scale vessel.

Foam control can be achieved by a mechanical foam breaker attached to the shaft or with chemical antifoaming agents by reducing surface tension at the air/liquid interface. The antifoaming liquids are composed of oils, fatty acids, esters, poly-glycols, and silicones. At low salt concentration, the foam stability increases due to the increase in protein stability, whereas addition of pure salts decreases foaming. Short-chain alcohols increase foaminess at concentrations of 1 to 2 % (v/v). Hence, alcohol concentration should either be increased or decreased from this critical level to reduce foaming. Chemical foam prevention by the addition of surface-active substance (e.g., silicone oil) influences coalescence behavior, mass transfer, mode of reaction, purity of products, and finally the down-stream processing (Blenke, 1987). The foam level can also be controlled by high-speed impellers or centrifugal devices that physically break up the foam. Foam control is automated through regulated addition of an antifoam chemical into the broth.

Increasing temperature may decrease foaming due to the increase in drainage by reduced viscosity. However, at higher temperatures, protein denaturation can influence foaming. The protein solution exhibits maximum foaming at the isoelectric point. Therefore, the pH of the broth should be controlled.

A Teflon™-coated electrode is suitable to measure foam level, in which foam completes the circuit between the probe tip and the reactor wall, activating a solenoid valve controlled by a timer (Armiger, 1985). A foam level controller (Kontes model 1000 FL; Mandel Scientific Co., Canada) is a dual-purpose lab and process instrument used with control of either fluid or foam levels under process conditions. When liquid or foam makes contact with the probe, a circuit is completed, which activates an output control port. In the foam control mode, an external pump or valve is activated to add antifoam reagent to the reactor. In the level control mode an external pump or valve is deactivated to stop the addition of liquid to the reactor vessel. The control may be reversed to add liquid, as is the case of the antifoam mode.

## 9.9.1.3 Dissolved Oxygen

Dissolved oxygen probes are used to measure the oxygen partial pressure. Oxygen sensor controls dissolved oxygen by means of agitation rate, aeration, or both. The rate of oxygen uptake within a fermenter can be obtained by measuring oxygen concentrations in the outlet and inlet gas streams using a mass spectrometer or a paramagnetic oxygen analyzer.

## 9.9.1.4 pH

For optimum microbial growth and product formation, the pH must be kept within a narrow range. Base or acid is supplied to the fermenter to control the pH over a wide range. The pH measurement requires a pH-sensitive glass membrane electrode, a reference electrode, and an automatic temperature compensator. The temperature compensator consists of an element in contact with the process solution, which changes its resistance with temperature.

An aqueous buffer solution resists the changes in its pH when either an acid or base is added to the process solution. All solutions have some buffering capacity, but process solutions are buffered to a great extent when containing organic salts, salts of strong acids and weak bases, or salts of weak acids and strong bases (Liptak and Venczel, 1982).

An on-off control is adequate when control point is below pH 5 or above pH 9. A relatively long hold-up time in the reaction vessel should be provided, and thorough mixing is needed. Multimode controls require proportional, integral, derivative, or any combination of these modes.

Junker et al. (1988) developed a method for measuring dissolved oxygen and pH using fiber-optic probes. A fluorescent molecule possesses a characteristic excitation wavelength corresponding to its spectral absorption peak. When exposed to this wavelength, the molecule moves to an electronically excited state. It then returns to its initial state by re-emitting this energy as fluorescent light of longer wavelength or lower energy. The wavelength of peak fluorescence is usually 50 to 100 nm higher than the excitation wavelength. The fluorescence intensity of oxygen-sensitive fluorophors such as pyrine butyric acid (PBA) decreases with an increase in the concentration. Similarly, fluorophors sensitive to pH such as 8-hydroxy-1,3,6-pyrene-tri-sulfonic-acid-trisodium salt (HPTS) possess a dissociating hydrogen atom. Their absorbance spectra contain two highly distinct peaks, corresponding to the associated and dissociated species. The Stern-Volmer equation relates the dissociated oxygen to a ratio of fluorescence intensities:

$$I_o / I_{Do} = 1 + K_{sv}$$

where

$I_o$ = fluorescence intensity in the absence of quencher oxygen

$I_{Do}$ = fluorescence intensity in the presence of oxygen

$K_{sv}$ = Stern-Volmer quenching constant

Similarly, the Handerson-Hasselbalch equation provides a relationship between the solution pH, the $pK_a$ of the indicator's dissociating hydrogen atom, and the ratio of the undissociated (*HA*) to dissociated (*A⁻*) indicator species.

$$pH = pK_a - \log (HA)/(A^-)$$

The fluorescence intensity at the wavelength of 405 nm was proportional to "*HA*" and at 460 nm to "*A⁻*." Therefore, the above equation was written as:

$$pH = pK_a - \log (I^{405}/I^{460})$$

These relationships can be used to measure the dissolved oxygen and pH.

## 9.9.1.5 Pressure

To measure pressure, generally diaphragm gauges or pressure transducers are used, if contamination can be avoided (Brown et al., 1987). A linear variable differential transformer (LVDT) protected by a diaphragm is one of pressure transducers used to measure fermenter head pressure.

## 9.9.1.6 Temperature

This is the most common and fundamental physical parameter that needs to be measured. It can be measured by:

(1) *Thermocouples:* The Seebeck thermal e.m.f. is the voltage produced between the two junctions in the circuit. The voltage depends on the temperatures of the two junctions and on the junctions' material. One temperature must be a reference known temperature because the Seebeck voltage is proportional to the difference between junction temperatures. Certain standard configurations of thermocouples using specific metals (or alloys of metals) have been adopted and given letter designations. Type E is ideally suited for low temperature measurements because of its high Seebeck coefficient, low thermal conductivity, and corrosion resistance. Types J, T, and K are also popular because of their high Seebeck coefficients and low price.

In some cases, a thermocouple is sheathed in a protective covering or even sealed in glass to protect the unit from a hostile environment. A thermocouple constitutes an excellent antenna for pickup of noise from electromagnetic radiation in the radio, TV, and microwave bands. The following noise reduction techniques are commonly used:

- Extension or lead wires from the thermocouple to the reference junction or measurement system are twisted and then wrapped with a grounded foil sheath.
- The measurement junction is grounded at the point of measurement.

Thermocouples are economical and rugged; they have reasonably good long-term stability. Because of their small size, they respond quickly and are good choices where fast response is important.

(2) *Resistance temperature detector (RTD):* It is an electrical circuit element consisting of a metal conductor, characterized by a positive coefficient of resistivity. Platinum and nickel are more commonly used. It has a temperature coefficient of about 0.4%/°C. Platinum RTDs are also available in the form of thin film on a ceramic substrate, for reduced size, increased ruggedness, and lower cost. These platinum films may be taped, cemented, or imbedded.

(3) *Thermistor:* A thermistor is also a temperature-sensitive resistor, and is composed of semiconductor materials. Most thermistors have a negative temperature coefficient, and their output is highly nonlinear. These are low in cost and have the highest sensitivity among common temperature transducers. These are generally quite small and response is fast.

(4) *Solid state temperature sensor:* This sensor is an integrated circuit temperature transducer. These are available in both voltage and current output configurations. Both supply an output that is linearly proportional to absolute temperature. Typical values are 1 $\mu A/°C$ and 10 mV/°C.

Optimum temperature is necessary for an efficient fermentation process. Thus, an accurate and reproducible temperature controller is required. Temperature is usually controlled by a heater, a heat exchanger, and a temperature sensor (Wernerspach, 1986).

## 9.9.1.7 Viscosity

Various equipment is available for the measurement of rheological properties. Picque and Corrieu (1988) used a vibrating rod sensor mounted on the fermenter to measure apparent viscosity on-line. Its operating principle involves the excitation of a vibrating rod (3 mm dia., 7 cm length) held at constant frequency (350 Hz) by an electromagnet. The change in the damping of the rod vibration corresponded to the apparent viscosity of the medium. Vibration amplitude decreased with an increase in the apparent viscosity of the medium. The sensor drift was very low (0.03% of measured value per hour).

## 9.9.1.8 Other Parameters

(1) *Agitator speed:* This can be measured using a tachometer on the agitator shaft.

(2) *Power input to an agitator:* An in-line torque measuring external to the vessel is used. The power losses due to friction, etc., can be determined under a no-load condition.

(3) *Liquid and gas flow rates:* Rotameters, magnetic flow, thermal mass flow, and turbine flow meters are used.

(4) *Volume:* In-line sensors for volume measurements include liquid level sensors, differential pressure sensors, and weight-measuring devices (load cells).

(5) *Redox probe:* This probe measures the oxidation-reduction potential in the reactor. The measurement can provide correlations for product formation (Armiger, 1985).

(6) *Specific ion probes:* A number of ion sensors are available to measure ammonium, calcium, chlorine, magnesium, nitrate, potassium, sodium, phosphate, sulphate, etc., in the reactors.

## 9.9.1.9 Microbial Growth

Microorganism counting can be accomplished by microscopic, electronic, or biological techniques.

(1) *Microscopic methods:* First the sample is diluted and properly stained, then cells are counted in a known volume after examining them under a suitable microscope. The Petroff-Hausser counting chamber is generally used, which is 20 $\mu$m deep with a cross-sectional area of $2.5 \times 10^{-9}$ m$^2$. A cell concentration lower than $3 \times 10^{14}$ cells/m$^3$ is recommended (Koch, 1981). Microscopic examination can be performed under phase contrast or using fluorescent stains. Cell motility and growth can be prevented by using a diluent containing 0.5 M sucrose, 0.2% formalin, and 0.1% anionic detergent neutralized with disodium hydrogen phosphate (Gerhardt, 1981). A dilution with 0.1 M hydrochloric acid provides stationary cells without aggregation. Electron microscopy is used for counting virus particles.

(2) *Electronic methods:* For automatic cell counting, the coulter counter or

the flow microfluorometer is used. The coulter counter records the variations in the resistance when microorganisms in saline solution are drawn through a small orifice (30−100 μm dia.) (Agar, 1985). For accurate results, solid particles and bubbles should be eliminated from the suspension, and it should be treated with ultrasonic equipment to separate cell aggregations.

The flow microfluorometer measures the fluorescence resulting when stained cells flow past an exciting laser beam (Agar, 1985). Non-biological material can be eliminated by using special stains, e.g., fluorescein isothiocyanate for proteins and propidium iodide for nucleic acids. The laser beam diffraction at various angles can be used to measure cell size and shape.

(3) *Biological methods:* In a commonly used method, samples of a ten-fold serial dilution are spread on 10 cm agar plates with 1.5 to 2% agar. Cells will provide visible colony after incubation for 24 to 72 h. Generally, plates with 30 to 300 colonies are used for counting (Agar, 1985).

## 9.9.1.10 Biomass

(1) *Dry or wet weight:* To measure the dry weight of the cell mass, cells are first removed by centrifugation or filtration, then washed with an appropriate solution or solvent to remove residual traces of the medium. Cell mass is then dried, usually at 80°C for 24 h or 110°C for 8 h. The drying conditions should prevent biomass loss through volatilization or oxidation (Pringle and Mor, 1975). The wet weight technique is faster than the dry weight technique, in which the drying is not required. The values obtained are generally four times the dry weights (Gerhardt, 1981).

(2) *Turbidimetry and nephelometry:* In these techniques, the optical density of microbial suspensions is correlated with the amount of biomass. The following equation for the absorbance of a microbial suspension at 420 nm wavelength ($A_{420}$) was derived by Koch (1981).

$$A_{420} = (7.114 \times 10^{-3})X - (7.702 \times 10^{-6})X^2$$

where $X$ is the concentration of the suspension in g/m³; and is valid as long as $A_{420} < 0.3$. Dilution of the suspension is required for higher values of $A_{420}$. A number of instruments are available to measure absorbance. The Klett-Summerson unit is used to measure optical density directly in a shake flask (Agar, 1985). This unit contains a parallel beam arrangement with a filter and a CdS detector, and 420 or 660 nm wavelengths are used. This is most suitable for microorganisms in the $4 \times 10^{-19}$ to $2 \times 10^{-18}$ m³ size range, and cell shape does not affect the accuracy (Koch, 1981).

Nephelometry measures the light scattered at a 90° angle to the incident beam by the cell suspension. This method can be used for low cell concentrations ($10^{10}$ cells/m$^3$) and viruses, and can provide information on cell shape and structure (Agar, 1985).

## 9.9.1.11 Biomass Components

First the biomass is separated by filtration or centrifugation. Then, the following components are generally measured.

(1) *Protein:* Commonly used techniques for microbial growth determination are the Biuret, and the Folin-Ciocaltean (or Lowry) (Cooney, 1982). The Biuret technique determines $\alpha$-peptide linkages. On the other hand, in the Lowry method, the aromatic amino acids tryptophan and tyrosine are measured after alkaline hydrolysis of the protein. This method is about 600 times more sensitive than the Biuret (Agar, 1985).

(2) *DNA:* The intercalating fluorescent dyes ethidium bromide and propidium iodide are reacted quantitatively with double-stranded nucleic acids to yield highly fluorescent and conjugated forms. First, the RNA from the cells is removed by treating with RNAase or barium hydroxide (LePecq and Paleotti, 1967).

(3) *Physical properties:* Cell mass concentration can be determined by the indirect measurement of related parameters, such as carbon dioxide generation, electrical conductivity, heat generation, oxygen use, and viscosity. The viscosity of the cell suspension may provide the relative concentration of the biomass. A number of equipment are available to measure viscosity. Some of them are also suitable for continuous on-line measurements. The capacitance of the cell suspension can also provide the relative microbial cell concentrations (Genar and Mutharasan, 1979).

## 9.9.2 Process Control

Successful fermentation process control should maintain optimal environmental conditions for microbial growth. Myer et al. (1985) and VanBrunt (1985) suggested that off-gas analysis used to measure variables such as oxygen uptake, carbon dioxide evolution rate, and respiratory quotient may provide the best general on-line cell growth measurements. Other variables such as pH (influenced by substrate concentration), temperature (monitoring heat production), and glucose concentration have also been implemented.

A good fermentation control system is a comprehensive process model that relates all important process inputs to the process outputs. Real time adaptive control methods based on the system model can then provide on-line control in a fermenter. The most commonly used model in the design of control strategies for a fermentation process is the homogeneous single organism form. A set of state-space equations for a fed batch fermentation under conditions of substrate feed and continual volume increase limiting substrate was described by Gerson et al. (1988). Williams et al. (1986) proposed a dynamic model based on material balance and fermentation kinetics for baker's yeast production.

Fermentation is a multi-variable process, exhibiting nonlinear and time-varying properties, and subjected to unexpected disturbances. Thus, key process variables may not be directly measured. Bastin and Dochain (1986) implemented an on-line specific growth rate estimation using continuous time estimation algorithms as part of adaptive control strategies, where controller parameters were identified and varied on-line.

Fermenter parameters such as aeration, agitation, temperature, substrate feeding, etc., should be optimized to increase product yields. The fermenter optimization involves maximizing product purity, quantity, and quality in the broth, and minimizing the use of raw materials and energy. For this, the descriptive mathematical model of the process can be used.

## 9.10 **OTHER CONSIDERATIONS**

### 9.10.1 **Inoculation**

In most fermenters the culture is grown as inoculum for each fermentation, and discarded at the end. In other cases a part of the culture is reused as inoculum of the next fermentation (Brown et al., 1987). The stock culture is a carefully maintained collection of a particular microbial strain. Stock cultures are kept in freeze-dried form, deep frozen in liquid nitrogen on silica gel pellets to maintain their viability. A small sample is used to check its viability, and cultures are then grown in larger amounts for the inoculum of the fermenter. Preparation of an inoculum requires careful proliferation of a few cells to a dense suspension from 1 to 20% of the fermenter volume.

Recycling of cells or nutrients to a fermenter is practiced commercially. For single-cell protein and ethanol fermentations, it was found advantageous to return spent broth from product recovery to the fermenter (Bungay, 1984).

### 9.10.2 **Nutrient**

A proper amount of nutrient is required to meet organism metabolic demand for optimum growth and/or product formation. The basic nutritional requirements of heterotrophic microorganisms are a carbon or a nitrogen source.

## 9.10.3 Pipes and Fittings

Various types of valves are required for an efficient and safe fermentation. A safety valve is needed to maintain a safe pressure within the vessel. To control fluid flow, diaphragm valves are most frequently used. These are convenient for both manual and automatic operations. Piping should normally be welded, with smooth joints, but at locations where periodic disassembling is required, bolted joints are used (Brown et al., 1987). A proper gas-collecting system is required when potentially useful gases (e.g., methane) are collected during fermentation.

The fermenter system should be arranged to allow independent sterilization of its components. All valves in the fermenter should be easy to maintain, clean, and sterilize. Positive pressure is maintained in the fermenter to ensure that leakage will be outward.

## 9.11 **DESIGN**

Fermentation process design requires a product's annual capacity, the yield of product isolation, and the fermenter productivity. The maximum fermenter capacity is determined by constraints such as down-stream capacity, air supply (increased height increases static pressure), and heat transfer (becomes limiting at about 4.5 m diameter) (Naveh, 1985).

The basic fermenter design criteria are (Smith, 1985) (1) microbiological characteristics, (2) cell growth and product formation kinetics, (3) heat and mass transfer, (4) aseptic design, (5) control of fermentation parameters, (6) economics, and (7) scale-up potential.

The bioreactor should be designed to satisfy the particular requirements and limitations of a process. Generally, there is a single rate-limiting step controlling the reaction, as well as a few secondary limitations. These rate limitations provide a basis for process design. Bioreactor productivity can be increased either by genetically improving the rate-limiting step in the cell or increasing the cell concentration. The bioreactor design is based on minimizing process constraints such as heat and mass transfer, and permitting optimal control of biocatalytic activity while minimizing total process cost. The overall objective is to maximize the yield of product per unit of biocatalyst. The product concentration in a bioreactor can be improved by reducing the flow rate or increasing the activity and concentration of the biocatalyst (Cooney, 1983).

## 9.11.1 Material Selection

The material selection for the fermenter construction is based on the factors that influence corrosion. These are pH, temperature, oxidizing agents, and film accumulation on the surface. The material should also be non-toxic, and able to withstand internal pressure. Stainless steel is the most preferred material. Type 304 stainless steel, containing 18% chromium and 8% nickel, is used to some extent in the food industry. Type 316 stainless steel, alloyed with 2% molybdenum for improved pitting and crevice corrosion, is used widely in the food industry.

## 9.11.2 Heat Transfer

The non-Newtonian nature of the broth effects bulk flow, and heat and mass transfer. Apparent viscosity may create problems in stirring and mixing.

The heat transfer coefficient can be estimated by using an empirical equation of the form:

$$\text{Nu} = C1*(\text{Re}**C2) * (\text{Pr}**C3) * (\mu/\mu_w)**C4$$

where $C1$ to $C4$ are constants depending on the flow type, geometry, and surface properties; Nu, Re, and Pr are the Nusselt, Reynolds, and Prandtl numbers, respectively; $\mu$ is viscosity at the average temperature, and $\mu_w$ is the viscosity at the wall temperature. For detailed heat transfer calculations, a heat transfer book should be consulted.

When designing the external cooling system, the organisms should not be overcooled, and care should be taken that they are not damaged by the shear action of the pump.

## 9.11.3 Kinetics

The fermentation kinetics information is necessary to size the fermenter and its associated equipment, and to predict bioreactor performance. This information is normally obtained from laboratory experimentation with a 1 to 3 L fermenter. The optimum fermenter size and fermentation conditions can then be calculated for the highest or most economic product yield by modeling the fermenter system.

In general, packed-bed reactors have kinetic advantages over continuous-stirred tank reactors for most types of reactions. In a continuous-stirred reactor, the average reaction rate is lower than in a packed-bed reactor due to the different operational concentration of substrates. However, stirred reactors are more suitable for reactions requiring high rates of oxygen transfer or the addition of base or acid for pH control (Smith, 1985).

Batch operation is still used for better control of the final product. In large operations, the use of continuous fermentation results in significant savings in capital, labor, and operating costs compared to batch operation.

The microbial specific growth rate in a fermenter ($u$) as a function of the substrate concentration ($S$) is given by the Monod equation:

$$u = u_{max} \cdot S/(k_s + S)$$

where $u_{max}$ is the maximum microbial specific growth rate, and $k_s$ represents a saturation constant. The rate of organism concentration increase ($dx/dt$) is the growth rate, while the specific growth rate is the rate of increase/unit of organism concentration, i.e., $(1/x)\cdot(dx/dt)$ (Patel, 1985). In many simple systems, growth rate is a constant fraction ($y$) of the substrate utilization rate ($dS/dt$), or $dx/dt = -y\cdot dS/dt$; where yield constant ($y$) over any finite growth period is the ratio of the mass of cells formed to the mass of substrate used. The net increase of the organisms is growth minus outflow ($D_x$), i.e.,

$$dx/dt = u_x - D_x$$

The net rate of change of substrate concentration is input-output-consumption, where consumption is the growth/yield constant. Thus, for the maximum output of cells or biomass, the dilution rate should be high but it cannot exceed $u_{max}$ (Smith, 1985).

The oxygen uptake rate ($OU$) of the culture is met with an oxygen transfer rate ($OT$). $OU$ [mol/(m³·s)] is given by (Naveh, 1985):

$$OU = Cx \cdot [(1/Y_{sx} - k_1)\mu + M_s]$$

where $Y_{sx}$ is the yield of biomass on substrate, $Cx$ is the cell density (mol/m³), $M_s$ is the maintenance coefficient (1/s), $\mu$ is the specific growth rate (1/s), and $k_1$ is a factor depending on the medium composition. The $OT$ is (Aiba et al., 1973) $K_La \cdot (C_G/M - C_L)$, where $C_G$ is the equilibrium oxygen concentration at the gas/liquid interface (mol/m³), $M$ is the partition coefficient between gas and liquid, $C_L$ is the liquid phase oxygen concentration (mol/m³), $K_L$ is the liquid side mass transfer coefficient (m/s), and "$a$" is the air/liquid interface area (m²/m³). Empirical correlations relating "$K_La$" to the gassed power and air velocity can be used for scale-up on a constant $K_La$ basis for Newtonian broths (Aiba et al., 1973): $K_La = C\,(Pg/V)^\alpha \cdot Vs^\beta$, where $C$ is a constant, $Pg$ is the gassed power draw of the agitator, $V$ is fermenter volume (m³), and $Vs$ is the superficial gas velocity (m/s). Exponents $\alpha$ and $\beta$ change with fermenter size.

## 9.11.4 Fermentation Products

The nature of the fermented product and its relationship to the microorganism growth also affect fermenter design. The product may be intracellular (contained within the cell) or extracellular (secreted from the cell) or secondary metabolite.

(1) *Intracellular products:* The majority of fermentation products are retained within the cell. In the production of intracellular product, the cells may be separated from the broth in non-sterile equipment following the fermentation (Russell, 1987).

(2) *Extracellular products:* Some fermented products are secreted by the organism. In such cases, the product may be recovered from the broth after removal of the cells, and the opportunity exists to recycle cells back to the fermenter.

(3) *Secondary metabolites:* These are not needed for growth of the micro-organism and are usually produced after the growth phase of the organism.

# CHAPTER 10

# Scale-Up Techniques

## 10.1 **INTRODUCTION**

*Scale-up* of biotechnological processes is not easy, but it can be viewed as challenging and exciting. The scale-up ratio is the relationship between the size of the contemplated commercial unit and the small-scale unit (Bisio and Kabel, 1985). Some scale-up problems are (1) unit shape affecting differences in agitation, fluid short-circuiting, or stagnation zones; (2) mode of operation resulting in different residence time distributions; (3) surface-to-volume ratios, flow patterns, heat removal, and geometry generating various gradients of concentration and temperature; and (4) wall, edge, and end effects. Rather simple operations in the laboratory, like transferring solutions, become quite complex in the plant. They require complicated design and selection of pumping equipment, piping, agitation, etc. Temperature increases of 15 to 20°C can develop during large-scale breakage of microbial cells, if temperature is not controlled properly (Patel, 1985).

A common goal in scale-up is to convert a batch process to a continuous operation. Often the continuous reactor selected is a tubular reactor. The advantages of a continuous operation are the reduction in reactor volume, a savings in labor costs, and improved quality control (Kabel, 1985). Not all equipment used in the laboratory is readily available, adaptable, or economical on a large scale. For example, the breakage of microbial cells using a sonifier is not suitable for large-scale operation (Patel, 1985).

The scale-up concept is usually convenient in a qualitative sense. If the reaction occurs at constant mass density, the residence time in a plug flow reactor is indeed the same as the batch reaction time, and this scale-up method gives accurate results. However, many tubular reactor calculations are not so simple (Kabel, 1985).

Most of the equipment involved, e.g., fermenters, sterilizers, separators, and crystallizers, can be designed and operated properly only if the scale-up technology can be applied correctly. The required data for scale-up are generally obtained from a properly equipped pilot plant (Kampen, 1983). Process conditions may affect the performance of the scale-up unit. For instance, although concentration of enzyme solution by ultrafiltration is easy to scale up, the total surface area available for ultrafiltration, concentration

**283**

rate, temperature, pH of the solution, concentration polarization, and shear will affect the process.

Time is one of the most important factors influencing the effectiveness of any scale-up attempt. On the other hand, handling problems result from the volume of material involved, the necessity for containment of biological material, and the type of equipment to be used (Patel, 1985). For a given flow rate, the pressure drop can be reduced by increasing the particle size. The drawbacks are due to flow pattern and chemical kinetics. If the ratio of tube diameter to particle diameter is too small, a large fraction of the reaction mixture will bypass the catalyst. With relatively fast reactions, important concentration gradients develop inside the catalyst particles (Froment, 1985).

## 10.2 SCALE-UP

Scale-up is not simply a matter of multiplication. The parameters that govern growth of microorganisms in a 1 L shake flask are not necessarily the same parameters that apply at a 100,000 L fermenter. To start with, at least lab and/or pilot plant scale data on the reactions involved must be available. All factors cannot be kept constant during scale-up, and generally the assumption is made that factors that are not held constant will not change enough to affect a large-scale fermentation deleteriously.

Scale-up involving process transformation can be done by setting up proper process models based on mass and energy balances, transport processes, and chemical kinetics. The usefulness of a given model for scale-up of a reactor will depend on the validity of the assumptions made for different reactor sizes. Mock-ups or homologous models are quite helpful for investigating hydro-dynamic aspects of the processes.

### 10.2.1 Required Data

For scale-up of any fermentation system, the following information is generally required:

(1) Kinetics, including reaction paths, rate constants, and heat of reaction for each path over a complete range of anticipated operating conditions, that can be estimated or measured

(2) Physical properties of the reaction mixture including densities, viscosities, surface tension of various phases, and thermal diffusivity

(3) Thermodynamic data including the solubility of various reactants and products; phase and reaction equilibrium constants

(4) Residence time distribution

(5) Non-adjustable quantities for multiphase reactor including phase hold-

ups, interfacial areas, heat and mass transfer properties, and dispersion coefficients or mixing parameters

(6) Others such as production rate, reactor geometry and type, and process data; these parameters control adjustable operating conditions such as velocities of materials in different phases, temperature, pressure, and the direction of flows

This data will permit a choice of the scale-up reactor geometry, dimensions, and the nature of the phase distributors (Froment, 1985). In multiphase systems, process data can seldom be measured independently and they are always coupled with interphase transport of heat and mass.

The information required can be obtained from investigations on a smaller scale by experimental simulation applying the rules of similarity, and mathematical simulation based on a more detailed model. Preliminary results from the smaller pilot-scale reactors yield important information for scale-up on new processes and process improvements.

In fermentation, both the organism and bioreactor are important in design and scale-up. Since many process parameters change with scale, the results of the first large-scale (or pilot-scale) experiments will often provide further modifications to the organism, the reactor, or other process parameters (Van Brunt, 1985).

## 10.2.2 **Geometric Similarity**

Scaling up a fermenter involves a considerable number of compromises; in general, the environment at a large scale cannot be the same as that at the small scale. The scale-up algorithm has been well-known for a long time. In these algorithms, complete geometrical similarity between the large- and small-scale units was maintained; and all the relevant dimensionless groups were required to be the same for the two units (Fair, 1985). If geometric similarity is to be maintained, then the general forms of correlation for heat and mass transfer, power input, and oxygen maintenance should be essentially scale-invariant. The neglect of biological features of biotechnological processes frequently creates problems of simultaneous maintenance of the values of many dimensionless groups of physical parameters.

Highly empirical methods based on the concepts of similitude and dimensionless groups have been developed. Two systems are geometrically similar when the ratios of corresponding dimensions in one system are equal to those in the other. Kinematic similarity exists between two systems of different sizes when they are not only geometrically similar but when the ratios of velocities between corresponding points in each system are also the same. Dynamic similarity exists between two systems when, in addition to being geometrically and kinematically similar, the ratios of forces between corresponding points in each system are equal (Himmelblau, 1985). Thus, for similarity to exist, the values of all of the dimensionless groups should remain constant during the scale-up. These dimensionless groups can be derived by means of the Buckingham Pi theorem. A number of such groups have been developed (Froment, 1985). This principle in practice frequently remains impossible to fulfill. Complete similarity is not possible between two non-isothermal and non-isobaric reactors, and the scaling-up will involve some compromise and some risks. Under these situations, a ''weight'' factor can be introduced.

Geometric similarity is not necessary to scale-up, and there are many cases where it is impractical to maintain it. Greater flexibility is possible if geometric similarity is not a priority. With non-geometric scale-up, $K_La$ and shear at the impeller can be kept constant. However, the correlations will be scale-dependent if geometric similarity is abandoned. One of the most frequently used methods is based on the maintenance of constant $K_La$ along with, in some cases, constant shear (Charles, 1985).

Consider the scale-up of a reaction in a pilot reactor 3 m in diameter to a process vessel 9 m in diameter, keeping the same geometrical shape and keeping the same heat transfer rate per unit mass in the large reactor as in the small reactor. The lengths are scaled up 3 times, the areas are scaled up 9 times, and the volumes 27 times. Thus, it is 3 times easier to heat the reaction mass in the small unit (Himmelblau, 1985; Kampen, 1983). Increasing the degree of agitation by doubling the tip speed increases the surface heat transfer coefficient by 10%. It is difficult to satisfy all or even a majority of the various similarity criteria in scale-up, particularly those related to the dimensionless thermal and chemical groups.

## 10.2.3 Relationships

It is possible to examine various relationships that might be relevant to a fermenter scale-up:

Power input $(P), \propto N^3 \cdot D^5$
Head $(H) \propto N^2 \cdot D^2$
Circulation capacity $(Q_c) \propto N \cdot D^3$
Circulation time $(tc) \propto Q_c / T^3$
Mixing time $(tm) \propto tc$
Impeller tip speed $(Nt) \propto N \cdot D$

where $N$ is impeller speed, $D$ is impeller diameter and $T$ is absolute temperature (Einsele, 1978). Various schemes can be considered:

(1) Scale-up at constant agitator power per unit broth volume $(P/V)$ and with geometric similarity. Circulation time, mixing time and impeller

tip speed will increase with the increase in head ($H$). Thus, circulation capacity has to increase substantially.

(2) Maintaining the geometrical similarity, tip speed, and head constant; $P/V$ will be decreased, which might seriously affect the aeration rates.

(3) Using two impellers, while keeping the impeller geometrically similar, will maintain constant $P/V$, head, and impeller tip speed on scale-up.

(4) Increasing impeller diameter ratio on scale-up will maintain $H$, $Nt$, $P/V$, and circulation rate constant.

A more common form of scale-up in laminar flow processes is to change the pipe dimensions but to maintain geometric similarity, e.g., $L_1/R_1 = L_2/R_2$, where $L$ is pipe length and $R$ is pipe radius. This form of scale-up provides similar pressure drop and residence time ($\bar{t}$); volumetric flow rate ($Q$) increases, i.e., $Q_2/Q_1 = R_2^3/R_1^3$ for Newtonian fluids, and $Q_2/Q_1 = (R_2/R_1)^{(2n+1)/n}$ for power law fluids at constant pressure drop; also $\Delta P_2/\Delta P_1 = (R_2/R_1)^{(n-1)/n}$ at constant "$\bar{t}$," where "$n$" is power law index (Nauman, 1985). Such a scale-up can be more desirable for the non-Newtonian than that for Newtonian fluids. Larger-sized equipment can operate at lower pressures and thus remain mechanically feasible to construct.

Scale-up at constant "$\bar{t}$" generally requires a longer length of the same diameter tube, or multiple tubes in parallel. A single-tube scale-up with geometric similarities gives $Q_2/Q_1 = L_2/L_1 = R_2/R_1$. Laminar systems frequently show large differences in composition or temperature which are eliminated in turbulent systems by eddy diffusion. Complete similarity of velocity, temperature and composition is difficult to achieve. One must choose to retain important aspects of the process that are required for its economic or technical success (Nauman, 1985).

The impeller tip speed determines the maximum shear rate and shear stress. On scale-up, $K_L a$ and impeller tip speed are generally kept constant by slight adjustments to the impeller diameter per fermenter ratio. The $P/V$ determines the Reynolds number, mass transfer coefficient and particulate sizes (Kampen, 1983).

Thus, in fermentations, power consumed per unit volume ($P/V$), $K_L a$, and the average velocity at a point in the vessel are the three most commonly used criteria for scale-up (Aiba, 1973).

## 10.2.4 Mixing, Agitation, and Aeration

Good bulk mixing is essential for better control, and for heat and mass transfer in a fermenter. Mixing time is a function of fermenter geometry, impeller design, operating conditions, and fluid rheological properties. There are several correlations available for Newtonian fluids; however, they give widely divergent estimates of mixing time. Non-Newtonian fluids usually are difficult to mix well and this causes many significant problems, particularly for large-scale fermentations.

In some instances, the scale-up of agitators is difficult due to the interrelation of various factors such as the mixing of particles. The particles stay the same size while the impellers are made larger in scale-up procedures (Garrison, 1983).

The mixing time increases as the size of the fermenter increases. Mixing time values of 29 s for 1800 L, 67 s for 60,000 L, and 140 s for 120,000 L were reported. Manfredini et al. (1983) found that for a 112,000 L fermenter, long mixing time had little effect on temperature in different parts of the vessel, even for viscous broth. Dissolved oxygen concentration, however, varied between 65 % air saturation at a point 1 m from the bottom to 46 % air saturation at a point 7.5 m from the bottom of the fermenter. However, the dissolved oxygen concentration measured at the same height, but at different distances from the axis, was more or less constant.

Factors affecting the liquid behavior in an agitated fermenter vessel are the power requirements of agitation, rotational speed of the impeller, and the pumping rate of the impeller. If scale-up is attempted on the basis of equal $P/V$, then an increase in the tip velocity of the impeller will result in the decrease of the shear rate (force per unit volume, $F/V$) of the liquid. With a decrease of $F/V$, the mixing time is expected to increase.

Constant values of the specific power input ($P/V$), the Reynolds number, the circumferential speed of the impeller, or other factors are generally used in scale-up of aeration and mixing devices.

In general, practical approaches to scale-up usually vary between using constant $P/V$ as a conservative estimate and using constant tip speed as a very unconservative estimate. For gas-liquid processes, the superficial gas velocity and $P/V$ are effective in correlating the gas-liquid mass transfer coefficient, $K_L a$. These relationships are relatively independent of scale. In blending process, to maintain pumping capacity from the impeller per unit tank volume constant on scale-up, the $P/V$ must increase proportionally with the square of the tank diameter. This is not normally practical. Therefore, the circulation time tends to increase with increase in tank diameter (Oldshue, 1985). In case of free-settling solids, constant $P/V$ is an approximate indicator of similarity on scale-up to larger tank diameter. When the slurries are very viscous, pseudoplastic, and have 40 % or more solids, $P/V$ generally decreases with an increase in tank diameter.

Generally, it is recommended that designers apply equal torque per unit volume of suspension by keeping geometric similarity and constant tip speed. Thus, the $P/V$ varies inversely with agitator diameter. However, Zwietering (1958) indicated that the stirrer speed can be reduced on scale-up proportional to the increase in agitator diameter by maintaining geometric similarity. All the impellers are more effective for complete suspension when placed nearer to the tank bottom, and a clearance of about 25% of vessel diameter is used (McCabe et al., 1985)., The optimum ratio of impeller diameter to vessel diameter for a given power input is an important factor in scale-up. It is impractical to maintain a similar blending time in the scale-up unit; however, a small increase in blending time in the scale-up unit will reduce power requirements. Sometimes horizontal baffles are used in a scale-up unit for better and uniform mixing (Figure 10.1).

## 10.2.5 Residence Time Distribution

One can never maintain all forms of similarity (geometric, dynamic, thermal, etc.) during a scale-up. The art of scale-up is knowing which similarities to keep and which to sacrifice (Nauman, 1985). The disadvantage of scaling-up with geometric similarity and constant "$\bar{t}$" is that the pressure drop will increase when going from the pilot plant to the production unit.

For the following situations, the scale-up of residence time distribution can be done with a high degree of confidence (Nauman, 1985):

(1) The pilot system is an open tube or packed bed with a residence time distribution that approximates piston flow.

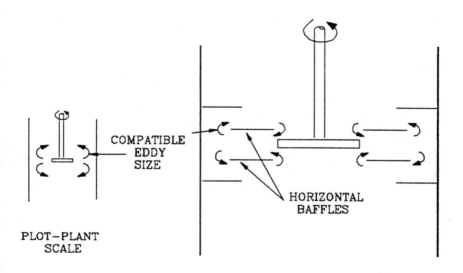

**FIGURE 10.1.** Use of baffles for uniform mixing in a scale-up unit.

(2) The pilot system is in laminar flow and molecular diffusivity is negligible.

(3) The pilot system is a stirred tank with an exponential distribution of residence times.

In these situations, the desired scale-up will maintain "$\bar{t}$."

## 10.2.6 Physical and Chemical Properties

The effect of physical and chemical parameters on scale-up are as follows (Buckland, 1984).

### 10.2.6.1 Oxygen Transfer

Most of the common aerobic fermenters have been found to be limited by the supply of oxygen from the gas phase into the liquid. After measuring the specific growth rate from small-scale experiments, the specific oxygen uptake rate can be determined. A knowledge of the interrelationships between $K_L a$ (oxygen transfer coefficient, $s^{-1}$) in the fully grown culture and the geometry of the unit, energy input, gas rate, etc., is required in order to create a detailed fermenter design. A number of expressions were developed to calculate $K_L a$ (Charles, 1985) for Newtonian and non-Newtonian fluids (see Chapter 9 for details). The results predicted by these correlations are highly variable. It is difficult to predict $K_L a$ with accuracy due to the effect of broth rheology on bubble size and because of the pseudoplastic properties of many of these broths.

Agitator speed, airflow rate or back pressure are adjusted to maintain the dissolved oxygen levels. The problem of foaming generally gets worse as the scale is increased due to the increase in superficial air velocity. As scale is increased, the $P/V$ ratio required to provide a certain $K_L a$ generally decreases. The opposite is true for the $P/V$ of air supply, which increases with scale because of greater hydrostatic head in the vessel. Scaling on the basis of constant $P/V$, therefore, is misleading. Also, scaling-up on the basis of constant $K_L a$ is also problematic. The $K_L a$ in a typical fermentation changes throughout the cycle and is a function of the broth rheology and chemical composition.

The dissolved oxygen level can be maintained above that of air saturation at all times to prevent foaming. In practice, the concentration of the starting medium is reduced, resulting in a lowered oxygen demand.

Commonly used methods of scale-up are based on the maintenance of constant $K_L a$ along with constant shear due to agitation. Thus, reasonable reliable methods are required to determine $K_L a$, power input required to reach the desirable $K_L a$ and shear, the effects of increased pressure, surface aeration, etc.; and spatial distributions of important variables (Charles, 1985). Control system design should be made an integral part of the overall design process.

## 10.2.6.2 Heat Transfer

A classic mistake is to install a more powerful agitator into a fermenter to enrich the medium, and then find that the fermenter has become heat transfer limited.

## 10.2.6.3 Media Preparation and Sterilization

As scale is increased, the heat stress on a batch sterilized medium increases. The heat-up and cool-down periods of a 100,000 L fermenter are typically a few hours, compared to 20 min for an 800 L fermenter. For sterilization, larger volumes require higher temperatures for longer times.

In laboratory-scale experiments, laboratory-grade material and deionized water are used, where in the pilot plant or industrial scale, technical-grade material and tap water are generally used. Also, low-cost ingredients used at the production level show a significant lot-to-lot variation.

The large-scale purification process could be designed to minimize the number of steps while maintaining high yields and product purity.

## 10.2.6.4 Culture Stability and Seed Preparation

Seed transfer often becomes critical as scale is increased. Transferring seeds on a scheduled age basis is a mistake because of the variability in the site of inoculation into the seed tank.

## 10.3 **RULES OF THUMB**

A number of important rules of thumb used in scale-up are listed below (Aerstin et al., 1985; Buckland, 1984):

(1) Reaction rates may double every 10°C, and vapor pressures may double every 20°C.

(2) Heat transfer coefficients in jacketed reactors may run 225 W/(m²·K) for metal and 170 W/(m²·K) for glass-lined equipment; for condensers this value may be 540 W/(m²·K), and 1130 W/(m²·K) for thermo-siphon reboilers.

(3) Column diameters should be at least 8 times the packing size to minimize large voids near the wall. Column length and throughput per cross-sectional area should be held constant to get identical results.

(4) If two liquids are immiscible, then the infinite dilution activity co-efficient is >8. Liquid distribution is the problem with a packed tower in many cases.

(5) Solid particles tend to collect at the liquid/liquid interfaces.

(6) Freezing points may be suppressed 1°C for every 1.5 mol% impurity present.

(7) The relative separation factor in a liquid-liquid extraction should be at least 5. Liquid-liquid extraction column volume can be reduced with mechanical agitation if the interfacial tension is high.

(8) The vessel diameter is generally limited to 4 m to avoid fabrication problems and also because of heat transfer requirements. The vessel size is typically in the range of 100,000 to 150,000 L. Several vessels in a factory are preferred to one tank because of the typical batch mode operation.

(9) A typical ratio of impeller diameter to tank diameter is 0.4 for shear-sensitive cultures.

(10) The motor, which is typically sized to give about 19.7 W/L (100 hp/1000 gallons) for a production-scale vessel, should have a variable speed or at least two speeds.

Sometimes it is useful to compare various process time constants during scale-up. These time constants are related to transport processes and mixing. Some of these are (Kossen, 1985):

- heat transfer, (mass)(specific heat)/[(surface area)(heat transfer coefficient)]
- oxygen transfer, $1/(K_L a)$
- diffusion, (length)$^2$/diffusivity
- flow, length/velocity
- microbial growth, 1/specific growth velocity
- chemical reaction, concentration of substrate/rate of conversion

These are also useful tools for regime analysis.

## 10.4 REACTOR SCALE-UP

The design of a commercial-size reactor based upon information gathered on a smaller scale involves many facets and several stages of evolution. Decisions concerning the reactor type and its configuration, or about the desirability of recycling, have to be made at an early stage to avoid costly detailed studies of unfruitful side tracks (Froment, 1985). The choice of reactor type is dependent upon the peculiarities of a particular process. The reactor should achieve a certain conversion with optimal selectivity at a given temperature and pressure, with minimum downtime and hazard, and maximum profit.

To control the temperature change in *adiabatic* reactors, a multi-stage reactor is used, and intermediate cooling can be achieved by means of internal or external heat exchangers, by injection of cold reacting fluid, or by a combination of both means.

The *micro-carrier* system, the *fluidized bed* bioreactor, and the *solid bed* bioreactor are suitable for scaling-up (Werner et al., 1987). The major issues in homogeneous reactor scale-up are the determination of the size, shape, and performance of the reactor. In performance, such matters as conversion, selectivity, and stability are of primary concern (Bisio and Kabel, 1985).

*Non-stirred* and *non-aerated* reactors are used for the production of beer, wine, and fermented dairy products. These are anaerobic batch processes and do not create scale-up problems. *Bubble column* (non-stirred and aerated) reactors are used for the production of baker's yeast. Scaling-up creates mixing problems. *Stirred-aerated* reactors are used for the production of amino acids, citric acid, etc. Scaling-up of these reactors is a problem of transport.

Biotechnological processes should be scaled-up very carefully. At each stepwise scale-up, the process is carefully examined for changes in yield, foaming, production rate, etc.

Kossen (1985) summarized the changes in various parameters when a 10 L reactor is scaled-up to 10 m³ by using various criteria, as shown in Table 10.1. This indicates that different scale-up criteria are incompatible.

In scale-up, it is assumed that the residence time in a *tubular* reactor, operating at plug flow, would be identical to the batch reaction time without the down time. The volume of the tubular reactor would then be equal to the product of this residence time and the volumetric feed rate necessary to give the desired production rate (Kabel, 1985).

Klei et al. (1987) modelled the fermenter hydrodynamics by considering the fermenter as a series of compartments. The dissolved oxygen concentrations in various parts of the fermenter was estimated by calculating oxygen balances using correlations for mass transfer and impeller pumping capacity

TABLE 10.1. *Changes in various parameters due to different criteria.*

| Scale-Up Criteria | Power | Power/Volume | $N$ | $Nt$ | $Re$ |
|---|---|---|---|---|---|
| Equal power/volume | $10^3$ | 1.0 | 0.2 | 2.1 | 21.0 |
| Equal stirrer speed ($N$) | $10^5$ | 100 | 1.0 | 10.0 | 100 |
| Equal tip speed ($Nt$) | 100 | 0.1 | 0.1 | 1.0 | 10.0 |
| Equal Reynolds no. (Re) | 0.1 | $10^{-4}$ | $10^{-2}$ | 0.1 | 1.0 |

for each compartment. The basis of the scale-up was to calculate operating conditions, such as agitation speed and aeration rate, that minimized power consumption while keeping the dissolved oxygen concentration above a critical level.

Hubbard (1987) emphasized the need to consider the non-Newtonian properties of broths when making scale-up computations for *stirred tank reactors*. The rheological model parameters can be included in gas-liquid mass transfer and heat transfer correlations. Data for heat and mass transfer to non-Newtonian liquids in stirred tanks are available in literature.

*Fluid-fluid* reactions can be classified as liquid-liquid, gas-liquid, and gas-liquid-solid reactions. Scale-up of a fluid-fluid reactor is a complex task because in most cases reactor performance depends upon hydrodynamics, transport, and mixing characteristics of the reactor. Since the apparent reaction rate depends upon the various transport resistances, the controlling resistance also can depend upon the reactor scale. Thus, the reactor performance model used for a small-scale reactor may not be useful for the large-scale reactor. For proper design and scale-up of a fluid-fluid reactor, a good mathematical model of the reactor is required. It depends on the specification of reactor geometry, adjustable reactor operating conditions, process data, reaction kinetics, physical properties of the reaction mixture, and thermodynamic data. However, in the scale-up of a multiphase fluid-fluid reactor, non-adjustable parameters are also required. Knowledge of the controlling reaction regime can simplify the model. Values of intrinsic kinetic parameters do not depend on the shape, size, and internal design of the reactor or the various phases' velocities. However, the phenomenological coefficients such as gas-liquid, and liquid-solid mass transfer coefficients and phase hold-ups depend on the scale, and the size and nature of the reactor. Scale-up will change the relative importance of transport and kinetic resistances on the apparent reaction rate (Shah and Deckwer, 1985).

*Bubble column* reactor scale-up usually increases the reactor length, diameter, and phase velocities. These changes will affect inflow regime, phase hold-up characteristics, axial and radial mixing, and transport coefficients. In large-diameter bubble columns, the assumption of gas-phase plug flow is questionable. When reactor diameter and length are increased in the scale-up of a gas-liquid reactor, phase distribution devices should also be changed. The gas sparger design should uniformly distribute the gas (Shah and Deckwer, 1985).

Kossen (1985) provided a scale-up method by experimental simulation of large-scale conditions at the small scale. These conditions are optimized at the small scale and then implemented at the full scale.

To simulate a variety of free cell or immobilized cell biochemical reactors to allow sizing and scale-up, a software program was developed for use on microcomputers (Klei et al., 1987).

# Applications

## 11.1 **INTRODUCTION**

A number of biotechnological techniques discussed in Chapters 2 through 10 are being used at various stages of food processing. It is difficult to cover all of these applications in this book. However, a few of these applications are discussed in this chapter.

## 11.2 **DAIRY PROCESSING**

### 11.2.1 Starter Cultures

In the dairy industry, microbial cultures are used to ripen and flavor various milk products, to reduce the maturation time for dairy products, and to hydrolyze large-chain molecules to make them soluble and improve the consistency of dairy products. For these purposes, microbial rennin, lactase, proteases and lipases are used (Kosaric, 1984). Presently, more than 20 million tons of fermented milk products are annually produced worldwide with the help of lactic acid bacteria. The dairy industry also uses propionic acid bacteria, red smear cultures (*Brevibacterium linens*), mould, and yeast cultures. Table 11.1 summarizes various starter cultures with their compositions used for fermented milk products (Hunger, 1987). Starter cultures assist in manufacturing products at a high quality level, control the fermentation time, and reduce spoilage (Hammes, 1987).

Lactic acid bacteria produce lactic acid from lactose. Starter cultures are available in different types: liquid cultures (stock culture), freeze-dried cultures, and concentrated deep-frozen cultures. Liquid cultures are the most widely used.

The lactic acid genera include *Lactobacillus*, *Pediococcus*, *Leuconostoc*, *Sporolactobacillus*, *Bifidobacterium*, and *Streptococcus*. *Lactobacillus* and group *N-Streptococci* are the most commonly used genera in food processing. These are used for producing fermented milk products such as yogurt and cheese.

Yogurt is prepared from concentrated milk by the addition of nonfat dry milk to pasteurized skimmed milk so that total solids are about 10%. *Lac-*

TABLE 11.1. *Starter cultures for fermented milk products (modified after Hunger, 1987).*

| Product | Culture |
|---|---|
| Blue cheese | DL, L |
| Camembert | DL, L, MSK, *Penicillium candidum* |
| Cheddar/Chester | O |
| Cottage cheese | L, O |
| Cream cheese | D, DL, L |
| Cultured butter | DL |
| Cultured buttermilk | DL, L |
| Edam, Gouda | D, DL, L |
| Emmental, Gruyere | A mixture of: *S. thermophilus*, *Lb. helveticus*, *Lb. lactis*, *Lb. bulgaricus* |
| Feta, White cheese | L, O |
| Harzer, Mainzer | A mixture of: *S. thermophilus*, *Lb. helveticus*, *Lb. bulgaricus*, *Brevibact. linens*, *P. candidum*, *Geotrichum candidum* |
| Kefir | A mixture of: *S. cremoris*, *S. lactis*, *Lc. cremoris*, *Lb. kefir*, *Lb. brevis*, *Candida kefir* |
| Mild sour milk | MSK cultures |
| Roquefort | *P. roqueforti* |
| Romandur | DL, L, MSK, *Brevibact. linens* |
| Set curd yoghurt | *S. thermophilus* |
| Stirred yoghurt | *Lb. bulgaricus*, *Lb. yugurti* |
| Sour cream | D, DL, L |
| Sour milk | D, DL, L |

D: composed of *S. cremoris, S. lactis, S. lactis* susp. *diacetylactis.*
DL: composed of *S. cremoris, S. lactis, S. lactis* susp. *diacetylactis, Lc. cremoris.*
L: composed of *S. cremoris, S. lactis, Lc. cremoris.*
MSK: composed of *Biofidobact.* spec., *Lb. acidophilus, Lb. bulgaricus, S. thermophilus.*
O: composed of *S. cremoris, S. lactis.*
*Lb.* = *Lactobacillus, Lc.* = *Leuconostoc, S.* = *Streptococcus.*

*tobacillus bulgaricus* and *Streptococcus thermophilus* are commonly used species for fermentation (Brown et al., 1987).

Commercially important plasmid-linked elements are lactose fermentation ability, proteolytic activity, nisin production, mechanisms of phage resistance, and citrate utilization. Work is being carried out on plasmid mapping to determine the positions of important genes and markers (Barach, 1985). Moreover, industrially important recombinant DNA products are being developed from lactic bacteria.

Mixed strain dairy starter cultures are used for proper acid production and flavor development. DNA fingerprinting is used to distinguish between different strains of a number of bacterial species (Lazo et al., 1987). This method is based on the extraction of total cellular DNA, digestion of DNA with a restriction endonuclease, and separation of DNA fragments by agarose gel electrophoresis (Ramos and Harlander, 1990) (see Chapter 2 for details). This provides a banding pattern unique to that particular bacteria. The optimum results were obtained by suspending the washed cells in 10:1 TE and treating them with a minimum of 10 $\mu$L of a 10 mg/mL lysozyme solution.

## 11.2.1.1 Streptococci

In cheese, lactic *Streptococci* concentrate and stabilize the curd by coagulating the protein and expelling water, and contribute to the texture and the flavor (Sandine, 1985). Genetic material has been transferred to these bacteria using transduction (Allen et al., 1963), conjugation (McKay et al., 1980), protoplast fusion (Gasson, 1980), and transformation (Kondo and McKay, 1981) techniques.

Huggins (1984) reviewed the progress in the use of mesophilic lactic *Streptococci* as dairy starters, with emphasis on the methods for overcoming the bacteriophage inhibition problem. In the past, bacteriophage-insensitive mutants (BIM) were isolated by exposing a phage-sensitive strain to a collection of phages, permitting a time period for phage-mediated lysis of sensitive cells and partial outgrowth of the indigenous BIMs. The cloned BIMs were purified and characterized for comparison with the parent strain. Phase-mediated transduction is used to move genes between strains. McKay et al. (1973) transduced lactic *Streptococci* by packaging lactose plasmid DNA into defective phage particles.

The multitude of different plasmids in lactic *Streptococci* obscured the results of gene transfer and cloning work. With plasmid-free stains, Kok et al. (1984) reported that polyethylene glycol-treated protoplasts of *Streptococcus lactis* took up free plasmid DNA. The application of genetic engineering techniques to the lactic *Streptococci* was extended with the development of many vectors, derived from the cryptic *S. cremoris* Wg2 plasmid, which replicates in *S. lactis*, *Bacillus subtilis*, and *E. coli*. Kok et al. (1985) described the cloning in *B. subtilis* and *S. cremoris* Wg2 plasmid DNA, which encoded the gene responsible for the proteolysis in the strain, in one of these pWVO1-based vectors. The proteolytic activity of *S. cremoris* Wg2 was specified by a 4.3 Md (megadalton) HindIII fragment of pWVO5. This provided the possibility of using HindIII fragment "C" or its sub-fragments as proteinase gene probes.

McKay et al. (1973) described the ability of the temperate phage to transduce the lactose marker to lactose-negative (lac⁻) recipients of *S. lactis* C2. Ultraviolet (UV)-induced phage lysates, from lactose-positive (lac⁺) *S. lactis* C2, transduced lactose fermenting ability to lac⁻ recipient cells of this organism. The phage lysate from *S. lactis* C2 also transduced maltose and mannose metabolism to the respective negative recipient cells.

Kondo and McKay (1985) reviewed the functional properties of plasmid DNA of *N-Streptococci* and gene transfer systems including transduction, conjugation, transformation, and protoplast fusion. The *N-Streptococci* including *S. lactis*, *S. cremoris*, and *S. lactis* ssp. diacetylactis, are used as starter cultures in the production of cheese and other fermented dairy products.

Most studies of transduction in group *N-Streptococci* have used temperature phages. The transduction has also assisted in testing of a recombination-deficient (*Rec⁻*) mutant of *S. lactis* ML3. The *rec* genes are responsible for recombination and introduction of chromosomal genes into cells, while introduction of plasmids into cells is *Rec*-independent. In conjugal gene transfer, a donor cell, in close physical contact with a recipient cell, may transfer a replica of genetic information to the recipient cell. The conjugal transfer of many metabolic properties of *N-Streptococci* have been reported.

Transformation provides a link between the *in vitro* analysis of DNA and its *in vitro* function, and is required for the use of recombinant DNA techniques in *N-Streptococci*. More work is required on the physiology of protoplasts, as well as their formation and regeneration. Protoplast fusion among *S. lactis* 527 derivatives was reported by Okamoto et al. (1983). Recombination of chromosomal markers, adenine biosynthesis and Str$^r$, mal, galactose, tryptophan utilization were also described. More work is required to utilize this gene transfer mechanism.

Kondo and McKay (1984) cloned the *lac* genes from pLM2001, a 24 Md lactose plasmid, into *S. lactis* LM0230, a *Lac⁻* plasmid-cured derivative of C2. On the other hand, Harlander et al. (1984) cloned *S. lactis* genes in *S. sanguis* and *E. coli* where molecular cloning was more cost and time efficient. According to Kondo and McKay (1985), direct molecular cloning in *S. lactis* is most desirable for cloning plasmid-coded genes. The least desirable is the cloning in *S. sanguis*, since deletions in plasmids occur during transformation. In the future, recombinant DNA techniques should assist in the development of an appropriate vector and genetically engineered strains with better capabilities.

Potential applications of genetic engineering techniques, such as lending mutagenesis and recombinant DNA technology to the important dairy *Streptococci*, was reviewed by McKay (1986). The recombinant DNA technology to "tailor-make" a strain requires an advanced understanding of the genetic makeup and plasmid biology of the *Streptococci* as well as the development of gene transfer systems. The genetic engineering techniques could also be used (1) to stabilize the desired functions by integrating the plasmid or the desired gene into the chromosome; (2) to construct strains for the over production of β-galactosidase; (3) to link nisin producing ability to plasmid

DNA, which may lead to the super-nisin producing ability to plasmid DNA, which may in turn lead to the super-nisin producing derivative; (4) to develop phase-resistant mutants by conjugally transferring the appropriate plasmid to selected phase-sensitive strains of *Streptococci*, and other lactic acid bacteria; and (5) to develop strains for use in preparing frozen culture concentrate or freeze-dried or spray-dried culture concentrates.

A recombinant plasmid was isolated by Anderson and McKay (1984) containing a transfer region that correlates with high-frequency conjugative. However, an efficient plasmid transformation system and the use of recombinant DNA technology are needed for further genetic studies. Konde and McKay (1984) suggested polyethylene glycol-induced transformation of *S. lactis* protoplasts using the plasmid DNA. This was demonstrated by cloning a DNA fragment coding for the lactose-metabolizing genes into *S. lactis*.

McKay (1986) summarized the following applications of genetic engineering techniques to dairy starter cultures:

- developing naturally produced flavor compounds (such as banana, citrus, peach, pineapple, strawberry) through the isolation of the responsible genes and transfer of these genes to lactic acid bacteria
- developing strains to convert whey into a marketable end product through flavor and textural changes
- cloning genes coding for the production of proteins with intrinsic sweetness qualities
- developing food-grade organisms into hosts used to produce medicinals
- cloning genes beneficial to human or animal health

## 11.2.1.2 Lactobacilli

*Lactobacilli* are widely used in dairy, alcoholic beverages, pickling and silage processes. *Lactobacilli* produce lactic acid from the fermentation of lactose. They also produce volatile compounds such as acetaldehyde, diacetyl, and alcohol in milk. Acetaldehyde is a major flavor component of yogurt. An erythromycin resistance plasmid, isolated from *Streptococcus faecalis*, was introduced into *L. casei*, *L. acidolphilus*, *L. reuteri*, and *L. salivarius* by conjugation (Lee-Wickner and Chassy, 1985). They described the isolation and restriction mapping of some of the small cryptic plasmids isolated from *L. casei* and the molecular cloning of these plasmids in *E. coli* and *S. sanguis*. Four small cryptic plasmids were isolated from *Lactobacillus casei* strains, and restriction endonuclease maps of these plasmids were made. Three of the small plasmids (pLZ18C, pLZ19E, and pLZ19F1) were cloned into *E. coli* K-12 using pBR322, pACYC184, and pUC8 as vectors. Two of the plasmids (pLZ18C and pLZ19E) were also cloned into *S. sanguis* by using pVA1 as the vector. Hybridization using nick-translated [32]p-labeled *L. casei* plasmid DNA as the probe showed that none of the cryptic plasmids had appreciable DNA-DNA homology with the large lactose plasmids of *L. casei* strains. Partial homology was detected among several plasmids isolated from different strains, but not among cryptic plasmids isolated from the same strain.

Mutagenesis using chemical or irradiation on *Lactobacillus* has been investigated (Batt, 1986). However, plasmid biology of *Lactobacillus* has not been well-investigated. Plasmid DNA has been reported in *Lactobacillus* (Lin and Savage, 1985), and most of the plasmids are cryptic. The most researched plasmid is a 35 kb plasmid (pLZ64) of *L. casei* 64H which encodes for the enzyme β-D-phosphogalactoside galactohydrolase (p-β-gal) (Lee et al., 1982). This was expressed in *E. coli* and encodes for a 56 kd protein. A shotgun clone bank of chimeric plasmids, containing restriction-enzyme-digested fragments of pLZ64 DNA, was constructed in *E. coli* K-12. One clone contained the gene coding for β-D-phosphogalactoside galactohydrolase on a 7.9 kb PstI fragment cloned into the vector pBR322 in *E. coli* strain X1849. After compiling the restriction map of the recombinant plasmid, a series of subclones were constructed. The specific activity of β-D-phosphogalactoside galactohydrolase was stimulated 1.8-fold in *E. coli* by growing in media containing β-galactosides.

Conjugal transfers of pAM B1 from *Streptococcus* into *L. casei*, *L. reuteri*, *L. acidophilus*, and *L. salivarus* have been reported (Vescovo et al., 1983). The transformations of both *L. bulgaricus* and *L. helveticus* have been investigated using a chimeric vector pLBC104EM (Batt, 1986). Batt identified the plasmid in both *L. bulgaricus* and *L. helveticus* by Southern hybridization. Some possible alternative selectable markers include nisin resistance, lactose utilization, and other desired catalytic functions. The genetic manipulation is also desired for the introduction and expression of heterologous genes to produce new proteins or metabolites. The acidification control may be obtained by genetically altering the regulation of the glycolytic pathway.

Although plasmid DNA has been reported in *L. acidophilus*, *L. reuteri*, *L. casei*, *L. fermentum* and *L. helveticus*, detection has been limited to a few strains within each species. Klaenhammer (1984) discussed an effective general method for plasmid isolation in *Lactobacillus* species using either lysozyme or mutanolysin to achieve cell lysis. Using an alkaline-detergent lysis method, plasmid DNA was released and characterized from cells treated with the above-mentioned enzymes for 1 hour.

Chassey et al. (1983) also cloned some *Lactobacillus* plasmid genes in *E. coli*. The transfer of the conjugative plasmid pAMβ from *S. lactis* to *Lactobacillus* strains has also been successful (Gibson et al., 1979). The conjugal exchange of genetic information is also used for strain development. Chassy and Rokaw (1981) isolated a conjugal lactose plasmid from *Lactobacillus casei*. A gene can be cloned in an organism such as *E. coli*, moved into *Streptococcus sanguis* by transformation, and finally into the *Lactobacillus* strain by conjugation (Chassy, 1986). Many investigators (Klaenhammer, 1984) have isolated plasmids from various *Lactobacilli*. Some of these plasmids were quite stable.

Mutagenesis has been used (Morishita et al., 1974) to unblock cryptic amino acid biosynthetic pathways of *Lactobacilli* providing amino acid prototrophs. Daeschel et al. (1984) described the strains of *Lactobacillus plantarum* that do not generate carbon dioxide from malate and that are useful in reducing bloating in pickles.

Thus, lactose plasmid DNA can be introduced into *Lactobacilli* for the ability to metabolize lactose by genetic transformation (Chassy, 1986). A suitable transformation system will be available in the near future, which will assist in developing innovative new strains for *Lactobacilli*.

## 11.2.1.3 Cheese Culture

Cheese preserves milk protein without dangerous proteolytic microbial spoilage. Various *Streptococcus* and *Lactobacillus* species are used in the production of most cheeses. A specially selected protease was blended with the lipase to enhance cheddar cheese textural development (Andres, 1986). The lipases break down the lipids, usually found in cheddar cheese, during aging.

Many investigators have developed recombinant DNA techniques to introduce the prochymosin gene from calf stomach cells into microorganisms such as *E. coli* or *Saccharomyces cerevisiae*. Efforts are being made to develop a fermentation-based source of microbial calf rennet. The gene for calf rennin has been isolated and cloned into yeast (Mellor et al., 1983; Moir et al., 1985), and fungi (Cullen et al., 1987).

Mellor et al. (1983) constructed a high-efficiency expression vector to direct the synthesis of heterologous polypeptides in yeast. The gene was inserted between the 5' and 3' control regions of the efficiently expressed yeast PGK gene. This vector was used to direct the expression of three derivatives of the calf chymosin gene—preprochymosin, prochymosin, and chymosin. The prochymosin was synthesized to about 5% of the total yeast-cell protein and activated to produce an enzyme that had milk-clotting activity.

The recombinant calf prochymosin synthesized in *E. coli* was shown to accumulate in the form of insoluble inclusion bodies. A process for the isolation of the inclusion bodies was described (Marston et al., 1984). Denaturation was essential for the solubilization of the recombinant product, and specific renaturation conditions were needed to recover a significant yield of soluble prochymosin which yielded active chymosin.

Prochymosin from *A. nidulans* was expressed and secreted by construction of plasmids containing a selectable marker (pyr4) and various expression units composed of either prochymosin or preprochymosin cDNA flanked by transcriptional, translational, and secretory control regions of the *A. niger* glucoamylase gene (Cullen et al., 1987). The glucoamylase production by these organisms was induced by the presence of starch or maltose in the culture medium. The expression and regulation of several glucoamylase-prochymosin fusions were described when integrated into the *A. nidulans* genome. To test the ability of the filamentous fungus *Aspergillus nidulans* to secrete bovine prochymosin, four plasmids were constructed in which the transcriptional, translational, and secretory control regions of the *A. niger* glucoamylase gene were functionally coupled to either prochymosin or preprochymosin cDNA. The secretion of polypeptides enzymatically and immunologically indistinguishable from bovine chymosin was achieved following transformation of *A. nidulans* with each of these plasmids.

## 11.2.2 Milk Proteins

Applications of genetic engineering techniques will provide the background information to induce the dairy cow to produce milk proteins with desirable functionality. Bovine casein cDNAs coding for $\alpha_{s1}$- and $\varkappa$-caseins have been cloned (Stewart et al., 1984), and the resultant plasmids were used to transform *E. coli* (Kang and Richardson, 1985). The cDNA fragments coding for portions of the caseins have also been cloned (Maki et al., 1983).

The oligonucleotide site-directed mutagenesis techniques will assist in producing the primary sequence of the caseins after incorporating a cDNA expression vector into a host cell. Later on casein structural genes may be incorporated into bovine genomes, and then it may be transmitted to the progeny via the germ cells (Kang and Richardson, 1985).

These techniques can also be used to modify chymosin to accelerate the rate of textural development in cheese, and to promote the faster ripening of the cheese.

## 11.2.3 Whey Processing

About 50% of world whey production is being processed for further utilization. The whey can be treated to transform lactose into lactic acid, ethanol, galactose, or a mixture of glucose and galactose. Whey and permeates contain all the minerals, vitamins, and trace elements required for the growth of microorganisms. Thus, it is an excellent basal medium for any type of culture.

Production of yeast from whey is possible by (1) the transformation of lactose, making it available to the yeast strain chosen, and (2) direct use of lactose by a yeast strain able to metabolize this substrate (Moulin and Galzy, 1984). In the first process, lactic bacteria are used to transform lactose into lactic acid, without aeration at pH 4.5 and 44°C. Then, lactic acid is used by *Candida utilis* and *Candida krusei* in an aerated fermenter. The second process could be achieved by several yeast strains; however, *Kluyveromyces fragilis* was used most often. The improvement in the strain can be made possible by an increase in growth rate; an increase in protein and nucleic acid content; a decrease in storage compound contents; and a loss of sporulation and sexual characteristics.

The production of protein-enriched whey was possible with the growth of microorganisms such as yeast and filamentous fungi. The Devos process consisted of growing *Saccharomyces cerevisiae* on the lactic acid in whey; the enriched whey was concentrated and dried. The dried powder contained 15 to 18% protein. Another process involved a two-stage fermentation followed by evaporation and drying. The substrate in the first fermenter was lactose, and lactic acid was used in the second fermenter (Moulin and Galzy, 1984).

Using genetic engineering techniques, yeast strains capable of fermenting lactose were developed. The lactose utilizing genes from enzymes were cloned into *Saccharomyces cerevisiae* (Sreekrishna and Dickson, 1985; Farahnak et al., 1986) and *Xanthomonas campestris* (Walsh et al., 1984) to convert lactose in whey into ethanol or xanthan gum. Schoutens et al. (1984) utilized whey permeate to produce butanol.

## 11.2.4 Lactic Acid from Lactose Fermentation

In the studies on lactose metabolism, plasmids were implicated in certain lactose specific enzymes such as the phosphotransferase lactose uptake mechanism (Collins-Thompson et al., 1985). Lactose phosphoenol pyruvate-phosphotransferase system and P-$\beta$-galactosidase are also controlled by plasmids.

Hydrolyzed whey and whey containing hydrolyzed lactose are useful in the food industry, as is hydrolyzed lactose syrup converted from whey permeates. Lactose was converted by *Lactobacillus thermophilus* into lactic acid and galactose (Moulin and Galzy, 1984). The production of lactulose from lactose with the enzyme $\beta$-galactosidase was also studied.

The following steps summarize the conventional catalysis to produce lactone from whey (Cayle et al., 1986). Whey-protein-concentrate and per-meate are separated from whey through ultrafiltration. Permeate is hydrolyzed in an immobilized lactase reactor. The glucose is then converted to ethanol in a fermenter containing a mutant and flocculating strain of *Saccharomyces cerevisiae*. Then alcohol is recovered by distillation, and galactose is recovered by ion exclusion chromatography. The dried galactose is then oxidized to D-galacturonic acid, and the acid is reduced to L-galactonic acid by inorganic catalysts. Then, L-galactonic acid is concentrated to form L-galactono-1,4-lactone. The L-galactono-1,4-lactone is then converted to L-ascorbic acid via a mutant strain of *Candida norvegensis*.

Okos (in Anon, 1988a) determined the feasibility of using *T. ethanolicus* to produce ethanol from lactose. Diacetyl was produced as a secondary metabolite during the lactic acid fermentation of lactose by *L. casei*. Okos also studied the fermentation of delactosed whey permeate. Delactosed whey permeate (DLP) is the mother liquor remaining from the production of crystallized lactose from concentrated whey permeate.

Using the protoplast fusion techniques, hybrids were constructed between auxotrophic strains of *S. cerevisiae* having high ethanol tolerance and an auxotrophic strain of lactose-fermenting *K. fragilis* isolated by ethyl methane-sulfonate mutagenesis. These hybrids were protrophic and capable of as-similating lactose and producing ethanol in excess of 13% (v/v) (Farahnak et al., 1986).

## 11.2.5 Enzymes

The following enzymes are commonly used in the dairy industry: (1) proteases for milk coagulation, flavor, and texture of cheese; (2) lipase for cheese flavor and texture; (3) lactase for low-lactose milk; (4) catalase for pasteurization, texture, and flavor of milk and milk products; and (5) glucose oxidase as an oxygen scavenger of cheese and milk powder (Collins-Thompson et al., 1985).

Lipases and esterase enzymes are used to develop flavors in enzyme-modified products. In the past, many patents were issued for the applications of different lipases to milk-based ingredients for flavor development (Dziezak, 1986). These include yogurt, various cheeses, milk fat, cream, and dry milk powder. Submerged mycelial cultures are used to produce blue cheese flavor concentrates from enzyme-modified butterfat. Enzyme-modified creams are used in coffee whiteners, soups, candies, dips, and baked goods. The biotechnological techniques will assist in developing enzyme-modified flavors for use in formulated foods.

To decrease the cheese aging time, enzymes are being used to develop the cheese flavor rapidly (Dziezak, 1986).

Greenberg and Mahoney (1981) reviewed the rationale for hydrolyzing lactose, and procedures and supports for immobilization of lactase. They also discussed the operating characteristics that would affect behavior in a commercial reactor system for milk and whey processing.

## 11.2.6 Membrane Processing

Dairy industry uses membrane processes for the following purposes (Hammes, 1987):

- reverse osmosis for milk concentration, and separation of protein and lactose
- ultrafiltration for milk concentration, and separation of protein and lactose from milk and sewage
- microfiltration for separation of protein from whey or milk, and removal of microorganisms from cheese brine
- electrodialysis for the separation of lactic acid from sewage
- dialysis for the concentration of milk

## 11.3 **MEAT PROCESSING**

Starter culture in meat fermentation provides lactic acid bacteria to produce the lactic acid required for taste and low pH. The lower pH increases the juiciness of the product and denatures the meat protein, contributing to the characteristic firm texture (Bacus and Brown, 1987). It also antagonizes some pathogens such as *Salmonella* and *Staphylococcus aureus*. Starter culture organisms such as *Lactobacillus plantarum* and *Pediococcus cerevisiae* provide both the acidity and antimicrobial agents (Table 11.2). Starter culture addition also lowers histamine levels and extends the shelf life of meats. The success of starter culture has also led to its application to nonfermented meats such as frankfurter and poultry meats (Collins-Thompson et al., 1985).

Most of these cultures are natural isolates of lactic acid bacteria, *Micrococci* and *Staphylococci*, which are able to control the fermentation processes. Cavoski et al. (1988) found the mixed cultures of *Streptococcus lactis* (AK60) and *Micrococcus* (M104) to be superior in dry-fermented sausages, compared to single cultures of *Micrococcus* (M104). On the other hand, Coventry et al. (1988) studied the performance of different starter cultures in salami manufacturing. *P. pentosaceus*, *L. plantarum* and *L. sake* showed similar characteristics over the pH range 4.7 to 6.3, while *S. carnosus* and *M. varians* were sensitive to the lower pH. Also, *P. pentosaceus* and *L. sake* showed greater psychrotrophic growth than *L. plantarum*. Salami made with *P. pentosaceus* maintained higher viable numbers in the product over 6 weeks than did *L. plantarum*. Further, the proliferation of *P. pentosaceus* and *L. plantarum* did not prevent the development of non-starter flora.

Meat fermentation technology is also being applied to traditionally non-fermented products to enhance flavor development, extend shelf-life, and provide control over food pathogens and toxic chemicals.

TABLE 11.2. *Microorganisms in starter preparations used for the production of raw sausages (modified after Hammes, 1987). Reprinted by permission of Gustav Fischer Verlag, Stuttgart, Germany.*

| Microbial Class | Species | Uses |
|---|---|---|
| Fungi | *Penicillium chrysogenum*<br>*Penicillium nalgiovense* | 1, 2, 4 |
| Lactic acid bacteria | *Lactobacillus curvatus*<br>*Lactobacillus plantarum*<br>*Lactobacillus sake*<br>*Pediococcus acidilactici*<br>*Pediococcus pentosaceus* | 1, 2, 3, 4, 5 |
| Micro-Coccaceae | *Micrococcus varians*<br>*Staphylococcus carnosus*<br>*Staphylococcus xylosus* | 1, 2 |
| Streptomycetaceae | *Streptomyces griseus* | 1 |
| Yeasts | *Candida famata*<br>*Debaryomyces hansenii* | 1, 2 |

1. flavor, 2. color, 3. shelf life, 4. reduction of hygienic risk, and 5. sliceability.

Meat fermentation research in the U.S. has focused on rapid acid-producing strains of both *Pediococci* and *Lactobacilli* to shorten fermentation schedules at a wide range of temperatures (21 to 46°C) without sacrificing product safety (Bacus, 1984). Many flavor-producing microorganisms are not assisted in fermentation and thus are used with lactic acid microorganisms. Current research using genetic engineering techniques provides the potential to selectively eliminate those undesirable fermentation end-products while retaining the desirable flavor compounds. *Staphylococcus carnosus*, used to produce dry sausage, is apathogenic and is being examined as a host strain for recombinant DNA, and to study the genetic organization of certain *Staphylococci*.

Proteolytic enzymes are used for tenderizing meat and poultry products (Kosaric, 1984).

Ultrafiltration is used to concentrate blood and plasma, and to separate fat and protein from wastewater (Hammes, 1987).

## 11.4 **BEVERAGE PROCESSING**

Beer is produced by fermentation of an extract of malted cereals, preferably barley, with aromatic herbs and hops for additional flavoring. The yeast population grows approximately 8-fold during the fermentation, which is limited by the falling pH, the rising ethanol concentration, and the yeast's inability to grow indefinitely under anaerobic conditions (Brown et al., 1987). Beer of the ale type is traditionally fermented by the "top" strain of *S. cerevisiae*.

During the fermentation of wine, a succession of yeasts develop. Species of *Saccharomyces*, including *S. cerevisiae*, are involved, but so are various species of the other yeast genera, e.g., *Kloeckera*, *Kluyveromyces*, *Torulaspora*, and *Zygosaccharomyces*. Each yeast contributes its own spectrum of flavor compounds (Brown et al., 1987).

In beer production, ultrafiltration is used to separate carbohydrate oligomers and polymers from wort. Alcohol is separated using reverse osmosis. In wine processing, ultrafiltration is used for the stabilization of tartrate and protein, and for clarification. Reverse osmosis can be used to separate alcohol and to concentrate must and wine. Reverse osmosis is used in the alcohol industry to increase yield with membrane bioreactors (Hammes, 1987). Many other separation techniques are employed in beverage processing.

## 11.4.1 Brewer's Yeast

Brewer's and baker's yeasts belong to the genus *Saccharomyces*, and exist in either diploid or haploid phases. Their life cycles usually change between the haploid and diploid states. Genetically, hybrid characteristics can be controlled by controlling the mating of haploids.

Genetic engineering techniques such as spheroplast fusion and transformation are being used to modify yeast strains to enable them to ferment at a rapid rate, change their flocculation properties, tolerate high ethanol concentration during brewing, increase the number of metabolizable sugars, adjust the metabolic by-products, etc. Transformation with the aid of recombinant DNA is a desirable technique for introducing single gene traits into strains.

Current research on brewer's yeast is being conducted in order to improve its capacity and efficiency to produce ethanol and carbon dioxide from a wide variety of substrates, and in order to employ yeast to synthesize and excrete a wide variety of proteins or peptides of commercial value (Russell et al., 1986).

The genetic methods used in the development of brewer's yeast strains are hybridization, mutation and selection, rare mating, spheroplast fusion, and transformation (see Chapters 2 and 5). Transformation can be carried out using native DNA, recombinant DNA, or by liposome-mediated DNA transfer. Rare mating has been used in conjunction with the *kar* (Karyogamy defective) strains to introduce zymocidal (killer) activity into brewing strains. Mutation and selection has been used to induce auxotrophs and to select such strains as spheroplast fusion partners and as recipients for transformation experiments; and to isolate depressed mutants of brewing strains so that these can metabolize maltose in the presence of glucose and thus have increased wort fermentation rates. Spheroplast fusion has been used to fuse strains constructed by hybridization with brewing strains to introduce the novel capabilities of the hybridized strain into the brewing strains. Transformation has been used to introduce genes from non-*Saccharomyces* yeast strains into brewing strains (Russell et al., 1986).

Stewart et al. (1983) fused non-flocculent ale and lager strains with a flocculent haploid strain that was unable to ferment either maltose or malto-triose. They also fused flocculent ale and lager strains with a non-flocculent haploid strain. These were able to out-ferment the wort in static culture, but they produced unpalatable beer. Thus, in these studies, the fusion was not specific enough to genetically alter existing brewing strains in a controllable manner. A number of ale strains (*S. cerevisiae*) showed plasmid DNA in many configurations under ultraviolet light.

Brewing efficiency would be increased by improving the ability of the yeasts to ferment starch and dextrins. Attempts are being made to transform a brewing yeast with glucoamylase genes from *S. castellii* (Panchal et al., 1984). Federoff et al. (1982) identified a structural gene, MAL6, which is able to control the maltose/maltotriose uptake. A similar system has been noticed with sucrose genes such as SUC1 to SUC5.

Spheroplast fusion was used by Stewart et al. (1982) for intergeneric fusion of *Saccharomyces* strains. The spheroplasts of the lager-brewing strain-21 *S. uvarum* were fused with spheroplasts of a diploid *S. diastaticus* strain-1384 to obtain *Saccharomyces* sp. strain-1400. This was found suitable for ethanol production.

Plasmid-mediated transformation is being used to transfer genes of interest into the brewing yeast. Russell and Stewart (1980) used this technique to transform the maltotriose uptake system into a strain of *Saccharomyces*, by treating a haploid maltose-utilizing but non-maltotriose-utilizing strain with DNA isolate from *S. cerevisiae*. This technique has been used to construct many chimeric plasmids containing yeast genes such as TRP1, SUC2, LEU2, URA3, and HIS3 by transforming them into suitable recipients (Botstein and Davis, 1982).

Figure 11.1 shows the schematics of the rare mating. When non-mating strains are mixed together at a high cell density, a few true hybrids with fused nuclei form, which can usually be selectively isolated. When the strain, which harbors a specific nuclear gene mutation (*kar*), hybridizes with another strain, the nuclei will not fuse and this permits the formation of cell lines with mixed cytoplasmic contents (heteroplasmons). This has been used to transfer "zymocidal" or "killer" factor from haploid strains to brewing yeast strains (Russell et al., 1986). The *kar* mutation prevents nuclear fusion, and hybrids selected contain only the brewing strain nucleus. It permits the introduction of cytoplasmically transmitted characteristics of the brewing strain.

## 11.4.2 **Wine Yeast**

Genetic engineering techniques are being applied to improve wine yeast strains by using intergeneric protoplast fusion, single-chromosome transfer, transformation with native DNA or plasmid vectors, and mutagenesis. Mutagenesis modifies yeast phenotypes for temperature and pressure tolerances, flavor improvement, alcohol tolerance and production. Clonal selection is used after mating desirable genotypes or recombining alleles that are naturally occurring or obtained through mutations. Mitotic recombination has not provided a useful industrial wine yeast. Cell fusion techniques are suitable for strains that sporulate or mate poorly or when interspecific recombination is required (Subden, 1987).

Barney et al. (1980) transformed a non-flocculent strain and then transformed *S. cerevisiae* into a dextrin-utilizing strain with native DNA from *S. diastaticus*. Efforts have been made to clone the gene for malate carboxylase

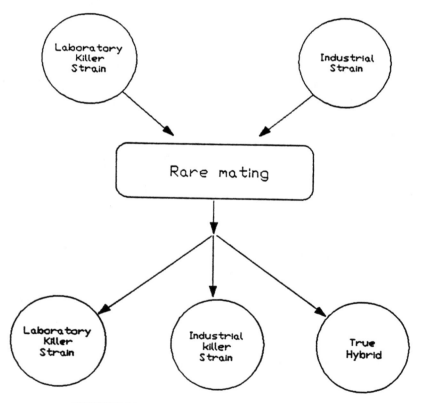

**FIGURE 11.1.** Rare mating process for brewer's yeast.

into *E. coli*, and then transform a wine yeast to construct a strain that would perform a malolactic and ethanolic fermentation.

To obtain commercial wine yeasts resistant to killer toxins, the dsRNA particles have to be transferred into the yeast. This involves the transfer of large (L) and medium (M) particles—"M" for the toxin and "L" for encapsulation. Hara et al. (1981) mated a killer sake yeast to a wine yeast then back-crossed the progeny many times to the parent while retaining the killer character. Seki et al.(1985) applied cytoduction with a killer strain and the montrachet wine yeast.

In wine making, primary fermentation converts sugar to alcohol, and secondary fermentation involves the production of L-lactic acid and $CO_2$ from L-malic acid. Three genera of bacteria are involved in the malolactic fermentation -*Leuconostoc, Lactobacillus*, and *Pediococcus*. Efforts have been made to introduce the required gene into wine yeast using recombinant DNA to provide malolactic capability in wine yeast, so that primary and secondary fermentations are controlled by a single yeast (Snow, 1985).

Snow (1985) sliced purified DNA from *Lactobacillus delbrueckii* CUC-1 with the SalI, and the fragments were ligated into plasmid pBR322. The ligated DNA was used to transform *E. coli* K-12 RR1, and ampicillin-resistant, tetracycline-sensitive transformants were selected. After screening 400 *E. coli* transformants, two produced L-lactate. These produced 30% as much L-lactate as *L. delbrueckii* under the same conditions. The strain carried the pBR322 plasmid with a 5 kb DNA insert at the SalI restriction site. Later a 5 kb fragment was transferred into a shuttle vector that could also replicate in yeast. This plasmid was called pRC3. Insertion of the malolactic fragment into the SalI restriction site produced a derivative (pHW2). This was transformed into a laboratory yeast strain that carried the trp1 mutation. Yeast strains with pRC3 and pHW2 produced 3.5 times more L-lactate than the untransformed yeast.

Boone et al. (1990) integrated the yeast K1 killer toxin gene into a genome of laboratory and commercial wine yeasts by gene replacement technology. It provided recombinants containing only yeast DNA. They also integrated K1 killer gene into two K2 wine yeasts to obtain stable K1/K2 double killer strains. This was found to have a wide spectrum of killing, and advantages over other strains of yeast or *Saccharomyces cerevisiae* in wine fermentations.

## 11.5 **VEGETABLES AND FRUITS PROCESSING**

Biotechnological and genetic engineering techniques are being used to improve the post-harvest quality of fruits and vegetables. The genes for two ripening enzymes, polygalacturonase from tomato and cellulase from avocado, have already been cloned (Della-Penna et al., 1986; Christoffersen et al., 1984). The gene for thaumatin has been cloned and expressed

(Ledeboer et al., 1984; Edens et al., 1984) and its 3-dimensional structure has been described (deVos et al., 1985). Genetic engineering techniques will also help to increase the nutritional value of plant-derived foods.

Tomato explants from leaves, stems, and hypocotyls have all been successfully grown in cultures. Somaclonal process reduces the development time for the new variety by about half over conventional techniques (Nevins, 1987). The potential for protoplast fusions from diverse genomes provides the opportunity for micro-injection of DNA directly into tomato cells. Somaclonal variation can be used to develop new variants that retain all the favorable qualities of the existing variety, while adding an extra trait. Fillatti et al. (1987) has introduced and expressed a mutant aroA gene in tomato plants for herbicide tolerance. Increasing fruit solids is considered a high priority for the improvement of tomato processing characteristics using recombinant DNA techniques.

Addy and Stuart (1986) discussed a more economical and efficient cloning system for improved potato by eradicating the pathogen and increasing the solid content. Tissue culture, mutagenesis, and regeneration techniques are also used. To effect mutagenesis, Hamrike-Wagner et al. (1982) introduced a callus from potato. On the other hand, Bright et al. (1982) mutagenized calli from tuber discs, and regenerants were taken to plants through organogenesis.

Potato pathogens are mostly RNA viruses. After isolating these viruses, their protein coat is used to raise antibody for detecting assays. The RNA of the virus has been isolated and used for genetic engineering techniques. The RNA virus can be converted into a cDNA copy using *in vitro* cDNA cloning techniques. This cDNA is then inserted into the plant nuclear DNA by using a suitable vector system (Bialy and Klausner, 1986). The second technique uses the plant's cross-protection response to fight incoming viruses. The gene for coat protein from the virus is inserted into the plant using a vector system. This has been successfully used by Abel et al. (1986) for tobacco mosaic virus in tobacco plants.

A lower cDNA library was constructed from poly(A)$^+$ RNA of a ripe avocado fruit (Christoffersen et al., 1984). Colony hybridization indicated many ripening specific clones of which one, pAV5, was found specific for cellulase. Hybrid selection with pAV5 gave a message from ripe fruit, from which *in vitro* translation yielded a polypeptide of 53 kd, comigrating with purified avocado cellulase on SDS polyacrylamide gel electrophoresis. Hybridization of pAV5 to poly(A)$^+$ RNA from unripe and ripe avocado fruit showed that there is at least a 50-fold increase in the cellulase message concentration during ripening.

Genetic engineering can help to prevent rancid flavor due to the action of the enzyme lipoxygenase in frozen vegetables. This can be accomplished by isolating and understanding the lipoxygenase gene, and then limiting its expression.

For frozen corn production, blanching is required to inactivate lipoxy-genase. The blanching requirement could be eliminated if corn mutants devoid of lipoxygenase activity are developed. This will save process and energy costs, and will improve the texture of the product. Genetic engineering techniques will assist in developing these mutants (Moshy, 1985).

Vegetables such as cucumbers and cabbages are preserved by fermentation. In pickling, microbial cultures and enzymes improve preservation for large stocks of vegetables awaiting processing (Kosaric, 1984). Bacteria such as *Lactobacilli*, *Pediococci*, and *Leuconostocs* have been used for the fermentation of cucumbers, cabbages, olives, and other vegetables. Genetic engineering techniques have been applied to improve *L. plantarum* and *P. pentosaceus*. Conjugal plasmid transfer has been reported between *Streptococci* and *Pediococci* (Gonzalez and Kunka, 1983) and plasmids have been observed in *L. Plantarum* (Klaenhammer, 1984).

The fermentation of ground soybean provides flavor and texture suitable for use in artificial cream cheese (Hofmann and Marshall, 1985).

Enzymes, such as amylases, glucose oxidase, proteases, pectinases and pectolytic, are used in preparing vegetable juices both in maceration and liquefaction. These solubilize and clarify fruit pulps. The pectinases are divided into two groups containing (1) polymethyl galacturonases and pectinyl-ases, and (2) pectin esterases (Collins-Thompson et al., 1985).

Ultrafiltration is used to concentrate and clarify juices, to separate citrus oil from emulsions, and to clarify and provide enzyme separation from pickling brine. Reverse osmosis can be used to separate sugars from waste-water (Hammes, 1987).

Micropropagation of fruit trees is a commercial application of tissue culture and *in vitro* propagation can now be extended to pome and stone fruits (Mullins, 1987). In grapevines, *in vitro* plant generation is well-developed for somaclonal variation and its contribution to clonal selection.

Many tissue and cell culture techniques are now also available for vegetable breeding. Carrots appeared to be an excellent model plant for the development of synthetic seeds (Dore, 1987).

To prevent the frost damage to fruits and vegetables, genetically modified *Pseudomonas syringae* spray can be developed, by deleting one of its proteins that promotes ice nucleation, which will compete with natural flora.

## 11.6 CEREALS PROCESSING

The deficiency of essential amino acids in grains and legumes can be eliminated by amplifying genes coding for proteins containing low levels of these amino acids for increased protein production.

Whereas amylases produce only linear maltodextrins, CGTase produces a cyclic oligomer molecule. It may be possible to alter the product specificity of CGTase to minimize production of linear maltodextrins through modification of its binding structure (Tao, in Anon, 1988a). He plans to explore the effects of deletion and substitution of amino acid sequences in *Bacillus macerars* CGTase upon its binding and catalytic properties, targeted at developing an understanding of the functional and structural features of this biocatalyst.

Sucrose and starch were reacted in the presence of different enzymes and CGTase to produce branched cyclodextrins (Tao, in Anon, 1988a).

## 11.6.1 High-Fructose Corn Syrup

High-fructose corn syrup (HFCS) has a sweetening power greater than that of sucrose. It is derived from corn starch by using a microbial enzyme as a catalyst for the conversion of starch to glucose and glucose to fructose.

The first step in HFCS production is corn wet-milling to separate corn starch (a portion of endosperm) from other fractions. This process differs from plant to plant but principally consists of cleaning, steeping, degerming, separation, grinding (milling), washing, and centrifugal separation. The steeping is carried out for about 40 h. Water in the steep tank contains 0.1 to 0.2% sulphur dioxide with a pH of 3 to 4 and is maintained at about 54°C. The sulphur dioxide inhibits fermentation and facilitates softening of the kernel, which provides optimum starch yields. After steeping, corn kernels are conveyed to degerminating mills, where their components are separated. To separate the germ, the corn slurry enters a continuous liquid cyclone. Starch and hull are separated by grinding and screening. Starch and gluten are separated by using a continuous centrifuge. The hull is used as an animal feed. The dried gluten, containing 60 to 70% protein (dry basis) is used as gluten meal or animal feed. Thus, wet milling yields corn oil, corn-germ meal, animal corn feed, and common cornstarch (Collins-Thompson et al., 1985; Joglekar et al., 1983).

The starch from wet milling is converted to HFCS by using various processes such as hydrolysis, saccharification, filtration, ion-exchange refining, isomerization, evaporation, and packaging. The corn hydrolysis yields 42% of fructose. Glucose isomerase enzymes are used to convert glucose to fructose. About 65 different microorganisms have been identified as producers of this enzyme. Corn hydrolysis and refining produce corn syrup, dextrose, HFCS, and maltodextrins.

HFCS is available in liquid form and crystallizes on storage. The beverage industry uses HFCS as a liquid sweetener.

## 11.6.2 Baker's Yeast

The most economical Baker's yeast strains are those that grow rapidly and have high yields on sugar. The genes involved in baking properties have emphasis on glycolytic activity. The sucrose-non-fermenter (SNF) genes should be present in the yeast for invertase production. Mutants in many of the genes encoding glycolytic enzymes have been isolated. Enzymes PFK1 and PFK2 (subunits of phospho-fructokinase) and PYK1 (pyruvate kinase) were found to be important in the control of this pathway. New yeast strains are being developed by using hybridization, spheroplast fusion, electrofusion, transformation, and other recombinant DNA techniques. Spontaneous petite mutants can be isolated or they can be induced with ultraviolet light, manganous ion, and ethidium bromide dye (Trivedi, 1986).

## 11.7 **OIL AND FAT PROCESSING**

Rattray (1984) reviewed the biotechnology of fat and oils of plant and microbial sources. Tissue culture techniques can provide improved crops and unique germ plasm. These techniques are also being used to increase the productivity of soybean, rapeseed, palm and sunflower. In soybean plants, efforts are being made to decrease linolenic acid, regulate oleic acid, and to desaturate and synthesize triacylglycerol in oil. This is done through the use of mutagens such as ethylmethane sulfonate. In rapeseed, research is being done to increase linoleic and erucic acids, and to decrease chlorophyll and glucosinolate contamination. One such product, canola oil, has been developed in Canada containing < 5 % erucic acid as a percentage of the total fatty acids and < 30 $\mu$moles of glucosinolates per g in the oil-free meal. Similarly, in sunflower, efforts are required to increase oleic acid and decrease wax.

To increase the extractability of oil from seeds, non-lipolytic enzymes are used. Release of fatty acid from triglycerides can be achieved enzymatically. Enzymes are also used in the production of different mono- and diglycerides with the release of fatty acid. Reaction of fatty acid in the presence of glycerol and lipase yielded mono- and diglycerides. Lipase is also used to modify the fatty acid composition of triglycerides through acyl exchange (Rattray, 1984). Immobilized soybean lipoxygenase has been used in the continuous production of 13-hydroperoxylinoleic acid and 15-hydro-peroxyarachidonic acid. Cocoa butter substitutes can be obtained from the growth of certain yeasts using enzymes.

### 11.7.1 Interesterification

Interesterification is used to alter the composition, and therefore the physical properties, of triacylglycerol mixtures in oils and fats. In this process, chemical catalysis by sodium or sodium alkoxide promotes the migration of fatty acyl groups between glycerol molecules, so that the oils and fats consist of acyl glycerol mixtures. In the mixture, the fatty acyl groups are randomly distributed among the glycerol molecules. To provide a large area for interesterification, macroporous inorganic particulate materials coated with hydrated extracellular microbial lipase are used as catalysts. Suitable support materials are kieselguhr, hydroxylapatite, and alumina. These catalysts are generally prepared by adding solvent (such as acetone, methanol, or ethanol) to a slurry of the particles in buffered lipase solution. The solvent-precipitated enzyme coats the particles, and the lipase-coated particles can be collected by filtration, dried, and stored. An alternative method of preparing the catalysts is to mix microbial lipase solution thoroughly with an excess of inorganic particles and then to dry the resulting wetted powder (Macrae, 1984).

Interesterification reactions are performed either batchwise in stirred tank reactors or continuously using packed bed reactors. For batch interesterification reactions, the catalyst material was activated by hydration with a small amount of water, and stirred with a reactant mixture. At the end of the reaction, the catalyst particles were removed by filtration, and the interesterified triacylglycerols were separated by conventional fat fractionation techniques. The catalyst was reused in subsequent reactions. The continuous enzymatic interesterification process was operated by pumping water-saturated feed stock through a packed bed of lipase catalyst (Macrae, 1984).

## 11.7.2 Soybean Oil

Recurrent selection methodology was applied to study genetic variability for unsaturated fatty acid composition in soybeans (Wilson, 1984). Selection for higher levels of oleic acid has resulted in lower concentrations of linoleic and linolenic acids. The genetic changes in lipid metabolism were not only a function of unsaturated fatty acid biosynthesis, but also were mediated by the mechanisms of triacylglycerol synthesis. Different soybean lines were developed through mutation breeding (Hammond and Fehr, 1984), containing lower linolenic and higher stearic acids.

Other applications are described in Chapter 4.

## 11.8 **IMMOBILIZED CELL APPLICATIONS**

Some of the immobilized cell applications are discussed below.

### 11.8.1 Beer Production

Kolpackni and Isaeva (1976) immobilized S. carlsbergensis yeasts by entrapment into various polymers — caprone, cellophane, phoroplast, and polyethylene. The immobilized cells were packed in a 7 L column reactor through which wort at pH 3.5 was passed. The rate of beer formation was 24 h faster than that obtained with batchwise brewing due to higher yeast cell concentration. Supports retained up to 55 to 75% of the cells, which could be regenerated and reused.

Navarro and Durand (1977) immobilized *S. carlsbergensis* by adsorption to polyvinyl chloride, porous brick, and kieselguhr for beer production. The brewing wort (pH 5.5) through the column reactor at a dilution rate of 250 mL/h and at 25 to 30°C. It took 90 min to ferment wort to beer.

### 11.8.2 β-Galactosidase Enzyme Production

Kolot (1988) immobilized *Saccharomyces lactis* by entrapment into acrylamide gel at optimum conditions — 3 g cells/10 mL of phosphate buffer, 50% acrylamide solution, and 15 min polymerization time. The immobilized cells retained 61% activity while only 29% was obtained for immobilized enzymes. pH profile or thermo-stability and Michaelis-Menten reaction constant ($k_m$) of the immobilized cells were not affected.

### 11.8.3 Sucrose Inversion

Sugar products are sweetened by this process. Kolot (1988) placed immobilized cells into a continuous flow fluidized bed reactor, and 58.4 mM sucrose solution at pH 5 was passed through the reactor at 47.5°C. Under these conditions 70% of sucrose hydrolysis and 100 h stable operation were achieved.

### 11.8.4 Ethanol Production

Navarro et al. (1976) (quoted by Kolot, 1988) produced ethanol in a continuous column reactor using immobilized *S. cerevisiae* by adsorption to brick and polyvinyl chloride chips. No new cells were required for 30 days.

Wada et al. (1981) immobilized *S. cerevisiae* into $\varkappa$-carrageenan at an initial concentration of $9 \times 10^6$ cells/mL, and immobilized cell in beads were packed in column reactors. The largest cell concentration ($5 \times 10^8$ cells/mL) obtained at the lowest glucose concentration (10%). Retention times for various substrate concentrations (15 to 25%) were 0.8 to 2.5 h.

### 11.8.5 2,3-Butanediol Production

Chua et al. (1980) used *Enterobacter aerogenes* for butanediol production, immobilized by entrapment into $\varkappa$-carrageenan gel. Cell concentration in the beads was increased 200-fold within 12 to 24 h after incubation in a nutrient medium. These beads (3 to 4 mm dia.) were used to produce butanediol in batch and continuous reactors. In batch operation, immobilized cells formed 6 times less acetone (a by-product) than free cells. In continuous reactor, product production remained at a steady level (3 g/L) for 10 days. The product synthesis was 1/3 that of free cells; however, the immobilized system had many advantages.

## 11.8.6 L-Aspartic Acid Production

Aspartic acid is used as a food additive, and is also used to produce the sweetener aspartame. Kolot (1988) described a procedure developed in Japan to produce aspartic acid using immobilized cells. *E. coli* cells were immobilized by entrapment into acrylamide gel. The gel beads were packed in a column reactor to produce the acid.

## 11.9 **PLANT GENETIC ENGINEERING APPLICATIONS**

Genetic engineering techniques are being used to modify plants to include genetic traits to improve processing characteristics such as solid content, color, flavor, size, sweetness, etc. Somaclonal and gametoclonal developed varieties for many crops are already in the field, and others are being developed. Considerable progress has been made with tomato, potato, banana, celery, wheat, corn, palm oil and sugarcane (Sharp et al., 1984). Other techniques include (Potrykus et al., 1987) gene transfer into various crops, gene transfer into chloroplasts and mitochondria, gene isolation and inactivation, analysis of regulatory DNA sequences, gene transfer across cell walls, and identification and analysis of genes important in the regulation of plant development.

DNA-mediated stable transformations of cultured cells of wheat, corn, and rice have been achieved by transforming protoplasts with isolated DNA (Lorz et al., 1987). High-efficiency transformation of corn by a mixture of pollen and exogenous DNA has been reported. Transformation of rye has been obtained by DNA-injection into plants prior to meiosis.

## 11.10 **MISCELLANEOUS APPLICATIONS**

Tissue and cell culture techniques will assist in developing natural sources of many high-value food ingredients such as flavors, pigments, antioxidants, and sweeteners. Sahai and Knuth (1985) provided the possibilities of producing the following products using above-mentioned techniques: colors (Anthocyanins, betacyanins, saffron), flavors (asparagus, celery, grape, strawberry, tomato, vanilla), oils (garlic, lemon, mint, onion, rose, vetiver), and sweeteners (miraculin, monellin, stevioside, thaumatin). Naturally occurring antioxidants have been isolated from the spice rosemary (Houlihan et al., 1984, 1985).

Mutagenesis has been employed to obtain bacteria which produce high concentrations of amino acids (Shiio, 1982), ferment milk rapidly (Singh and Chopra, 1982), and produce high levels of flavors and acetoin.

## 11.10.1 Pigment

Ilker (1987) discussed the plant tissue culture for *in vitro* color production for non-food use such as anthocyanins and betalains. Anthocyanins produce the hues of flowers and fruits, and betalains provide pigments in red beets. Only two hormones, auxin and cytokinin are needed for growth and secondary product formation. Correct spectral requirements, including the duration of illumination, is experimentally determined for maximum color production. Tissue cultured cells have been optimized to produce 845 times more shikonin than is obtainable from plant roots.

## 11.10.2 Sweetener

Thaumatin is a sweet-tasting plant protein, which can be isolated from the fruits of the West African shrub *Thaumatococcus deniellii* Benth. In these fruits, five different thaumatin forms occur, which are designated thaumatins I, II, III, b, and c. On a mass basis they are 2000 times sweeter than sucrose. Nucleotide sequence studies showed that mature thaumatin II consists of a single polypeptide chain 207 amino acid residues long. Various maturation forms of the thaumatin were expressed in yeast, using a promoter fragment of the glyceraldehyde-3P-dehydrogenase (GAPDH) gene. Plasmids encoding preprothaumatin were shown to direct the synthesis of a processed form of the plant protein. Nucleotide sequence analysis of the 843 nucleotide GAPDH promoter fragment showed a characteristic structure with two regions of dyad symmetry containing translational starts of GAPDH and a putative 38 amino acid peptide (Evans et al., 1984).

Five different clones, homologous to the structural gene for thaumatin, were isolated from leaf DNA of *Thaumatococcus deniellii* Benth. Restriction maps, hybridization studies, and S1-nuclease mapping showed that the isolated thaumatin genes belonged to one multi-gene family, and had two very small introns situated at different positions in the various structural genes (Ledeboer et al., 1984). DNA was isolated from the freeze-dried leaves by specific precipitation of DNA and RNA with cetyl-trimethylammonium bromide at low salt concentration. DNA was purified further by CSCl-Et Br density gradient centrifugation. Hybridizations were performed in 50% formamide, 3XSSC at 42°C. Restriction maps were made for the restriction enzymes – AluI, HhaI, and TaqI. RNA purified twice on oligo(dT) cellulose, and nick-translated inserts of the isolated genomic clones were used.

## 11.10.3 **Amino Acids**

Microbial synthesis by hyper-producing organisms can be used to produce amino acids. Synthesis of lysine and glutamate have been improved by using conventional genetic approaches (Haas, 1984). Their production can be further improved by directed modification of the genes of the producing organisms. rDNA techniques are being used to increase the methionine biosynthesis.

Tsuchida and Momose (1986) investigated the genetic improvement of L-leucine productivity in strain 218, an ile$^{-2}$-thiazolealanine-resistant mutant of *Brevibacterium lactofermentum* 2256. In strain 218, which produced 28 mg of L-leucine per mL from 13% glucose, $\alpha$-isopropylmalate synthetase was genetically desensitized and derepressed to the effect of L-leucine, whereas $\alpha$-acetohydroxy acid synthetase remained unaltered. As a result, 103 mutants resistant to $\beta$-hydroxyleucine were isolated. Among these, three produced only L-leucine. The improved strains produced 34 mg of L-leucine per mL.

## 11.10.4 **Enzymes**

Microbial production of enzymes (amylases, glucose oxidase, lipases, cellulases, proteases, glucose isomerase, and invertase) can be obtained by the transfer of their genes to bacteria. The genes for rennin and $\alpha$-amylase have been cloned (Palva, 1982).

New enzymes can be designed and produced using recombinant DNA technology by knowing the relations between amino acid sequences, three-dimensional protein structures, and catalytic action. Site-specific mutagenesis modifies the enzymes by incorporating a specific mutation in the corresponding gene. The enzymes can also be modified using oligonucleotide-directed mutagenesis. This technique involves synthesizing a short section of DNA (the oligonucleotide) that corresponds to the gene sequence to be mutated. When this oligonucleotide is introduced into suitable cells, this acts as a primer for DNA replication, which incorporates the altered sequence into the gene (Simon et al., 1989).

To enhance their thermostability, enzymes can be chemically modified by using (1) polymer attachment, (2) cross-linking with bifunctional reagents, and (3) low-molecular weight mono-functional reagents. Genetic engineering techniques can also be used to improve the thermostability of enzymes. Mutagenesis, transformation, and recombinant DNA techniques are found to be suitable for this. Mutants can be developed by exposing cultures either to chemical mutagens, such as *N*-methyl-*N*-nitro-*N*-nitrosoguanidine, or to ultraviolet or gamma radiation. Hyper-production colonies are then screened by incorporating a suitable indicator into the growth medium (Wasserman,

1984). This has been used for α-amylase, cellulase, lactase, and other enzymes. Hyper-producing α-amylase strains were developed by transferring DNA between wild-type and mutant strains of related species. Hitotsuyangi et al. (1979) developed a strain of *B. subtilis* that produced 1500 to 2000 times greater α-amylase activity than the parent strain. Similarly, Shinomiya et al. (1982) transformed DNA from an unidentified thermophilic bacterium into an α-amylase-deficient *B. subtilis strain*. This produced thermostable enzyme similar to the donor strain.

Inserting genes of one organism into another will permit the transplanting of thermophilic enzymes into organisms such as *B. subtilis* and yeast. Other techniques such as gene splicing and gene amplification are also used for this purpose. The molecular study of genes coding for α-amylase in *Bacillus* has been successful (Heslot, 1985). Cornelis et al. (1982) cloned the α-amylase gene from *B. coagulans* into *E. coli* by phage infection. The α-amylase retained the thermostability of the *B. coagulans* form. Similarly, Fujii et al. (1983) cloned the gene for *B. stearothermophilus* CU-21 neutral protease in protease-deficient strains of *B. subtilis* and *B. stearothermophilus*. Both expressed high activities and retained thermostability of *B. stearothermophilus* CU-21. Thus, thermostable enzymes can be produced by using genetic engineering techniques coupled with the three-dimensional structure of enzymes.

A disulfide bond has been introduced into T4 lysozyme to enhance the thermostability of a protein (Perry and Wetzel, 1984). Computer graphic analysis of the X-ray crystal structure assisted the location where the presence of two cysteine residues should make possible the formation of a disulfide bridge. After mutagenic removal of an unpaired cysteine in the same molecule (Wetzel, 1986), a cross-linked lysozyme was generated, which was considerably more stable than the wild type to irreversible thermal inactivation. This procedure may be applicable to other proteins, increasing their thermal stability.

The overproduction of the food processing enzymes is achieved by cloning and expression for endogenous (such as amino acids, xanthan gum) and exogenous (such as rennin) products. Genetically blocked mutants of *Bacillus subtilis* were used to clone the IMP dehydrogenase gene (Miyagawa et al., 1986). This increased the yield of guanosine by shifting the rate-limiting step. Similarly, many higher eukaryotic genes have been cloned and produced in bacterial and fungal hosts. On the other hand, improvement in the production of enzymes is being achieved by recombinant DNA techniques applied to prokaryotic or lower eukaryotic organisms; *B. subtilis* and *Saccharomyces cerevisiae* are the most commonly used host vector systems for the expression of cloned genes (Lin, 1986).

Site-directed mutagenesis can be used to change activity-critical residues in enzymes that are susceptible to chemical oxidation. For example, methionine is a primary site for oxidation inactivation of subtilisin. Estell et al. (1985) used cassette mutagenesis methods to express all 19 amino acid substitutions at this site in the cloned subtilisin gene.

White et al. (1984) described the cDNA cloning and characterization of the genes of three enzymes (glucoamylase, exocellobiohydrolaseI, and pyranose-2-oxidase) isolated from filamentous fungi. Highly enriched and specific cDNA probes for each enzyme were prepared from messenger RNA isolated from fungal cultures that were grown under induced or uninduced conditions. This cDNA was used for screening genomic libraries and for making cDNA libraries. All three genes were found to contain introns that were required to obtain cDNA clones for each enzyme. The screening method used to identify these cloned genes will be applicable to the cloning of other enzymes (White et al., 1984).

Protein engineering techniques are being used to engineer a specificity change into a microbial based enzyme to maintain its primary desired activity but to discard its secondary activities. These techniques can also enhance the productivity of the enzyme, or can change or introduce other properties such as stability to heat, pH, or shear forces, desired kinetic parameters, etc.

Immobilized proteolytic enzymes such as bromelain or papain could be used to avoid chill-haze in beer (Finley et al., 1977). Similarly, immobilized pectin for clarifying fruit juices, naringinase for removing the bitter principle from grapefruit juice, and invertase enzymes for the production of invert sugar from sucrose, have been suggested.

## 11.10.5 Butanol

Improvement of *C. acetobutylicum* using genetic engineering techniques should allow for improvement in the metabolic pathway, end product tolerance, and catabolic enzyme systems of this microorganism. The food industry will be able to use this microorganism for disposing of food wastes and processing by-products such as mycotoxin-contaminated corn or cheese whey (Blaschek, 1986).

An artificial protoplast-based DNA uptake process involves the use of natural conjugation mechanisms using cell-to-cell contact (Blaschek, 1986). Conjugal transfer of the broad-host-range plasmid pAM $\beta$1 from *S. lactis* to *C. acetobutylicum* was investigated by Oultram and Young (1985). Although the cloning of *Clostridium* genes in *E. coli* has now been successful, Zappe et al. (1986) cloned *C. acetobutylicum* DNA.

## 11.10.6 Ethanol

Using genetic engineering techniques, yeasts for ethanol production can be modified to have high osmotolerance; to degrade cellulose, starch, and xylose; and to have high tolerance to ethanol. Fusion between *S. cerevisiae* and *S. diastaticus* was successful in improving ethanol production.

Ethanol yields from xylose may be improved by inserting the gene for xylose isomerase into these yeasts such as *P. tannophilus* (Alexander, 1986). A number of clones were identified containing the region of DNA coding for xylose isomerase activity. The transformed yeast cells were capable of expressing the xylose isomerase gene.

Johannsen et al. (1985) demonstrated the protoplast fusion of *C. shehatae*, which resulted in an increase in ploidy level and a small increase in ethanol production from xylose.

## 11.10.7 Suitable Host

Recombinant DNA technology for *E. coli* has little applicability to the food processing industry due to its associated endotoxins. *Corynebacterium glutamicum* seems to be a substitute for *E. coli*, and has been used in the biosynthesis of a variety of L-amino acids and flavor-enhancing nucleotides. Follettie and Sinskey (1986) have developed an rDNA system for *C. glutamicum* and utilized the techniques to clone various amino acid biosynthetic genes. These genes have been isolated from a genomic library constructed directly in *C. glutamicum* by the complementation of auxotrophic variants. Cloning of the *C. glutamicum* pheA gene onto a multicopy plasmid (pWS124) increased the specific activity of its protein product prephenate dehydratase 6-fold. Cloning of the *C. glutamicum* ThrA gene encoding homoserine dehydrogenase increased its specific activity 20-fold. Thus, the introduction of novel pathways into this organism permits the development of new product design and synthesis.

$\alpha$   volume fraction of hydrocarbon

$\alpha_{AB}$   separation factor for materials A and B

$\beta$   selectivity coefficient

$\gamma$   shear rate

$\Delta F$   change in fundamental frequency of the coated crystal

$\Delta m$   the change in mass due to the deposited coating

$\Delta P$   pressure drop

$\mu$   specific growth rate or viscosity

$\mu_a$   apparent viscosity

$\tau$   shear stress

$A$   area of the filter surface on which the cake forms, or surface area, or a constant, or adenine

$a$   bubble surface area, or surface area per unit volume

BOD   biological oxygen demand

$b$   consistency coefficient

$C$   actual dissolved oxygen concentration in the bulk liquid phase, or number of components, or cytosine

$C^*$   dissolved oxygen concentration in equilibrium

$C1$ to $C4$   constants

cDNA   complementary DNA

CE   capillary electrophoresis

$CL_o$   zero oxygen desorption achieved by sparging nitrogen

$CL_\infty$   dissolved oxygen at equilibrium

$C_n$   sugar concentration

CPC   centrifugal partition chromatography

CSTR   continuous stirred tank reactor

$Cx$   cell density

$D$   vessel diameter, or impeller diameter, or oxygen diffusion coefficient

$d$   stirrer diameter

Da   Damkohler number

DNA   deoxyribonucleic acid

DNAaseI   deoxyribonucleaseI

**323**

dNTPs  deoxynucleotide triphosphates
DOT  dissolved oxygen tension
$E$  activation energy, or extract, or electric field strength
ENFET  enzyme FET
$F$  degree of freedom, or feed, or force, or resonant frequency of the crystal
FET  field effect transistor
$G$  specific growth rate, or guanine
GE  gel electrophoresis
$Gm$  maximum $G$
$H$  head or partition coefficient
$h$  vessel height
HETS  height equivalent to a theoretical stage
HPCE  high-performance capillary electrophoresis
HPCPC  high-performance centrifugal partition chromatography
HPLC  high-performance liquid chromatography
$I_{Do}$  fluorescence intensity in the presence of oxygen
IMFET  immunochemical FET
$I_o$  fluorescence intensity in the absence of quencher oxygen
ISFET  ion-selective FET
$K$  equilibrium coefficient
KGB  potassium glutamate buffer
$K_L$  liquid film mass transfer coefficient
$K_L a$  volumetric oxygen transfer rate
$K_m$  Michaelis-Manten constant
$K_s$  reaction rate constant
$k_s$  mass transfer coefficient
$Ksv$  Stern-Volmer quenching constant
$L$  underflow, or pipe length, or equivalent thickness of the filter
LC  liquid chromatography
LEM  liquid emulsion membrane
LLM  liquid-liquid membrane
$M$  cell or molar mass
$m$  distribution coefficient
MF  microfiltration
mRNA  messenger RNA
$Ms$  maintenance coefficient
$Mw$  molecular weight
$n$  flow behavior index
$N$  rotational speed of impeller/agitation
$N$  any nucleotide or impeller speed
$Nt$  impeller tip speed
Nu  Nusselt number
$N_w$  oxygen transfer rate in water

$OU$   oxygen uptake rate
$OT$   oxygen transfer rate
$P$   number of phases or power input
$Pg$   gassed power draw of the agitator
Pr   Prandtl number
Pu   purine
PV   pervaporation
Py   pyrimidine
$Q$   heat generation rate or volumetric flow rate
$Q_c$   circulation coefficient
QCM   quartz crystal micro-balance
$Q_o$   volumetric flow rate of outlet air before seeding
$R$   raffinate, or pipe radius, or gas constant
$r$   specific resistance of the filter cake
Re   Reynolds number
RNA   ribonucleic acid
RO   reverse osmosis
rRNA   ribosomal RNA
RTD   resistance temperature detector
$S$   solvent or substrate concentration in bulk solution
$s$   radial length of the baffle
SCP   single cell protein
$T$   absolute temperature, or thymine
$t$   residence time, or filtration time
$tc$   circulation time
$tm$   mixing time
tRNA   transfer RNA
U   uracil
$u$   electrophoretic mobility, or microbial specific growth rate
UF   ultrafiltration
$V$   reaction rate, or overflow, or volume
$Vep$   electrophoretic velocity of a solute
$V_m$   molar volume
$V_{max}$   maximum reaction rate
$V_s$   linear velocity of aeration based on cross-sectional area of the stirred tank, or reaction rate per unit surface area
$Vs$   superficial gas velocity
$V_w$   volume of water
$W$   solid content per unit volume of liquid
$w$   solute free phase ratio
$x$   concentration
$X$   initial concentration of mixture
$y$   concentration of components after passing through the membrane

# GLOSSARY

**Activation Energy**   The energy required to initiate the transformation of reactants to products.

**Activity**   The effective concentration of a biocatalyst.

**Adsorption**   The taking up of molecules of gases, liquids, or dissolved substances at the surfaces of solids or liquids.

**Aerobe**   Microorganism that requires molecular oxygen for growth.

**Aerobic**   Active only in the presence of free oxygen.

**Affinity Chromatography**   Separation and purification of compounds by capturing them by an immobilized matrix of other compounds.

**Agar**   A natural polymeric complex polysaccharide extracted from various species of marine macroalgae.

**Airlift Fermenter**   A fermenter in which sparged gas is used for agitation through draft tubes.

**Amino Acids**   The building blocks of proteins.

**Amylase**   An enzyme that catalyzes the starch hydrolysis.

**Anaerobe**   Microorganism that grows in the absence of molecular oxygen.

**Anaerobic**   Active without free oxygen.

**Anaerobic Digestion**   Metabolic breakdown of organic compounds by anaerobic microorganisms.

**Annealing**   The process in which individual strands of DNA come close to form a double helix.

**Antibody**   An immunoglobulin (protein) produced in human or animal bodies in response to exposure to an antigen.

**Anticodon**   Three nucleotides, complementary to the codon in mRNA, that are present in the tRNA at a constant position.

**Antigen**   Any foreign substance (usually a protein molecule) that, when injected into an animal, provokes synthesis of a specific antibody or antibodies.

**Asepsis**   The prevention of infection by avoiding the contact of microorganisms.

**Aseptic**   The exclusion of unwanted microorganisms.

**Assay**   A biological response measuring technique.

***att* sites**   The sequences on a phage and the bacterial chromosome whose recombination results in integration of the phage DNA into the bacterial genome.

**Autoradiography**    The method of detecting radioactively labeled molecules by obtaining an image on an X-ray film.

**Auxins**    Plant growth regulators affecting the organization of developing tissues.

**Axial Dispersion**    Mixing in the direction of the fluid flow.

**Bacterial Transformation**    The process by which a bacterial host acquires new genetic markers by incorporation of the added DNA molecule.

**Bacteriocins**    Antibacterial substances produced by many bacterial species.

**Bacteriophages**    Viruses that infect bacteria.

**Base Pair (bp)**    In a DNA double helix, adenine (A) pairs with thymine (I), and guanine (G) with cytosine (C).

**Biocatalyst**    A biological catalyst such as enzyme, cell, tissue.

**Bioconversion**    Conversion of reactants to products using biocatalyst.

**Biodegradation**    The degradation or breakdown of substances by microorganisms.

**Biological Oxygen Demand (BOD)**    The oxygen requirements of aerobic microorganisms in breaking down organic compounds dissolved or suspended in water.

**Biomass**    Organic material from a biological source such as plant, or microorganism.

**Biopolymers**    Naturally available macromolecules such as protein, polysaccharide.

**Bioprocess**    A process requiring cells or their components to provide required chemical or physical changes.

**Bioreactor**    Containment system for a bioreaction, e.g., fermentation, microbial growth.

**Biosensor**    An electric or electronic device detects specific compounds (e.g., enzymes) using biological molecules.

**Biosurfactant**    A compound from microorganisms, which assists in solubilizing compounds by surface tension between the compound and liquid.

**Biosynthesis**    Production of a compound by microorganisms by synthesis or degradation.

**Buffer**    A solution containing a mixture of a weak acid and its conjugate weak base, which resists large changes in pH upon the addition of small amounts of acidic or basic substances.

**Callus**    Plant cells capable of repeated cell division.

**Cap**    A methylguanosine triphosphate residue that is present at the 5' end of eukaryotic mRNA.

**cDNA**    DNA complementary to an mRNA, synthesized from it by the action of the enzyme reverse transcriptase.

**cDNA Cloning**    The cloning of double-stranded DNA representing an mRNA in a cloning vector.

**Cell** The smallest structural unit of living material capable of functioning independently.

**Cell Culture** *In vitro* growth of cell(s).

**Cell Fusion** Joining two or more cells together into a single cell.

**Cellulase** An enzyme that catalyzes the conversion of cellulose to sugars.

**Cellulose** A polymer of six carbon sugars found in all plants.

**Chemical DNA Synthesis** The synthesis of fragments of DNA with defined sequences by chemical methods.

**Chitin** A homopolysaccharide of *N*-acetyl-D-glucosamine from the exoskeleton of insects and crustaceans.

**Chromatography** Separation and purification of gases, liquids or solids in a mixture or solution by adsorption on an adsorbent medium.

**Chromosome** A discrete segment of the genomic DNA to which a number of proteins are attached.

**Clone** A group of genetically identical organisms or cells or nucleic acids.

**Cloning Vector** A DNA cloning vehicle.

**Coagulation** The process that reduces the forces tending to keep the suspended particles apart.

**Codon** A triplet of nucleotides that represents an amino acid.

**Cofactor** A molecule that is required for the enzyme's activity.

**Complementary** In a double-stranded DNA molecule, each strand represents the nucleotide sequence in the other strand.

**Complementation** The ability of two independent genes to produce wild phenotype when the products of these mutant genes are mixed together.

**Concatamer** A tandem repeat of the same DNA segment.

**Cosmid** A plasmid with bacteriophage lambda *cos* sites.

**Culture** A population of a cell or cells.

**Cytokinins** Plant growth regulators affecting the organization of developing tissues.

**Deletion** The removal of a portion of DNA sequence.

**Denaturation** The process in which double-stranded nucleic acid is converted into single-stranded state.

**Diacetyl** 2,3-butane dione provides buttery flavor.

**Dialysis** The separation of macromolecules from ions and low-molecular weight compounds using a membrane that is permeable to crystalloids and liquid but is impermeable to macromolecules.

**Dilution Rate** Reciprocal of the residence time of a culture in a bioreactor.

**Distribution Coefficient** A numerical value indicating the ability of the solvent to dissolve the solute.

**DNA (Deoxyribonucleic Acid)** A sequence of deoxyribonucleotides usually in a double-stranded helical form.

**DNAase** Any enzyme that cleaves DNA.

**DNA Clone**   A large number of DNA molecules identical to a single parental DNA molecule.

**DNA Cloning**   The process of isolating a DNA molecule and amplifying it in a cloning vector.

**DNA Hybridization**   The process in which pairing of complementary DNA strands takes place to form double-stranded DNA molecule.

**DNA Probe**   Usually a radioactively labeled DNA sequence that binds to a specific region of a denatured DNA molecule.

**DNA Sequencing**   Determination of the order of nucleotides in a DNA fragment or gene.

**DNA Synthesis**   DNA separation by the sequential addition of nucleotide bases.

**Double Helix**   A three-dimensional DNA structure where two strands of molecule are twisted together in a helical structure.

**Downstream Processing**   Separation and purification of product(s) from a reactor or a process.

**Duplex DNA**   A double-stranded DNA molecule.

**Electrophoresis**   The procedure of separating proteins or nucleic acids in an electric field.

**Embryogenesis**   Induction of embryos that directly develop into plants.

**Emulsification**   The process of making lipids, fats, and oils more soluble in water.

**End Labeling**   The process of adding a radioactively labeled nucleotide or phosphate group to 5' or 3' end of a DNA strand.

**Endoenzyme**   An enzyme that acts at random in cleaving molecules of substrate.

**Endonuclease**   Any enzyme that makes internal cuts in nucleic acid chains.

**Enzyme**   Protein-based catalyst produced by living cells.

**Enzyme Adsorption**   Physical adsorption of enzymes onto the surface of solid matrices.

**Enzyme Covalent Bonding**   Covalent bonding of enzymes to water-insoluble matrices.

**Enzyme Electrode**   A device that consists of an electrochemical sensor in contact with a layer of an immobilized enzyme.

**Enzyme Transistor**   Immobilized enzyme combined with a hydrogen and ammonia sensitive field-effect-transistor.

***Escherichia coli (E. coli)***   A bacteria species that inhabits the intestinal tract of most vertebrates.

**Eukaryote**   A microorganism with membrane-bound, structurally discrete nuclei and well-developed cell organelles.

**Exoenzyme**   An enzyme that acts by cleaving the ends of molecular chains in a substrate.

**Exonuclease**    Any enzyme that removes nucleotides from 3' or 5' or both ends of a polynucleotide chain.

**Explant**    Pieces of leaves, shoots, roots, or immature fruits.

**Fatty Acids**    Organic acids with long carbon chains.

**Fermentation**    An enzymatic or microbial bioprocess of converting large molecular weight organic substrates or feed stocks to lower molecular weight products.

**Fermenter**    A vessel in which a bioconversion process takes place.

**Filtrate**    The solution that passes through the filter or ultrafilter.

**Flocculating Agent**    A reagent used to disperse solids in a liquid.

**Flocculation**    The agglomeration of suspended material.

**Foam-Fractionation**    Selective adsorption of proteins at the gas/liquid interface generated by a rising ensemble of bubbles through the solution.

**Forced Convection**    Fluid motion that is induced by external forces in a reactor.

**Foreign DNA**    Any DNA fragment other than the vector DNA used for its cloning.

**Free Convection**    Fluid motion induced by density differences or temperature gradient in a reactor.

**Gellangum**    An extracellular polysaccharide secreted by the microorganism.

**Gene**    A segment of DNA responsible for the mRNA sequence during the process of transcription. It also includes the regulatory element.

**Gene Expression**    The way and efficiency by which a gene shows its phenotype.

**Genetic Engineering**    Gene manipulative techniques, e.g., recombinant DNA.

**Genomic DNA**    The total chromosomal DNA of any organism.

**Gram Negative**    A microorganism that does not retain the initial Gram stain but retains the counter-stain.

**Gram Positive**    A microorganism that retains the initial Gram stain and is not stained by the counter-stain.

**Growth Curve**    A plot of the growth of an organism w.r.t. time in a nutrient medium.

**Growth Hormone**    A group of peptides regulating growth in higher animals.

**Hemicellulose**    A polymer of D-xylose that contains side chains of other sugars.

**Hereditary Material**    The material (DNA) containing genetic information that is transferred from one generation to another.

**Heteroduplex DNA**    Partially double-stranded DNA molecule.

**High-Performance Liquid Chromatography (HPLC)**    A type of chromatography used for the separation and purification of compounds.

**Hormone**    A cellular product that circulates in the blood and induces a reaction at some distance from its point of origin.

**Hydrolysis**    A chemical reaction involving addition of water to break bonds.

**Immobilization**    A process of converting enzymes or cells from the free mobile state to the immobile state.

**Induction**    To increase the rate of synthesis of an enzyme using an inducer.

**Insert DNA**    See FOREIGN DNA.

**Integration**    The process of inserting a DNA fragment into a host genome.

**Intron**    A segment of DNA that is transcribed, but is excised during mRNA processing.

**Inverted Repeats**    Two copies of the same DNA sequence repeated in opposite orientation on the same DNA molecule.

***In vitro***    "In glass"; used for a biological reaction taking place in a vessel or container.

***In vivo***    "In life"; used for a biological reaction taking place in a microorganism.

**Ion Exchange**    Involves the reversible interchange of ions between a functionalized insoluble resin and an ionizable substance in a solution.

**Ionic Strength**    A measure of the ionic concentration of a solution.

**Isoelectric Point**    The pH at which a molecule has a net zero charge.

**Katal**    Amount of enzyme activity that converts one mole of substrate per second.

**Kb**    Kilo (1000) base pairs of DNA.

**$\varkappa$-Carrageenan**    A natural polymeric complex polysaccharide containing $\beta$-D-galactose sulphate and 3,6-anhydrogalactose.

**Laminar Flow**    Fluid motion in which steady fluid particle trajectories can exist; in pipe flow it exists when Reynolds number is less than 2100.

**Leaching**    The separation of a soluble compound from a solid mixture by washing or percolating.

**Library**    A set of cloned DNA or cDNA fragments that represents the entire genome.

**Ligation**    The process of forming a phosphodiester bond to join two adjacent nucleotides.

**Lignocellulose**    The composition of woody biomass, including lignin and cellulose.

**Lipase**    An enzyme that catalyzes fat hydrolysis.

**Lipids**    Water-insoluble fat-based organic molecules.

**Liquid Emulsion Membrane (LEM)**    An emulsion of two immiscible phases that is dispersed in a third phase.

**Mass Transfer**    The process of transfer of mass of a component from one material to another.

**Medium**    A mixture of nutrient substances.

**Meiosis**    Division of diploid cell to two haploid cells.

**Meristematic Cells**   Plant cells located at the tips of stems and roots, in leaf axils and in callus tissue, which initiate new growth.

**Messenger RNA (mRNA)**   RNA molecules produced by the process of transcription for DNA and used as templates for protein synthesis.

**Metabolite**   A reactant or product of biochemical activity, i.e., the reactions of metabolism.

**Microencapsulation**   The process of surrounding enzymes or cells with a permeable membrane for immobilization.

**Microorganisms**   Microscopic living entities.

**Mitosis**   A plant cell division that provides two cells with the same number of chromosomes as the original cells.

**Mixed Culture**   Culture containing two or more types of microorganisms.

**Multiple Cloning Site**   A segment of vector DNA containing a number of unique restriction enzyme sites for the purpose of cloning foreign DNA fragments.

**Mutagen**   Any substance that causes changes in DNA.

**Mutagenesis**   The process of introducing mutation in a DNA molecule.

**Mutation**   Any change in a DNA sequence.

**Mycotoxin**   Toxic fungal secondary metabolite.

**Natural Convection**   Similar to free convection.

**Newtonian Fluid**   A fluid in which the ratio of shear stress to shear rate is constant.

**Nicking**   The process of introducing single-stranded breaks in a double-stranded DNA molecule.

**Non-Newtonian Fluid**   A fluid in which the ratio of shear stress to shear rate is not constant.

**Northern Blotting**   The technique of transferring RNA from an agarose gel to a nitrocellulose filter.

**Nucleic Acid**   Molecules for alternating sugar and phosphate groups from which protrude certain organic bases attached to sugars, such as RNA, DNA.

**Nucleotide**   Building blocks of the nucleic acids.

**Oligonucleotide**   A short chain of nucleotides (usually 10 to 30 bases long).

**Oncogenicity**   Tumor-producing ability.

**Organelle**   A part of a cell that conducts a specific function, e.g., nuclei, chloroplasts.

*ori*   Origin of replication.

**Origin of Replication**   The initiation site for DNA replication.

**Osmosis**   Solvent mobility across a semipermeable membrane from low solute concentration to higher solute concentration.

**Osmotic Pressure**   The pressure that causes a solvent to move in osmosis.

**Oxygen Transfer Rate**   Oxygen mass transfer in a medium.

**Pathogen**   A disease-producing microorganism.

**Pectin**    A fruit-based polysaccharide.

**Peptide**    A linear polymer of amino acids.

**Permeate**    The solution that passes through the filter.

**Plaque**    A phage colony identified on a growth-free area on a bacterial agar plate.

**Plasmid**    An extrachromosomal circular DNA capable of replicating independently of the bacterial genome.

**Plug Flow**    Flow of materials with no mixing in the direction of flow.

**Point Mutation**    A single base pair change in DNA sequence.

**Polycloning Site**    See MULTIPLE CLONING SITE.

**Polymer**    A linear or branched molecule of repeating subunits.

**Polypeptide**    A chain of amino acids held together by peptide bonds.

**Polysaccharide**    A polymer of sugars.

**Primer**    A short oligonucleotide sequence that pairs with one of the DNA strands, and with the help of DNA polymerase allows synthesis of a deoxyribonucleotide chain.

**Prokaryote**    Organism or bacterium that does not have a nucleus.

**Promotor**    A region of DNA that has the site of RNA polymerase binding to initiate transcription.

**Protease**    Protein-digesting enzyme.

**Protein**    A polypeptide consisting of amino acids.

**Protein Domain**    A portion of the amino acid sequence (of a protein) having a specific function.

**Protein Engineering**    Rearranging amino acids on the protein molecule to create new proteins or change their properties.

**Protoplast Fusion**    The joining of two cells by first removing their cell walls.

**Pullulanases**    Enzymes that hydrolyze the 1-6 bonds in amylopectin molecules.

**Pyrazine**    Heterocyclic, nitrogen-containing compound that provides a nutty flavor upon heating.

**Receptor**    A site that recognizes and binds a specific intercellular messenger and initiates the messenger's effects, usually a protein on the cell's outer membrane.

**Recombinant DNA**    A DNA molecule synthesized by joining DNA fragments from two different sources.

**Regulatory Element**    A segment of DNA sequence (usually present upstream of the protein initiation start site of a gene) that controls the gene's activity.

**Reporter Gene**    Any gene, the expression of which can easily be monitored.

**Restriction Enzyme**    Any endonuclease that recognizes specific short sequences (recognition sequence) within a DNA molecule and makes double-stranded cleavage.

**Retentate**    The part of the unfiltered solution that does not pass through the filter.

**Ribosome**    The cellular organelle that is the site of mRNA translation.

**RNA (Ribonucleic Acid)**    A chain of ribonucleotides usually in a single-stranded form.

**RNA:DNA Hybrid**    The product of hybridization of a RNA molecule with its complementary DNA.

**Saccharification**    The degradation of polysaccharides to sugars.

**Scale-Up**    The transition of a process from small-scale to large-scale.

**Secondary Metabolite**    Non-essential metabolite for microbial life-support system.

**Semiconductor**    A material whose electrical conductivity is in between a good conductor and insulator.

**Single Cell Oil**    Oil produced by microorganisms.

**Single Cell Protein**    Cells or protein extracts of microorganisms.

**Sludge**    Precipitated solid matter produced by sewage treatment.

**Solid Substrate Fermentation**    Microbial growth on solid materials with no or little water.

**Somaclonal Variation**    Genetic variation produced from plant cell cultures from a pure breeding strain.

**Somatic Hybrid**    Non-sexual, vegetative hybrid.

**Southern Blotting**    The technique of transferring DNA from an agarose gel to a nitrocellulose filter.

**Species**    A taxonomic subdivision of a genus.

**Specific Growth Rate**    Microbial population growth rate per unit cell mass.

**Splicing Junctions**    The sequences present at the borders of exon-interon boundaries.

**Starch**    The major form of storage for carbohydrates in plants.

**Sterile**    Free from viable microorganisms.

**Sterilization**    The complete destruction of all viable microorganisms in a material.

**Stop Codon**    A triplet of nucleotides that represents a termination signal.

**Strain**    A group of microorganisms of the same species having distinctive characteristics but not usually considered a separate variety.

**Substrate**    A compound converted to a product by an enzyme.

**Tangential Flow Filtration**    Similar to cross-flow filtration.

**Template**    A single-stranded DNA or RNA molecule that can serve as a mold for the synthesis of a complementary strand.

**Terpenes**    Hydrocarbons that provide flavors of oils and grape aroma.

**Thermal Diffusivity**    A ratio of the thermal conductivity to the volumetric specific heat of a substance.

**Thermophile**    An organism that grows at high temperatures (45 to 70°C).

**Thermophilic**    Heat-loving microorganisms, which can survive at elevated temperatures.

**Tissue Culture**    Processes that permit the growth and maintenance of cell lines (tissue), generally used to regenerate plants.

**Toxicity**    The ability of a substance to produce a harmful effect on an organism if consumed.

**Transcription**    The process in which synthesis of mRNA takes place from a DNA template.

**Transduction**    Genetic material transfer between cells by viral vector and subsequent incorporation by recombination.

**Translation**    The process in which information present in the mRNA directs the synthesis of a polypeptide on ribosome.

**Trypsin**    An enzyme that cleaves bonds next to arginine and lysine residues.

**Turbulent Flow**    Fluid motion with particle trajectories varying randomly with time.

**Vector**    Transfers genes into cells, usually a plasmid or a viral DNA.

**Viscosity**    A ratio between shear stress and shear rate of a fluid, which is a measure of the fluid resistance to flow.

**Volatile Fatty Acids (VFA)**    A mixture of acids produced by acidogenic microorganisms during anaerobic digestion.

**Western Blotting**    The technique of transferring polypeptides from an acrylamide gel to a nitrocellulose filter.

**Xenoenzyme**    Engineered enzyme.

# REFERENCES

Abel, P. P., R. S. Nelsen, B. De, N. Hoffman, S. G. Rogers, R. T. Fraley and R. N. Breachy. 1986. "Delay of Disease Development in Transgenic Plants that Express Tobacco Mosaic Virus Coat Protein Gene," *Science*, 232:738.

Adams, J. M., L. A. Ash, A. J. Brown, R. James, D. B. Kell, G. J. Salter and R. P. Walter. 1988. "A Range of Ceramic Biosupports," *Am. Biotechnol. Lab.*, 6(10):34.

Addy, N. D. and D. A. Stuart. 1986. "Impact of Biotechnology on Vegetable Processing," *Food Technol.*, 40(10):64.

Adler, I. 1987. "Equipment and Bioreactors," in *Biochemical Engineering*. H. Chmiel, W. P. Hammes and J. E. Bailey, eds. New York: Gustav Fischer Verlag, pp. 342–354.

Adlercreutz, P. and B. Mattiasson. 1982. "Oxygen Supply to Immobilized Cells. III. Oxygen Supply by Hemoglobin or Emulsion of Perfluoro Chemicals," *Eur. J. App. Microb. Biotechnol.*, 16:165.

Agar, D. W. 1985. "Microbial Growth Rate Measurement Techniques," in *Comprehensive Biotechnology, Vol. 4*. M. Moo-Young, C. W. Robinson and J. A. Howell, eds. New York: Pergamon Press, pp. 305–328.

Aiba, S., A. E. Humphrey and N. F. Millis. 1973. *Biochemical Engineering*. New York: Academic Press, Inc.

Alexander, N. J. 1986. "Genetic Manipulation of Yeast for Ethanol Production from Xylose," *Food Technol.*, 40(10):99.

Alexander, P. W. and J. P. Joseph. 1981. "A Coated-Metal Enzyme Electrode for Urea Determinations," *Anal. Chim. Acta*, 131:103.

Allen, L. K., W. E. Sandine and P. R. Elliker. 1963. "Transduction in *Streptococcus lactis*," *J. Dairy Res.*, 30:351.

Amasino, R. M. 1986. "Acceleration of Nucleic Acid Hybridization Rate by Polyethylene Glycol," *Anal. Biochem.*, 152:304.

Amourache, L. and M. A. Vijayalakshmi. 1986. "Nylon Filters with Rennet Enzyme (Chymosin) for Continuous Milk Clotting," in *Proc. Int. Symp. on Immobilized Enzymes and Cells, Univ. of Waterloo*, Waterloo: GW Biotech.

Anderson, S. 1981. "Shotgun DNA Sequencing Using Cloned DNAaseI-Generated Fragments," *Nucleic Acids Res.*, 9(13):3015.

Andres, C. 1986. "Biotechnology and Enzyme Breakthroughs Benefit Cheese Producers and Consumers," *Food Process.* (Feb.): 31.

Anon. 1985a. "Biotechnology—Moving toward Commercial Reality," *Food Process.* (July):46.

Anon. 1985b. "Enzymes," in *Functional Properties of Food Components*. Y. Pumeranz, ed. Toronto: Academic Press.

Anon. 1986a. *Int. Sym. on Immobilized Enzymes and Cells, University of Waterloo*, Waterloo: GW Biotech.

**337**

Anon. 1986b. "Protein Eng.," *Biotechnology Canada/Canada Research* (Jan.):30.

Anon. 1988a. *Research Activities 1987*. Dept of Ag. Eng., Purdue University, West Lafayette, IN.

Anon. 1988b. *The Filter Book*. Rexdale, Canada: Gelman Sciences Inc.

Anon. 1988c. "Food Biotechnology," *Can. Inst. Food Sci. Technol. J.*, 21(4):334.

Anon. 1989. "Single Cell Oil: A Technology Looking for Applications," *Food Eng.*, 61(9):108.

Arcuri, E. J., R. M. Worden and S. E. Shumate. 1980. "Ethanol Production by Immobilized Cells of *Zymonas mobilis*," *Biotech. Letters*, 2:499.

Armiger, W. B. 1985. "Instrumentation for Monitoring and Controlling Bioreactors," in *Comprehensive Biotechnology, Vol. 2*. M. Moo-Young, ed. New York: Pergamon Press, pp. 133−148.

Arnold, M. A. and G. A. Rechnitz. 1982. "Optimization of a Tissue Based Membrane Electrode for Guanine," *Anal. Chem.*, 54:777.

Asenjo, J. A. and J. Hong, eds. 1986. *Separation, Recovery and Purification in Biotechnology*. Washington, DC: Am. Chem. Soc.

Aston, W. J. and A. P. F. Turner. 1984. "Biosensors and Biofuel Cells," in *Biotechnology and Genetic Engineering Reviews, I*. G. E. Russell, ed. Newcastle-upon-Tyne, UK: Intercept, pp. 89−120.

Atkinson, B. and F. Mavituna. 1983. *Biochemical Engineering and Biotechnology Handbook*. New York: The Nature Press.

Atkinson, B. and P. Sainte. 1982. "Development of Downstream Processing," *J. of Chem. Tech. & Biotech.*, 32:100.

Bacus, J. 1984. "Update: Meat Fermentation," *Food Technol.*, 38(6):59.

Bailey, J. E. and D. F. Ollis. 1986. *Biochemical Engineering Fundamentals*. New York: McGraw Hill.

Baker, K. 1978. "Design, Construction and Operation of Some All-Glass Pilot Plant Fermenters," *Biotechnol. Bioeng.*, 20:1345.

Banerjee, M., A. Chakrabarty and S. K. Majumdar. 1982. "Immobilization of Yeast Cells Containing β-Galactosidase," *Biotechnol. Bioeng.*, 24:1839.

Bankier, A. T., K. M. Weston and B. G. Barrell. 1987. "Random Cloning and Sequencing by the M13/Dideoxynucleotide Chain Termination Method," *Methods Enzymol.*, 155:51.

Barach, J. T. 1985. "What's New in Genetic Engineering of Dairy Starter Cultures and Dairy Enzymes," *Food Technol.*, 39(10):73.

Barney, M. C., G. P. Jansen and J. R. Helbert. 1980. "Use of Genetic Transformation for the Introduction of Flocculants into Yeast," *Am. Soc. Brew. Chem. Proc.*, 38:71.

Bastin, G. and D. Dochain. 1986. "On-line Estimation of Microbial Specific Growth Rates," *Automatica*, 22(6):707.

Batt, C. A. and A. J. Sinskey. 1984. "Use of Biotechnology in the Production of Single Cell Protein," *Food Technol.*, 38(2):108.

Batt, C. A. 1986. "Genetic Engineering of *Lactobacillus*," *Food Technol.*, 40(10):95.

Beggs, J. D. 1978. "Transformation of Yeast by a Replicating Hybrid Plasmid," *Nature*, 275:104.

Berk, A. J. and P. A. Sharp. 1977. "Sizing and Mapping of Early Adenovirus mRNAs by Gel Electrophoresis of S1 Endonuclease-Digested Hybrids," *Cell*, 12:721.

Bertrand, C., P. R. Coulet and D. C. Gautheron. 1981. "Multipurpose Electrode with Different Enzyme Systems Bound to Collagen Films," *Anal. Chim. Acta.*, 126:23.

Birch, J. R., C. R. Hill and A. C. Kenney. 1985. "Affinity Chromatography: Its Role in Industry," in *Biotechnology Applications and Research*. P. N. Cheremisinoff and R. P. Ouellette, eds. Lancaster, PA: Technomic Pub. Co., Inc., pp. 594–606.

Birnboim, H. C., and J. Doly. 1979. "A Rapid Alkaline Extraction Procedure for Screening Recombinant Plasmid DNA," *Nucleic Acids Res.*, 7:1513.

Bisio, A. and R. L. Kabel, eds. 1985. *Scale-Up of Chemical Processes*. New York: John Wiley & Sons.

Blackadder, D. A. and R. M. Nedderman. 1971. *A Handbook of Unit Operations*. New York: Academic Press.

Bland, R. R., H. C. Chen, W. J. Jewell, W. D. Bellamy and R. R. Zall. 1982. "Continuous High Rate Production of Ethanol by *Zymonas mobilis* in an Attached Film Expanded Bed Fermenter," *Biotech. Lett.*, 4:323.

Blaschek, H. P. 1986. "Genetic Manipulation of *Clostridium Acetobutylicum* for Production of Butanol," *Food Technol.*, 40(10):84.

Blenke, H. 1987. "Process Engineering Contribution to Bioreactor Design and Operation," in *Biochemical Engineering*. H. Chmiel, W. P. Hammes and J. E. Bailey, eds. New York: Gustav Fischer Verlag, pp. 69–91.

Bolivar, F., R. L. Rodriguez, P. J. Greene, M. C. Betlach, H. L. Heyneker, H. W. Boyer, J. H. Crosa and S. Falkow. 1977b. "Construction and Characterization of New Cloning Vehicles II. A Multipurpose Cloning System," *Gene*, 2:95.

Bolivar, F., R. L. Rodriguez, M. C. Betlach and H. W. Boyer. 1977a. "Construction and Characterization of New Cloning Vehicles. I. Ampicillin-Resistant Derivatives of the Plasmid pMB9," *Gene*, 2:75.

Bond, P., A. Scragg and M. Fowler. 1987. "Recovery and Disruption of Plant Cells," *Separations for Biotechnology*. M. S. Verrall and M. J. Hudson, eds. London: Ellis Horwood Ltd., pp. 80–89.

Boone, C., A. M. Sdicu, J. Wagner, R. Degre, C. Sanchez and H. Bussey. 1990. "Integration of the Yeast K1 Killer Toxin Gene into the Genome of Marked Wine Yeasts and Its Effector Vinification," *Amer. J. Enol. Vitic.*, 41(1):37.

Boss, F. C. 1983a. "Filtration," in *Fermentation and Biochemical Engineering Handbook*. H. C. Vogel, ed. Park Ridge, NJ: Noyes Pub., pp. 163–174.

Boss, F. C. 1983b. "Centrifugation," in *Fermentation and Biochemical Engineering Handbook*, H. C. Vogel, ed. Park Ridge, NJ: Noyes Pub. pp. 296–316.

Botstein, D. and R. M. Davis. 1982. "Principles and Practice of Recombinant DNA Research with Yeast," in *The Molecular Biology of the Yeast, Saccharomyces. Vol. 2*. J. N. Strathem, E. M. Jones and J. R. Broach, eds. New York: Cold Spring Harbor Lab., pp. 607–636.

Boyce, C. O. L. 1986. *Novo's Handbook of Practical Biotechnology*. Denmark: Novo Indstri A/S.

Bresser, J. and D. Gillespie. 1983. "Quantitative Binding of Covalently Closed Circular DNA to Nitrocellulose in NaI," *Anal. Biochem.*, 29:357.

Breuer, M. E., C. Y. Yoon, D. P. Jones and M. J. Nurry. 1977. "Countercurrent Controlled Cycle Liquid-Liquid Extraction," *Chem. Eng. Progress*, 73:95.

Bright, S. W. J., R. S. Nelson, A. Karp, V. A. Jarrett, G. P. Creisson, G. Ooms, B. J.

Miflin and E. Thomas. 1982. "Variation in Culture-Derived Potato Plants," *Proc. 5th Intl. Congress Plant Tissue and Cell Culture*, p. 413.

Brocklebank, M. P. 1985. "Usage and Role of Membranes in Bio-Processes," *Processing* (April):27.

Brocklebank, M. P. 1987. "Large-Scale Separation and Isolation of Proteins," in *Food Biotechnology I*. R. D. King and P. S. J. Cheetham, eds. London: Elsevier Applied Science, pp. 139–192.

Brodelius, P. 1984. "Immobilization of Cultured Plant Cells and Protoplasts," in *Cell Culture and Somatic Cell Genetics of Plants, Vol 1*. J. K. Vasil, ed. New York: Academic Press, Inc., pp. 535–546.

Brodelius, P. 1985. "Immobilized Plant Cells," in *Enzymes and Immobilized Cells in Biotechnology*. A.I. Laskin, ed. Don Mills, Ont., Canada: The Benjamin/Cummings Pub. Co., pp. 109–148.

Brodelius, P. 1988. "Immobilized Plant Cells as a Source of Biochemicals," in *Bioreactor Immobilized Enzymes and Cells—Fundamentals and Applications*. M. Moo-Young, ed. New York: Elsevier Applied Sci., pp. 167–196.

Brodelius, P., B. Deus, K. Mosbach and M. H. Zenk. 1980. *Enzyme Engineering, Vol. 5*. H. H. Weetall and G. P. Royer, eds. New York: Plenum Press, pp. 373–381.

Brodelius, P. and K. Nilsson. 1980. "Entrapment of Plant Cells in Different Matrices. A Comparative Study," *FEBS Lett.*, 122:312.

Brown, C. M., I. Campbell and F. G. Priest. 1987. "Introduction to Biotechnology." London: Blackwell Sci. Pub.

Brunner, K. H. 1987. "Extraction in Downstream Processing," in *Biochemical Engineering*. H. Chmiel, W. P. Hammes and J. E. Bailey, eds. New York: Gustav Fischer Verlag, pp. 272–287.

Buchanan, R. 1986. "Food Fermentation with Molds," in *Biotechnology in Food Processing*. S. K. Harlander and T. P. Labuza, eds. Park Ridge, NJ: Noyes Pub., pp. 209–221.

Buckland, B. C. 1984. "The Translation of Scale in Fermentation Processes: The Impact of Computer Process Control," *Bio/Technol.*, 2(10):875.

Bungay, H. R. 1984. "Recent Advances in Designs of Biological Reactors," in *Biotechnology for the Oils and Fats Industry*. C. Ratledge, P. Dawson and J. Rattray, eds. Am. Oil Chemists Soc., pp. 45–54.

Burnette, W. N. 1981. "Western Blotting: Electrophoretic Transfer of Proteins From Sodium Dodecyl Sulfate-Polyacrylamide Gels to Unmodified Nitrocellulose and Radiographic Detection with Antibodies and Radio-Iodinated Protein A," *Anal. Biochem.*, 112:195.

Busche, R. M. 1985. "Acetic Acid Manufacture—Fermentation Alternatives," in *Biotechnology Applications and Research*. P. N. Cheremisinoff and R. P. Ouellette, eds. Lancaster, PA: Technomic Pub. Co., Inc., pp. 88–102.

Cahn, R. P. and N. N. Li. 1974. "Separation of Phenol from Waste Water by the Liquid Membrane Technique," in *New Developments in Separation Methods*. E. Grushka, ed. Marcel Dekker, Inc., pp. 13–27.

Cairns, W. L., R. Rumble and N. Kosaric. 1984. "Chemical Species Contributing to the De-Emulsifying Ability of Bacterial Cell Surfaces," in *Biotechnology for the Oils and Fats Industry*. C. Ratledge, P. Dawson and J. Rattray, eds. Am. Oil Chemists Soc., pp. 223–240.

Campo, M. S. 1985. "Bovine Papillomavirus DNA: A Eukaryotic Cloning Vector," in

*DNA Cloning Volume II—A Practical Approach.* D. M. Glover, ed. Oxford: IRL Press.

Capecchi, M. R. 1980. "High Efficiency Transformation by Direct Microinfection of DNA into Cultured Mammalian Cells," *Cell*, 22:479.

Casey, J. and N. Davidson. 1977. "Rates of Formation and Thermal Stabilities of RNA:DNA and DNA:DNA Duplexes at High Concentrations of Formamide," *Nucleic Acids Res.*, 4:1539.

Cavoski, D., D. Obradovic, R. Radovanovic, M. Perunovic and T. Fridl. 1988. "The Effect of Starter Cultures on Sensory and Physicochemical Properties of Long Ripened Dry Sausage," *Proc. 34th Int. Congress of Meat Sci. & Technol.*, *Brisbane, Australia*, pp. 366–367.

Cayle, T. et al. 1986. "Production of L-Ascorbic Acid from Whey," in *Biotechnology in Food Processing.* S. K. Harlander and T. P. Labuza, eds. Park Ridge, NJ: Noyes Pub., pp. 157–169.

Cazes, J. 1988. "Centrifugal Partition Chromatography," *Bio/Technol.* ( Dec.)

Cazes, J. 1989. "High Performance CPC for Downstream Processing of Biomaterials," *Am. Biotech. Lab.*, 7(6):17.

Cazes, J. and K. Nunugaki. 1987. "Centrifugal Partition Chromatography," *Am. Lab. Feb.*

Chang, H. N., B. H. Chung and I. H. Kim. 1986. "Dual Hollow Fiber Bioreactor for Aerobic Whole Cell Immobilization," in *Separation, Recovery and Purification in Biotechnology.* J. A. Asenjo and J. Hong, eds. Washington, DC: *Am. Chem. Soc.*, pp. 32–42.

Chang, H. N., I. S. Joo and Y. S. Ghim. 1984. "Performance of Rotating Packed Disc Reactor with Immobilized Glucose Oxidase," *Biotechnol. Lett.*, 6(8):487.

Charles, M. 1985. "Fermenter Design and Scale Up," in *Comprehensive Biotechnol. Vol. 2.* M. Moo-Young, ed. New York: Pergamon Press, pp. 57–74.

Charm, S. E. 1978. "Fundamentals of Food Engineering." Westport, CT: AVI Pub. Co.

Chassy, B. 1986. "Lactobacilli in Food Fermentations in Biotechnology," in *Food Processing.* S. K. Harlander and T. P. Labuza, eds. Park Ridge, NJ: Noyes Pub., pp. 197–207.

Chassy, B. M., L. J. Lee, J. B. Hansen and G. Jagusztyn-Krynicka. 1983. "Molecular Cloning of *Lactobacillus casei* Lactose Metabolic Genes," in *Developments in Industrial Microbiology, 24.* C. H. Nash and L. A. Underkofler, eds. Arlington, VA: Soc. of Ind. Microbiology, pp. 71–77.

Chassy, B. M., L. J. Lee-Wickner and E. V. Porter. 1983. "Molecular Characterization of Plasmids and Selected Chromosomal Genes Isolated from *Lactobacillus* Species," *Lactic Acid Bacteria in Foods Symp., Wageningen, the Netherlands.*

Cheetham, P. S. J. 1985. "Principles of Industrial Enzymology: Basis of Utilization of Soluble and Immobilized Enzymes in Industrial Processes," in *Enzyme Biotechnology.* A. Wiseman, ed. UK: Ellis Horwood Ltd., pp. 74–86.

Chen, H. C. and R. R. Zall. 1982. "Continuous Fermentation of Whey into Alcohol Using an Attached Film Expanded Bed Reactor," *Process Biochem.*, 17(1):20.

Cheryan, M. and M. A. Mehaia. 1986. "Membrane Bioreactors," in *Membrane Separations in Biotechnology.* W. C. McGregor, ed. New York: Marcel Dekker, Inc., pp. 255–301.

Chibata, I. and T. Tosa. 1976. "Industrial Applications of Immobilized Enzymes and Immobilized Microbial Cells," in *Applied Biochemical and Bioengineering, Vol. 1,*

*Immobilized Enzyme Principles*. L. B. Wingard, E. Katchalski-Katzir and L. Goldstein, eds. New York: Academic Press, pp. 329–357.

Chibata, I. 1978. *Immobilized Enzymes*. New York: John Wiley.

Chilgwin, J. M., A. E. Przybyla, R. J. MacDonald and W. J. Rutter. 1978. "Isolation of Biologically Active Ribonucleic Acid from Sources Enriched in Ribonuclease," *Biochemistry*, 18:5294.

Chmiel, H. 1987. "Downstream Processing in Biotechnology," in *Biochemical Engineering*. H. Chmiel, W. P. Hammes and J. E. Bailey, eds. New York: Gustav Fischer Verlag, pp. 242–251.

Christoffersen, R. E., M. L. Tucker and G. G. Laties. 1984. "Cellulase Gene Expression in Ripening Avocado Fruit: The Accumulation of Cellulase mRNA and Protein as Demonstrated by cDNA Hybridization and Immunodetection," *Plant Molecular Biology*, 3:385.

Chua, J. M., A. Erarslan, S. Kinoshita and A. Taguchi. 1980. "2,3-Butanediol Production by Immobilized Enterobacter Aerogenes IAM1133 with *x*-Carrageenan," *J. Ferm. Technol.*, 58(2):123.

Cohen, S. N., A. C. Y. Chang and L. Hsu. 1972. "Nonchromosomal Antibiotic Resistance in Bacteria: Genetic Transformation of *Escherichia coli* by R-factor DNA," *Proc. Natl. Acad. Sci.*, 69:2110.

Colbère-Garapin, F., F. Horodniceanu, P. Kourilsky and A. C. Garapin. 1981. "A New Dominant Hybrid Selective Marker for Higher Eukaryotic Cells," *J. Mol. Biol.*, 150:1.

Collins-Thompson, D. L., J. D. Cunningham and J. T. Trevors. 1985. "Food Microbiology and Biotechnology – An Update," in *Biotechnology Applications and Research*. P. N. Cheremisinoff and R. P. Ouellette, eds. Lancaster, PA: Technomic Pub. Co., Inc., pp. 188–195.

Cooney, C. L. 1982. "Growth of Microorganisms," in *Biotechnology*. H. J. Rehm and G. Reed, eds. Weinheim: Verlag Chimie., pp. 73–97.

Cooney, C. L. 1983. "Bioreactors: Design and Operation," *Science*, 219:728.

Cooney, C. L., D. I. C. Wang and R. I. Mateles. 1969. "Measurement of Heat Evolution and Correlation with Oxygen Consumption During Microbial Growth," *Biotechnol. Bioeng.*, 6(3):269.

Cornelis, P., C. Digneffe and K. Willemot. 1982. "Cloning and Expression of a *Bacillus coagulans* Amylase Gene in *E. coli*," *Mol. Gen. Genet.*, 186:507.

Coventry, J., R. Carzino and M. W. Hickey. 1988. "Meat Starter Cultures in Salami Manufacture," *Proc. 34th Int. Congress of Meat Sci. & Technol.*, Brisbane, Australia, pp. 368–370.

Cowan, D. A., R. M. Daniel, A. M. Martin and H. W. Morgan. 1984. "Some Properties of a $\beta$-Galactosidase from an Extremely Thermophilic Bacterium," *Biotechnol. Bioeng.*, 26:1141.

Cullen, D., G. L. Gray, L. J. Wilson, K. J. Hayenga, M. H. Lamsa, M. W. Rey, S. Norton and R. M. Berka. 1987. "Controlled Expression and Secretion of Bovine Chymosin in *Aspergillus nidulans*," *Biotechnol.*, 5:369.

Daeschel, M. A. 1989. "Antimicrobial Substances from Lactic Acid Bacteria for Use as Food Preservatives," *Food Technol.*, 43(1):164.

Danielsson, B. 1985. "Enzyme Probes," in *Comprehensive Biotechnology*. M. Moo-Young, C. W. Robinson and J. A. Howell, eds. New York: Pergamon Press, pp. 395–422.

Danielsson, B., I. Lundstrom, K. Mosbach and L. Stiblert. 1979. "On a New Enzyme Transducer Combination: The Enzyme Transistor," *Anal. Lett.*, 12:1189.

Danziger, R. 1979. "Distillation Columns with Vapor Recompression," *Chem. Eng. Prog.*, pp. 58−64.

Davidson, V. 1990. "Quinoproteins: A New Class of Enzymes with Potential Use as Biosensors," *Am. Biotechnol. Lab.*, 8(2):32.

Dechow, F. J. 1983. "Ion Exchange," in *Fermentation and Biochemical Engineering Handbook*. H. C. Vogel, ed. Park Ridge, NJ: Noyes Pub., pp. 202−226.

Deininger, P. L. 1983. "Random Subcloning of Sonicated DNA: Applications to Shotgun DNA Sequence Analysis," *Anal. Biochem.*, 129:216.

deMan, J. C., M. Rogosa and M. E. Sharpe. 1960. "A Medium for the Cultivation of *Lactobacilii*," *J. Appl. Bacteriol.*, 23(1):130.

deRosnay, J. 1985. "From Biotechnology to Biotics: The Engineering of Molecular Machine," in *Biotechnology: Applications and Research*. P. N. Cheremisinoff and R. P. Ouellette, eds. Lancaster, PA: Technomic Pub. Co., Inc., pp. 1−8.

Devivo, J. F. 1983. "Nonadiabatic Drying," in *Fermentation and Biochemical Engineering Handbook*. H. C. Vogel, ed. Park Ridge, NJ: Noyes Pub., pp. 317−330.

Dick, J. E., M. C. Magli, R. A. Philips and A. Bernstein. 1986. "Genetic Manipulation of Hemotopoietic Stem Cells with Retroviral Vectors," *Trends Genet.*, 2:165.

Donofrio, D. M. 1989. "Use of Hollow Fiber Bioreactors to Facilitate Manufacture of Cell Products," *Am. Biotechnol. Lab.*, 7(9):33.

Dore, C. 1987. "Application of Tissue Culture to Vegetable Crop Improvement," in *Plant Tissue and Cell Culture*. C. E. Green, D. A. Somers, W. P. Hackett and D. D. Biesboer, eds. New York: Alan R. Liss, Inc., pp. 419−433.

Dretzen, G., M. Bellard, P. Sassone-Corsi and P. Chambon. 1981. "A Reliable Method for the Recovery of DNA Fragments from Agarose and Acrylamide Gels," *Anal. Biochem.*, 112:295.

Drioli, E. 1986. "Membrane Processes in the Separation, Purification, and Concentration of Bioactive Compounds from Fermentation Broths," in *Separation, Recovery and Purification in Biotechnology*. J. A. Asenjo and J. Hong, eds. Washington, DC: *Am. Chem. Soc.*, pp. 52−66.

Duncan, B. K., P. A. Rockstroh and H. R. Warner. 1978. "*Escherichia coli* K-12 Mutants Deficient in Uracil-DNA Glycosylase," *J. Bacteriol.*, 134:1039.

Dunlop, E. H., W. A. Feiler and M. J. Mattione. 1984. "Magnetic Separation in Biotechnology," *Biotech. Advs.*, 2:63.

Dziezak, J. D. 1986. "Biotechnology and Flavor Development−Enzyme Modification of Dairy Products," *Food Technol.*, 40(4):114.

Easson, D. D., Jr., O. P. Peoples and A. J. Sinskey. 1986. "Genetic Studies on *Zoogloea ramigera* and Chemical Characterization of $\beta$(1-6) Glucosidic Linkages." Quoted by Sinskey et al., *Biotech. Bioeng.*, 1986.

Edens, L., I. Bom, A. M. Ledeboer, J. Matt, M. Y. Toonen, C. Visser and C. T. Verrips. 1984. "Synthesis and Processing of the Protein Thaumatin in Yeast," *Cell*, 37:629.

Einsele, A. 1978. "Scaling Up Bioreactors," *Process Biochem.*, 13:13.

Elder, J. T., R. A. Spritz and S. M. Weissman. 1981. "Simian Virus 40 as a Eukaryotic Cloning Vehicle," *Annu. Rev. Genet.*, 15:295.

Enfors, S. O. 1981. "Oxygen-Stabilized Enzyme Electrode for D-Glucose Analysis in Fermentation Broths," *Enzyme Microb. Technol.*, 3:29.

Ervin, J. L., J. Geigert, S. L. Neidleman and J. Wadsworth. 1984. "Substrate Dependent and Growth Temperature Dependent Changes in the Wax Ester Compositions Produced by Acinetobacter sp. H01-N," in *Biotechnology for the Oils and Fats Industry*. C. Ratledge, P. Dawson and J. Rattray, eds. *Am. Oil Chemists Soc.*, pp. 217–222.

Estell, D. A., T. P. Graycar and J. A. Wells. 1985. "Engineering an Enzyme by Site-Directed Mutagenesis to be Resistant to Chemical Oxidation," *J. Biol. Chem.*, 260:6518.

Evans, D. A. 1988. "Applications of Somaclonal Variation," *Adv. In. Biotech. Processes*, 9:204.

Evans, D. and W. Sharp. 1986. "Potential Applications of Plant Cell Culture," in *Biotechnology in Food Processing*. S. K. Harlander and T. P. Labuza, eds. Park Ridge, NJ: Noyes Pub., pp. 133–143.

Evans, D. A., W. R. Sharp and C. E. Flick. 1981. "Growth and Behavior of Cell Cultures—Embryogenesis and Organogenesis," in *Plant Tissue Culture—Methods and Applications in Agriculture*. T. A. Thorpe, ed. New York: Academic Press, pp. 45–114.

Evans, D. A. and R. J. Whitaker. 1987. "Technology for the Development of New Breeding Lines and Plant Varieties for the Food Industry," in *Food Biotechnology*. D. Knorr, ed. New York: Marcel Dekker, p. 320.

Fair, J. R. 1985. "Continuous Mass Transfer Processes," in *Scale-Up of Chemical Processes*. A. Bisio and R. L. Kabel, eds. New York: John Wiley & Sons, pp. 504–548.

Farahnak, F., T. Seki, D. D. Y. Ryu and D. Ogrydziak. 1986. "Construction of Lactose-Assimilating and High Ethanol Producing Yeasts by Protoplast Fusion," *Appl. Environ. Microbiol.*, 51:362.

Federoff, H. J., J. D. Cohen, T. R. Eccleshall, R. B. Needleman, B. A. Buchferer, J. Giacalone and J. Marmur. 1982. "Isolation of a Maltose Structural Gene from *Saccharomyces carlsbergensis*," *J. Bacteriol.*, 149(3):1064

Fein, J. E., H. G. Lawford, G. R. Lawford, B. C. Zawadzki and R. C. Charley. 1983. "High Productivity Continuous Ethanol Fermentation with a Flocculating Mutant Strain of *Zymonas mobilis*," *Biotech. Letters*, 5:19.

Feinberg, A. P. and B. Vogelstein. 1983. "A Technique for Radiolabeling DNA Restriction Endonuclease Fragments to High Specific Activity," *Anal. Biochem.*, 132:6.

Felgner, P. L., T. R. Gadek, M. Holm, R. Roman, H. W. Chan, M. Wenz, J. P. Northrop, G. M. Ringold and M. Danielsen. 1987. "Lipofection: A Highly Efficient, Lipid-Mediated DNA-Transfection Procedure," *Proc. Natl. Acad. Sci.*, 84:7413.

Felix, H. R. and K. Mosbach. 1982. "Enhanced Stability of Enzymes in Permeabilized and Immobilized Cells," *Biotechnol. Lett.*, 4(3):181.

Fillatti, J. J., J. Kiser, B. Rose and L. Comai. 1987. "Efficient Transformation of Tomato and the Introduction and Expression of a Gene for Herbicide Tolerance," in *Tomato Biotechnology*. D. J. Nevins and R. A. Jones, eds. New York: Alan R. Liss, Inc., pp. 199–210.

Finley, J. W., W. L. Stanley and G. G. Watters. 1977. "Removal of Chill Haze from Beer with Papain Immobilized on Chitin," *Biotechnol. Bioeng.*, 19:1895.

Finn, R. K. 1967. "Agitation and Aeration," in *Biochemical and Biological Engineering Science, Vol 1*. N. Blakeborough, ed. London: Academic Press.

Finn, R. K. and E. Ercoli. 1986. "A Membrane Reactor for Simultaneous Production of

Anaerobic Single-Cell Protein and Methane," in *Separation, Recovery and Purification in Biotechnology.* J. A. Asenjo and J. Hong, eds. Washington, DC: Am. Chem. Soc., pp. 43–51.

Finnerty, W. R. 1984. "The Application of Hydrocarbon-Utilizing Microorganisms for Lipid Production," in *Biotechnology for the Oils and Fats Industry.* C. Ratledge, P. Dawson and J. Rattray, eds. Am. Oil Chemists Soc., pp. 199–215.

Fitts, R. 1986. "Biosensors for Biological Monitoring," in *Biotechnology in Food Processing.* S. K. Harlander and T. P. Labuza, eds. Park Ridge, NJ: Noyes Pub., pp. 271–278.

Follettie, M. T. and A. J. Sinskey. 1986. "Recombinant DNA Technology for *Corynebacterium glutamicum,*" *Food Technol.,* 40(10):89.

Fowler, M. W., P. Bond and A. H. Scragg. 1987. "Developments in Plant Cell Culture Technology," in *Biochemical Engineering.* H. Chmiel, W. P. Hammes and J. E. Bailey, eds. New York: Gustav Fischer Verlag, pp. 333–341.

Frankenfeld, J. W., R. P. Cahn and P. Zoybekian. 1969. "Coagulating Microbial Cells to Enhance Their Separation." U.S. patent 3,427,223.

Freese, H. L. 1983. "Evaporation," in *Fermentation and Biochemical Engineering Handbook.* H. C. Vogel, ed. Park Ridge, NJ: Noyes Pub., pp. 227–276.

Friend, B. A. and K. M. Shahani. 1982. "Characterization and Evaluation of Aspergillus Oryzae Lactase Coupled to a Regenerable Support," *Biotechnol. Bioeng.,* 24:329.

Froment, G. F. 1985. "Homogeneous Reaction Systems," in *Scale-Up of Chemical Processes.* A. Bisio and R. L. Kabel, eds. New York: John Wiley & Sons., pp. 167–200.

Gamborg, O. L. 1984. "Plant Cell Cultures: Nutrition and Media," in *Cell Culture and Somatic Cell Genetics of Plants V.1.* I. K. Vasil, ed. New York: Academic Press, Inc.

Gamborg, O. L., J. P. Shyluk and E. A. Shahin. 1981. "Isolation, Fusion, and Culture of Plant Protoplasts," in *Plant Tissue Culture—Methods and Applications in Agric.* T. A. Thorpe, ed. New York: Academic Press, pp. 115–154.

Garrison, C. M. 1983. "How to Design and Scale Mixing Pilot-Plants," *Chem. Eng. Progress,* 79(2):63.

Gary, K. 1989. "On-Line Electrochemical Sensors in Fermentation," *Am. Biotechnol. Lab.,* 7(2):26.

Gasson, M. J. 1980. "Production, Generation and Fusion of Protoplasts in Lactic *Streptococci,*" *FEMS Microbiol. Lett.,* 9:99.

Gatfield, I. L. 1988. "Production of Flavor and Aroma Compounds by Biotechnology," *Food. Technol.,* 42(10):110.

Geller, A. I. and X. O. Breakfield. 1988. "A Defective HSV-1 Vector Expresses *Escherichia coli* β-Galactosidase in Cultured Peripheral Neurons," *Science,* 241:1667.

Genar, M. A. and R. Mutharasan. 1979. "Determination of Biomass Concentration by Capacitance Measurement," *Biotech. Bioeng.,* 21:1097.

Gerhardt, P. 1981. "Diluents and Biomass Measurement," in *Manual of Methods for General Bacteriology.* P. Gerhardt, ed. Washington, DC: Am. Soc. Microbiol.

Gerson, D., M. Kole, B. Ozum and Oguztoreli. 1988. "Substrate Concentration Control in Bioreactors," *Biotechnol. Genetic Eng.,* 6:67.

Gestrelius, S. 1982. *Enzyme Engineering, Vol. VI.* I. Chibata, S. Fukui and L. B. Wingard, Jr., eds. New York: Plenum Press, pp. 245–250.

Gibson, E. M., N. M. Chance, S. B. London and J. London. 1979. "Transfer of Plasmid-Medicated Antibiotic Resistance from *Streptococci* to *Lactobacilli*," *J. Bacteriol.*, 137:614.

Gilbert, E. 1987. "Production of Single Cell Protein (SCP) with Reduced Nucleic Acid Content," in *Biochemical Engineering*. H. Chmiel, W. P. Hammes and J. E. Bailey, eds. New York: Gustav Fischer Verlag, pp. 464–468.

Gilboa, E., M. A. Eglitis, P. W. Kantoff and W. F. Anderson. 1986. "Transfer and Expression of Cloned Genes Using Retroviral Vectors," *Biotechniques*, 4:504.

Glasgow, S. M. 1983. "Crystallization," in *Fermentation and Biochemical Engineering Handbook*. H. C. Vogel, ed. Park Ridge, NJ: Noyes Pub., pp. 277–295.

Glatz, B. A., E. G. Hammond, K. H. Hsue, L. Baehman, N. Bati, W. Bednarski, D. Brown and M. Floetenmeyer. 1984. "Production and Modification of Fats and Oils by Yeast Fermentation," in *Biotechnology for the Oils and Fats Industry*. C. Ratledge, P. Dawson and J. Rattray, eds. Am. Oil Chemists Soc., pp. 163–176.

Gobie, W. A. and C. F. Ivory. 1986. "High Resolution, High Yield Continuous-Flow Electrophoresis," in *Separation, Recovery and Purification in Biotechnology*. J. A. Asenjo and J. Hong, eds. Washington, DC: Am. Chem. Soc., pp. 169–184.

Goldsworty, A. 1988. "Growth Control in Plant Tissue Cultures." *Adv. In Biotechnol. Process*, 9:36.

Gonzalez, C. F. and B. S. Kunka. 1983. "Plasmid Transfer in *Pediococcus* spp.: Intergeneric and Intrageneric Transfer of PIP501," *Appl. Environ. Microbiol.*, 46:81.

Gorton, L. and L. Ogren. 1981. "Flow Injection Analysis for Glucose and Urea with Enzyme Reactors and On-Line Dialysis," *Anal. Chim. Acta.*, 130:45.

Grabner, R. 1986. "Separation Technology for Bioprocesses," in *Biotechnology in Food Processing*. S. K. Harlander and T. P. Labuza, eds. Park Ridge, NJ: Noyes Pub., pp. 237–248.

Graham, A. and M. Moo-Young. 1985. "Biosensors: Recent Trends," *Biotechnol. Adv.*, 3:209.

Graham, F. L. and A. J. van der Eb. 1973. "A Technique for the Assay of Infectivity of Human Adenovirus 5 DNA," *Virology*, 52:456.

Greco, G., Jr., D. Albanesi, D. M. Cantaralla and V. Scardi. 1980. "Experimental Technique for Multilayer Enzyme Immobilization," *Biotechnol. Bioeng.*, 22:215.

Green, C. E., D. A. Somers, W. P. Hackett and D. D. Biesboer. 1987. "Plant Tissue and Cell Culture." New York: Alan R. Liss, Inc.

Green, J. H. and A. Kramer. 1979. "Food Processing Waste Management." Westport, CT: AVI Publishing Co. Inc., pp. 184–188, 394–396.

Greenberg, N. A. and R. R. Mahoney. 1981. "Immobilization of Lactose ($\beta$-Galactosidase) for Use in Dairy Processing—A Review," *Process Biochem.*, 16(2):2.

Greenberg, N. A. and R. R. Mahoney. 1982. "Production and Characterization of $\beta$-Galactosidase from *Streptococcus thermophilus*," *J. Food Sci.*, 47:1824.

Griffith, W. L., A. L. Compare, C. G. Westmoreland and J. S. Johnson, Jr. 1981. "Separation of Biopolymer from Fermentation Broths," in *Synthetic Membranes, Vol. 2: Hyper- and Ultra-Filtration Uses, ACS Symposium Series #154*. F. Turbak, ed. Washington, DC: *Am. Chem. Soc.*, pp. 171–192.

Groginsky, C. M. and R. A. Houghten. 1985. "Peptide Synthesis," in *Biotechnology Applications and Research*. P. N. Cheremisinoff and R. P. Ouellette, eds. Lancaster, PA: Technomic Pub. Co., Inc., pp. 487–518.

Gubler, U. and B. J. Hoffman. 1983. "A Simple and Very Efficient Method for Generating cDNA Libraries," *Gene*, 25:263.

Gudernatsch, W., K. Kimmerle, H. Strathmann and H. Chmiel. 1987. "Continuous Removal of Ethanol from Fermentation Broths by Pervaporation," in *Biochemical Engineering*. H. Chmiel, W. P. Hammes and J. E. Bailey, eds. New York: Gustav Fischer Verlag, pp. 409–415.

Guilbault, G. G. 1982. "Ion Selective Electrodes Applied to Enzyme Systems," *Ion Selective Electrode Reviews*, 4:187–231.

Guilbault, G. G. and G. J. Lubrano. 1974. "Amperometric Enzyme Electrodes II. Amino Acid Oxidase," *Anal. Chim. Acta.*, 69:183.

Guilbault, G. G. and F. Shu. 1971. "An Electrode for the Determination of Glutamine," *Anal. Chim. Acta.*, 56:333

Guilbault, G. G. and A. Suleiman. 1990. "Piezoelectric Crystal Biosensors," *Am. Biotechnol. Lab.*, 8(3):28.

Guilbault, G. G. and M. Tarp. 1974. "A Specific Enzyme Electrode for Urea," *Anal. Chim. Acta.*, 73:355

Haas, M. J. 1984. "Methods and Applications of Genetic Engineering," *Food Technol.*, 38(2):69.

Hall, E. A. H. 1988. "The Developing Biosensor Arena," *Enzyme & Microbiol. Technol.*, 8(11):651.

Hamamci, H. et al. 1987. "Performance of Tapered Column Packed Bed Bioreactor for Ethanol Production," *Biotechnol. and Bioeng.*, 29:994.

Hames, B. D. 1981. "An Introduction to Polyacrylamide Gel Electrophoresis," in *Gel Electrophoresis of Proteins*. B. D. Hames and D. Rickwook, eds. London: IRL Press.

Hammes, W. P. 1987. "Biotechnology, Biochemical Engineering and Food Technology," in *Biochemical Engineering*. H. Chmiel, W. P. Hammes and J. E. Bailey, eds. New York: Gustav Fischer Verlag, pp. 11–35.

Hammond, E. G. and W. R. Fehr. 1984. "Progress in Breeding for Low Linolenic Acid Soybean Oil," in *Biotechnology for the Oils and Fats Industry*. C. Ratledge, P. Dawson and J. Rattray, eds. Am. Oil Chemists Soc., pp. 89–96.

Hanahan, D. 1983. "Studies on Transformation of *Escherichia coli* with Plasmids," *J. Mol. Biol.*, 166:577.

Hanisch, W. 1986. "Cell Harvesting," in *Membrane Separations in Biotechnology*. W. C. McGregor, ed. New York: Marcel Dekker Inc., pp. 61–88.

Hanish, J. and M. McClelland. 1988. "Activity of DNA Modifications and Restriction Enzymes in KGB, a Potassium Glutamate Buffer," *Gene Anal. Tech.*, 5:105.

Hara, S., Y. Iimura, H. Oyama, T. Kozeki, K. Kitano and K. Otsuka. 1981. "The Breeding of Cryophilic Killer Wine Yeasts," *Agric. Biol. Chem.*, 45:1327.

Harlander, S. K. and L. L. McKay. 1984. "Transformation of *Streptococcus sanguis* Challis with *Streptococcus lactis* Plasmid Deoxyribonucleic Acid," *Appl. Environ. Microbiol.*, 48:342.

Hattori, R., T. Hattori and C. Furusaka. 1972. "Growth of Bacteria on the Surface of Anion-Exchange Resin. I. Experiment with Batch Culture," *J. Gen. and Appl. Microbiol.*, 18:271.

Havas, J. and G. Guilbault. 1982. "Tyrosine-Selective Enzyme Probe and Its Application," *Anal. Chem.*, 54:1991.

Heath, C. and G. Belfort. 1987. "Modeling of Substrate Mass Transfer and Uptake in Hollow Fiber and Microcapsule Bioreactors," in *Biochemical Engineering*.

H. Chmiel, W. P. Hammes and J. E. Bailey, eds. New York: Gustav Fischer Verlag, pp. 307–326.

Hemrike-Wagner, A., K. C. M. Kruek and L. H. W. Van der Plas. 1982. "Influence of Growth Temperature on Respiratory Characteristics of Mitochondria from Callus-Forming Potato Tuber Discs," *Plant Physiol.*, 70:602.

Henikoff, S. 1984. "Unidirectional Digestion with Exonuclease III Creates Targeted Breakpoints for DNA Sequencing," *Gene*, 28:351.

Hershfield, V., H. W. Boyer, C. Yanofsky, M. A. Lovett and D. R. Helinski. 1974. "Plasmid CoIEI as a Molecular Vehicle for Cloning and Amplification of DNA," *Proc. Natl. Acad. Sci.*, 71:3455.

Heslot, H. 1985. "Cloning and Heterologous Expression of Thermostable α-Amylase Genes from *Bacillus* sp.," in *Biotechnology and Bioprocess Engineering*. T. K. Ghose, ed. New Delhi: I.I.T., pp. 489–492.

Hikuma, M., T. Kubo, T. Yasuda, I. Karube and S. Suzuki. 1979. "Amperometric Determination of Acetic Acid with Immobilized *Trichosporon brassicae*," *Analytica Chemica Acta*, 109:33.

Hikuma, M., T. Kubo, T. Yasuda, I. Karube and S. Suzuki. 1979. "Microbial Electrode Sensor for Alcohols," *Biotech. Bioeng.*, 21:1845.

Hikuma, M., H. Obana, T. Yasuda, I. Karube and S. Suzuki. 1980. "A Potentiometric Microbial Sensor Based Immobilized *E. coli* for Glutamic Acid," *Analytica Chimica Acta*, 116:61.

Hill, D. E., A. R. Oliphant and K. Struhl. 1987. "Mutagenesis with Degenrate Oligonucleotides: An Efficient Method for Saturating a Def Region with Base Pair Substitutions," *Methods Enzymol.*, 155:558.

Hill, F. F. 1987. "Dry Living Micro-Organisms-Products for the Food Industry," in *Biochemical Engineering*. H. Chmiel, W. P. Hammes and J. E. Bailey, eds. New York: Gustav Fischer Verlag, pp. 199–215.

Himmelblau, D. M. 1985. "Mathematical Modelling," in *Scale-Up of Chemical Processes*. A. Bisio and R. L. Kabel, eds. New York: John Wiley & Sons, pp. 34–76.

Hinnen, A., J. Hick and G. R. Fink. 1978. "Transformation of Yeast," *Proc. Natl. Acad. Sci.*, 75:1929.

Hitotsuyanagi, K., K. Yamane and B. Maruo. 1979. "Stepwise Introduction of Regulatory Genes Stimulating Production of α-Amylase Extrahyper Producing Strain," *Agric. Biol. Chem.*, 43:2343.

Hochhauser, S. J. and B. Weiss. 1978. "*Escherichia coli* Mutants Deficient in Deoxyuridine Triphosphate," *J. Bacteriol.*, 134:157.

Hofschneider, P. H. 1963. "Untersuchungen Iiber 'Kleine' *E. coli* K12 Bakterio-phagen. 1 und 2. Mitteilung. Z. Naturforsch." *B. Chem. Biochem. Biophys. Biol.*, 186: 203.

Hong, Y. C., S. K. Harlander, T. P. Labuza and P. E. Read. 1989. "Development of Tissue Culture from Immature Strawberry Fruits," presented at the Am. Soc. Agric. Eng. Annual Meeting.

Horitsu, H., Y. Takahashi, S. Adachi, R. Xioa, T. Hayashi, K. Kawai and H. Kautola. 1988. "Production of Organic Acids by Immobilized Cells of Fungi," in *Bioreactor Immobilized Enzymes and Cells—Fundamental and Applications*. M. Moo-Young, ed. New York: Elsevier Applied Sci., pp. 287–300.

Huggins, A. R. 1984. "Progress in Dairy Starter Culture Technology," *Food Technol.*, 38(6):41.

Hultin, H. O. 1983. "Current and Potential Uses of Immobilized Enzymes," *Food Technol.*, 37(10):66

Hunger, W. 1987. "Starter Culture for Food Industry—New Developments in Dairy Industry," in *Biochemical Engineering.* H. Chmiel, W. P. Hammes and J. E. Bailey, eds. New York: Gustav Fischer Verlag, pp. 216—227.

Hutchison, C. A., III, S. Phillips, M. H. Edgell, S. Gillam, P. Jahnke and M. Smith. 1978. "Mutagenesis at a Specific Position in a DNA Sequence," *J. Biol. Chem.*, 253:6551.

Huynh, T. V., R. A. Young and R. W. Davis. 1985. "Constructing and Screening cDNA Libraries in λgt10 and λgt11," in *DNA Cloning: A Practical Approach, Vol. 1.* D. M. Glover, ed. Oxford: IRL Press.

Huysman, P., P. Van Meenen, P. Van Assche and W. Verstraete. 1983. "Factors Affecting the Colonization of Non-Porous and Porous Packing Materials in Model Upflow Methane Reactors," *Biotech. Letters*, 5:643.

Ichijo, H, T. Suehiro, J. Nagasawa, A. Yamauchi and M. Sagesaka. 1985. "Enzyme Reactor with Knitted Fabric Made of Poly(Vinyl Alcohol) Superfine Filaments," *Biotechnol. Bioeng.*, 27:1077.

Ilker, J. 1987. "*In vitro* Pigment Production: An Alternative to Color Synthesis," *Food Technol.*, 41(4):70.

Illanes, A., M. E. Zuniga, R. Chamy and M. P. Marchese. 1988. "Immobilization of Lactase and Invertase on Crosslinked Chitin," in *Bioreactor Immobilized Enzymes and Cells—Fundamental and Applications.* M. Moo-Young, ed. New York: Elsevier Applied Sci., pp. 233—250.

Innis, M. A., K. B. Myambo, D. H. Gelfand and M. A. D. Brow. 1988. "DNA Sequencing with *Thermus aquaticus* DNA Polymerase and Direct Sequencing of Polymerase Chain Reaction-Amplified DNA," *Proc. Natl. Acad. Sci.*, 85:9436.

Inouye, M. and R. Sarma, eds. 1986. "Protein Engineering." New York: Academic Press.

Ish-Horowicz, D. and J. F. Burke. 1981. "Rapid and Efficient Cosmid Cloning," *Nucleic Acids Res.*, 9:2989.

Ishimori, Y., I. Karube and S. Suzuki. 1981. "Determination of Microbial Population with Piezoelectric Membranes," *Applied & Environ. Microbio.*, 42:632.

Ito, H., Y. Fukada, K. Murata and A. Kimura. 1983. "Transformation of Intact Yeast Cells Treated with Alkali Cations," *J. Bacteriol.*, 153:163.

Jimenez-Flores, R. and T. Richardson. 1987. "Effects of Chemical, Genetic and Enzymatic Modifications on Protein Functionality," in *Food Biotechnology I.* R. D. King and P. S. J. Cheetham, eds. London: Elsevier Applied Science, pp. 87—137.

Jirku, V., T. Maeck, T. Vanek, V. Krumphanzl and V. Kubanek. 1981. "Continuous Production of Steroid Glyco-Alkaloids by Immobilized Plant Cells," *Biotechnol. Lett.*, 3(8):447.

Joglekar, R., R. J. Clerman, R. P. Ouellette and P. N. Cheremisinoff, eds. 1983. "Biotechnology in Industry." Ann Arbor, MI: Ann Arbor Sci. Pub.

Johannsen, E., L. Eagle and G. Bredenhann. 1985. "Protoplast Fusion Used for Construction of Presumptive Polyploids of the D-Xylose Fermenting Yeast *Candida shehatae*," *Curr. Genet.*, 9:313.

Johnson, D. E. and A. Ciegler. 1969. "Substrate Conversion by Fungal Spores Entrapped in Solid Matrices," *Archiv. Biochem. Biophys.*, 130:384.

Johnson, J. C. 1979. "Immobilized Enzymes—Preparation and Engineering—Recent Advances," Park Ridge, NJ: Noyes Data Corp.

Johnson, M., G. Andre, C. Cavarie and J. Archambault. 1990. "Oxygen Transfer Rates in Mammalian Cell Culture Bioreactor Equipped with a Cell-Lift Impeller," *Biotechnol. and Bioeng.*, 35:43.

Jones, K. 1990. "Affinity Chromatography: A Technology Update," *Amer. Biotech. Lab.* (Oct.):26.

Junker, B. H., D. I. C. Wang and T. A. Halton. 1988. "Fluorescence Sensing of Fermentation Parameters Using Fiber Optics," *Biotechn. and Bioeng.*, 32:55.

Kabel, R. L. 1985. " (i) Homogeneous reaction systems. (ii) Selection of Reactor Types," in *Scale-Up of Chemical Processes*. A. Bisio and R. L. Kabel, eds. New York: John Wiley & Sons, pp. 117–166 and pp. 253–274.

Kachholz, T. and M. Schlingmann. 1987. "Possible Food and Agricultural Application of Microbial Surfactants–An Assessment," in *Biosurfactants and Biotechnology*. N. Kosaric, W. L. Cairns and N. C. C. Gray, eds. New York: Marcel Dekker, Inc., pp. 183–210.

Kadonaga, J. T. and J. R. Knowles. 1985. "A Simple and Efficient Method for Chemical Mutagenesis of DNA," *Nucleic Acids Res.*, 13:1733.

Kamakura, M. and I. Kaetsu. 1984. "Encapsulation of Enzymes by Radiation Technique," *Biotechnol. Lett.*, 6(7):409.

Kameya, T. 1979. "Studies on Plant Cell Fusion: Effect of Dextran and PronaseE on Fusion," *Cytologia*, 44:449.

Kampen, W. H. 1983. "Industrial Pilot Plant," in *Fermentation and Biochemical Engineering Handbook*. H. C. Vogel, ed. Park Ridge, NJ: Noyes Pub., pp. 48–76.

Kang, Y. and T. Richardson. 1985. "Genetic Engineering of Caseins," *Foods Technol.*, 39(10):89.

Kangas, T. T., C. L. Cooney and R. F. Gomez. 1982. "Expression of a Proline Enriched Protein in *E. coli*," *Appl. Env. Microbiol.*, 43:629.

Karube, I. 1984. "Possible Developments in Microbial and Other Sensors for Fermentation Control," in *Biotechnology and Genetic Engineering Reviews II*. G. E. Russell, ed. Newcastle upon Tyne, UK: Intercept, pp. 313–340.

Karube, I. 1985. "Biosensors in Fermentation and Environmental Control," in *Biotechnology Applications and Research*. P. N. Cheremisinoff and R. P. Ouellette, eds. Lancaster, PA: Technomic Pub. Co., Inc., pp. 135–155.

Karube, I., S. Mitsuda, T. Matsunaga and S. Suzuki. 1977. "A Rapid Method for Estimation of BOD by Using Immobilized Microbial Cells," *J. Fermentation Technol.*, 55:243.

Karube, I., S. Mitsuda and S. Suzuki. 1979. "Glucose Sensor Using Immobilized Whole Cells of *Pseudomonas fluorescenes*," *European J. Applied Microbiol. & Biotech.*, 7:343.

Karube, I., I. Satoh, Y. Araki, S. Suzuki and H. Yamada. 1980. "Monoamine Oxidase Electrode in Freshness Testing of Meat," *Enzyme Microb. Technol.*, 2:117.

Kaul, R., S. F. D'Souza and G. B. Nadkarni. 1984. "Hydrolysis of Milk Lactose by Immobilized β-Galactosidase–Hen Egg White Powder," *Biotechnol. Bioeng.*, 26:901–904.

Kaup, E. C. 1973. "Design Factors in Reverse Osmosis," *Chem. Eng.* (April): 46–53.

Kennedy, J. F. 1985. "Principles of Immobilization of Enzymes," in *Enzyme Biotechnology*. A. Wiseman, ed. London: Ellis Horwood Ltd., pp. 147–200.

Kennedy, J. F., J. D. Humphreys, S. A. Barker and R. N. Greenshields. 1980. "Application of Living Immobilized Cells to the Acceleration of the Continuous Conversions of

Ethanol (Wort) to Acetic Acid (Vinegar)—Hydrous Titanium Oxide—Immobilized *Acetobacter* sp.,'' *Enzyme and Microbial Tech.*, 2:209.

Kennedy, J. F. and C. A. White. 1985. "Principles of Immobilization of Enzymes," in *Handbook of Enzyme Biotechnology*. A. Wiseman, ed. Chichester, UK: Ellis Horwood Ltd., pp. 147–207.

Kessler, C., P. S. Neumaier and W. Wolf. 1985. "Recognition Sequences of Restriction Endonucleases and Methylases—A Review," *Gene*, 33:1.

Keyes, M. H., F. E. Semersky and D. N. Gray. 1979. "Glucose Analysis Utilizing Immobilized Enzymes," *Enzyme Microb. Technol.*, 1:91.

King, R. D. and P. S. J. Cheetham. 1987. *Food Biotechnology I*. London: Elsevier Applied Science.

Klaenhammer, T. R. 1984. "A General Method for Plasmid Isolation in *Lactobacilli*," *Curr. Microbiol.*, 10:23.

Klein, T. M., E. D. Wolf, R. Wu and J. C. Sanford. 1987. "High-Velocity Microprojectiles for Delivering Nucleic Acid into Living Cells," *Nature*, 327:70.

Knight, P. 1989. "Downstream Processing," *Biotechnol.*, 7(8):777.

Knorr, D., M. D. Beaumont, C. S. Caster, H. Dornenburg, B. Gross, Y. Pandya and L. G. Romagnoli. 1990. "Plant Tissue Culture for the Production of Naturally Derived Food Ingredients," *Food Technol.*, 44(6):71.

Kobayashi, T., T. Yano, S. Shimiya. 1980. "Automatic Control of D. O. Concentration with a Microcomputer," in *Advances in Biotechnology, Vol. 1*. Toronto: Pergamon Press.

Koch, A. L. 1981. "Growth Measurement," in *Manual of Methods for General Bacteriology*. P. Gerhardt, ed. Washington, DC: *Am. Soc. Microbiol.*, pp. 179–207.

Kok, J., J. M. B. M. Van der Vossen and G. Venema. 1984. "Construction of Plasmid Cloning Vectors for Lactic *Streptococci* which Also Replicate in *Bacillus subtilis* and *E. coli*," *Appl. Environ. Microbiol.*, 48:726.

Kok, J., J. M. Van Dijl, J. M. B. M. van der Vossen and G. Venema. 1985. "Cloning and Expression of a *Streptococcus cremoris* Proteinase in *Bacillus subtilis* and *Streptococcus lactis*," *Appl. Environ. Microbiol.*, 50:94.

Kolot, F. B. 1988. "Immobilized Microbial Systems—Principles, Techniques and Industrial Applications." Malabar, FL: R. E. Krieger Pub. Co.

Kolpackni, A. P., V. S. Isaeva, A. Y. Zhvirblyanskaya, E. N. Kazantsev, E. N. Serova and N. N. Raltel. 1976. "Binding of Brewing Yeasts to Polymeric Materials," *Appl. Biochem. Micro.*, 12(6):703.

Kondo, J. K. and L. L. McKay. 1982. "Transformation of *Streptococcus lactis* Protoplasts by Plasmid DNA," *Appl. Environ. Microbiol.*, 43:1213.

Kondo, J. K. and L. L. McKay. 1984. "Plasmid Transformation of *Streptococcus lactis* Protoplasts: Optimization and Use in Molecular Cloning," *Appl. Environ. Microbiol.*, 48:252.

Kondo, J. K. and L. L. McKay. 1985. "Gene Transfer Systems and Molecular Cloning in Group *N. Streptococci*: A Review," *J. Dairy Sci.*, 68:2143.

Konrad, E. B. and I. R. Lehman. 1975. "Novel Mutants of *Escherichia coli* that Accumulate Very Small DNA Replicative Intermediates," *Proc. Natl. Acad. Sci.*, 72:2150.

Kosaric, N. 1984. "Possible Industrial Applications of Biotechnology," in *Biotechnology for the Oils and Fats Industry*. C. Ratledge, P. Dawson and J. Rattray, eds. Am. Oil Chemists Soc., pp. 55–69.

Kosaric, N., N. C. C. Gray and W. L. Cairns. 1987. "Introduction: Biotechnology and the Surfactant Industry," in *Biosurfactants and Biotechnology*. N. Kosaric, W. L. Cairns and N. C. C. Gray, eds. New York: Marcel Dekker, Inc.

Kossen, N. W. F. 1985. "Problems in the Design of Large Scale Bioreactors," in *Biotechnology and Bioprocess Engineering*. T. K. Ghose, ed. New Delhi: I.I.T., pp. 365–380.

Koyama, M., Y. Sato, M. Aizawa and S. Suzuki. 1980. "Improved Enzyme Sensor for Glucose with an Ultrafiltration Membrane and Immobilized Glucose Oxidase," *Anal. Chim. Acta*, 116:307.

Kulys, J. and K. Kadzianskiene. 1980. "Yeast BOD Sensor," *Biotech. Bioeng.*, 22:221.

Kunkel, T. A. 1985. "Rapid and Efficient Site-Specific Mutagenesis without Phenotypic Selection," *Proc. Natl. Acad. Sci.*, 82:488.

Kunkel, T. A., J. D. Roberts and R. A. Zakour. 1987. "Rapid and Efficient Site-Specific Mutagenesis without Phenotypic Selection," *Methods Enzymol.*, 154:367.

Kyte, L. 1983. *Plants from Test Tubes*. Portland, OR: Timber Press.

Laemmli, U. K. 1970. "Cleavage of Structural Proteins during the Assembly of the Head of Bacteriophage T4," *Nature*, 227:680.

Lahey, A. and R. L. St. Claire. 1990. "A Comparison of Ion-Pairing LC and MECC in the Separation of Nucleosides and Nucleotides," *Amer. Lab.* (Nov.):68.

Larson, A. 1974. "Various Approaches to the Separation Process for Harvesting the Products of Fermentation in the Field of Antibiotics," in *Advances in Microbial Engineering. Biotech. and Bioeng. Symposium #4*. B. Sikyata, A. Prokop and M. Novak, eds. New York: John Wiley, pp. 917–931.

Lasky, S. J. and D. A. Buttry. 1990. "Development of a Real Time Glucose Biosensor by Enzyme Immobilization on the Quartz Crystal Microbalance," *Am. Biotechnol. Lab.*, 8(2):8.

Ledeboer, A. M., C. T. Verrips and B. M. M. Dekker. 1984. "Cloning of the Natural Gene for the Sweet-Tasting Plant Protein Thaumatin," *Gene*, 30:23.

Lee, L. J., J. B. Hansen, E. K. Jagusztyn-Krynicka and B. M. Chassy. 1982. "Cloning and Expression of the $\beta$-D-Phosphogalactoside Galactohydrolase Gene of *Lactobacillus casei* in *E. coli* K-12," *J. Bacteriol.*, 152:1138.

Lee, T. S. 1979. "Membrane Processes for the Treatment of Process Streams in the Corn Wet Milling and Fermentation Industries." Ph.D. Dissertation, Columbia University, New York.

Lee-Wickner, L. J. and B. M. Chassy. 1985. "Characterization and Molecular Cloning of Cryptic Plasmids Isolated from *Lactobacillus casei*," *Appl. Environ. Microbiol.*, 49:1154.

Leonil, J., S. Sicsic, J. Braun and F. Goffic. 1984. "A New Photoactivable Support for Protein Immobilization: Fixation of $\beta$-D-Galactosidase on Modified Chitosan," *Enzyme Microb. Technol.*, 6:517.

LePecq, J. B. and C. Paleotti. 1967. "A Fluorescent Complex between Ethidium Bromide and Nucleic Acids," *J. Mol. Biol.*, 27:87.

Lichtenstein, C. and T. Draper. 1985. "Genetic Engineering of Plants," in *DNA Cloning, Vol II—A Practical Approach*. D. M. Glover, ed. Oxford: IRL Press.

Lin, J. H. C. and D. C. Savage. 1985. "Cryptic Plasmid in *Lactobacillus* Strains Isolated from the Murine Gastrointestinal Tract," *Appl. Environ. Microbiol.*, 49:1004.

Lin, Y. L. 1986. "Genetic Engineering and Process Development for Production of Food Processing Enzymes and Additives," *Food Technol.*, 40(10):104.

Lindahl, T. 1974. "An *N*-Glycosidase from *Escherichia coli* that Release Free Uracil from DNA Containing Delaminated Cytosine Residue," *Proc. Natl. Acad. Sci.*, 71:3649.

Lindblom, M. G. and H. L. Mogren. 1976. "Process for Preparing from a Microbial Cell Mass a Protein Concentrate Having a Low Nucleic Acid Content, and the Protein Concentrate Thus Obtained," U.S. patent 3,996,104.

Lindsay, K., M. M. Yeoman, G. M. Black and F. Mavituna. 1983. "A Novel Method for the Immobilization and Culture of Plant Cells," *FEBS Lett.*, 155(1):143.

Lindsey, K. and M. M. Yeoman. 1987. "Techniques for Immobilization of Plant Cells," in *Methods in Enzymology, Vol. 135*. New York: Academic Press, p. 140.

Linek, V., P. Benes and V. Vacek. 1988. "Dynamic Pressure Method for $K_La$ Measurement in Large Scale Bio-Reactors," *Biotechnol. Bioeng.*, 33:1406.

Linko, P. 1980. "Enzyme Engineering in Food Processing," in *Food Process Engineering, Vol. II*. New York: Applied Sci. Pub., pp. 27–39.

Linko, P. 1985. "Immobilized Lactic Acid Bacteria," in *Enzymes and Immobilized Cells in Biotechnology*. A. I. Laskin, ed. Don Mills, Ont., Canada: The Benjamin/Cummings Pub. Co., pp. 25–36.

Linko, Y. Y., H. Kautola, S. Votila and P. Linko. 1986. "Alcoholic Fermentation of D-Xylose by Immobilized *Pichia stipitis* Yeast," *Biotech. Lett.*, 8:47.

Liptak, B. G., and K. V. Venczel. 1982. *Instrument Engineers' Handbook: Process Measurement*. Radnor: Chilton Book Company.

Loh, V. Y., S. R. Richards and P. Richmond. 1986. "Particle Suspension in a Circulating Bed Fermenter," *Chem. Eng. J.*, 32:B39.

Lorz, H., B. Junker, J. Schell and A. dela Pena. 1987. "Gene Transfer in Cereals," in *Plant Tissue and Cell Culture*. C. E. Green, D. A. Somers, W. P. Hackett and D. D. Biesboer, eds. New York: Alan R. Liss, Inc., pp. 303–316.

Lowe, C. R., M. J. Goldfinch and R. T. Lias. 1983. "Some Novel Biomedical Biosensors," in *Biotech 83: Proc. of the Int. Conf. on the Commercial Applications of Biotechnology*. Northwood, London: Online Pub., pp. 633–641.

Lubrano, G. J. and G. G. Guilbault. 1978. "Glucose and L-Amino Acid Electrode Based on Enzyme Membranes," *Anal. Chim. Acta.*, 97:229

Ludwig, K. and K. O'Shaughnessey. 1989. "Tangential Flow Filtration—A Technical Review," *Am. Biotechnol. Lab.*, 7(6):41.

Lu-Kwang, J., C. S. Ho and R. F. Baddour. 1988. "Simultaneous Measurement of Oxygen Diffusion Coefficients and Solubilities in Fermentation Media with Plarographic Oxygen Electrodes," *Biotechnol. Bioeng.*, 31:995.

Lund, D. B. 1987. "Food Processing Techniques Tap Gains in Biotechnology," *Ag. Eng.*, 68(12):7.

Luong, J. H. T., K. B. Male and A. L. Nguyen. 1988. "Development of A Fish Freshness Sensor," *Am. Biotechnol. Lab.*, 6(6):38.

Mackett, M., G. L. Smith and B. Moss. 1984. "A General Method for the Production and Selection of Infectious Vaccinia Virus Recombinants Expressing Foreign Genes," *J. Virol.*, 49:856.

Macrae, A. R. 1984. "Microbial Lipases as Catalysts for the Inter-Esterification of Oils and Fats," in *Biotechnology for the Oils and Fats Industry*. C. Ratledge, P. Dawson and J. Rattray, eds. Am. Oil Chemists Soc., pp. 189–198.

Maki, M., M. Nagao, M. Hirose and H. Chiba. 1988. "Cloning of cDNA Sequence Coding for Bovine $\alpha_{s1}$-Casein," *Agric. Biol. Chem.*, 47:441.

Manfredini, R., V. Cavallera, L. Maini and R. Donati. 1983. "Mixing and Oxygen Transfer in Conventional Stirred Fermenters," *Biotech. Bioeng.*, 25(12):3115.

Maniatis, T., E. F. Fritsch and J. Sambrook. 1982. "Molecular Cloning—A Laboratory Manual." Cold Spring Harbor, New York: Cold Spring Harbor Laboratory.

Manjon, A., J. Bastida, C. Romero, A. Jimeno and J. L. Iborra. 1985. "Immobilization of Narginase on Glycophase-Coated Porous Glass," *Biotechnol. Lett.*, 7(7):477.

Mannino, R. J. and S. Gould-Fogerite. 1988. "Liposome-Mediated Gene Transfer," *Biotechniques*, 6:682.

Margaritis, A. and J. B. Wallace. 1982. "The Use of Immobilized Cells of *Zymomonals mobilis* in A Novel Fluidized Bioreactor to Produce Ethanol," in *Biotechnology and Bioengineering Symposium #12*. E. L. Gaden, ed. New York: John Wiley, pp. 147–159.

Margaritis, A. and J. B. Wallace. 1984. "Novel Bioreactor Systems and Their Applications," *Biotech.*, 2(5):447.

Margaritis, A. and C. R. Wilke. 1978. "The Roto-Fermenter. Description of the Apparatus, Power Requirement and Mass Transfer Characteristics," *Biotech. and Bioeng.*, 20:709.

Mark, H. F., D. F. Othmer, C. G. Overberger and S. T. Seaborg. 1980. "Encyclopedia of Chemical Technology, Vol. 9." New York: John Wiley and Sons.

Marston, F. A. O., P. A. Lowe, M. T. Doel, J. M. Schoemaker, S. White and S. Angel. 1984. "Purification of Calf Prochymosin (Prorennin) Synthesized in *E. coli*," *Biotechnol.*, 2:800.

Mateles, R. I. 1971. "Calculation of Oxygen Required for Cell Production," *Biotechnol. Bioeng.*, 13:581.

Matsumura, M., M. Obara, H. Yoshitome and J. Kobayashi. 1972. "Oxygen Equilibrium Distribution and Its Transfer in an Air-Water-Oil System," *J. Ferment. Technol.*, 50:742.

Matsunaga, T., I. Karube and S. Suzuki. 1978. "Rapid Determination of Nicotinic Acid by Immobilized *Lactobacillus arabinosus*," *Analytica Chimica Acta*, 99:233.

Matsunaga. T., I. Karube and S. Suzuki. 1979. "Electrode System for the Determination of Microbial Population," *Applied and Environ. Microbiol.*, 37:117.

Mattiasson, B. 1983. *Immobilized Cells and Organelles, Vols. 1 and 2*. Boca Raton: CRC Press.

Mattiasson, B. and R. Kaul. 1986. "Use of Aqueous Two Phase Systems for Recovery and Purification in Biotechnology," in *Separation, Recovery and Purification in Biotechnology*. J. A. Asenjo and J. Hong, eds. Washington, DC: *Am. Chem. Soc.*, pp. 78–92.

Mattingly, J. A., B. J. Robinson, A. Boehm and W. D. Gehle. 1985. "Use of Monoclonal Antibodies for the Detection of Salmonella in Foods," *Food Technol.*, 39:90.

McCabe, W. L., J. C. Smith and P. Harriott. 1985. *Unit Operations of Chemical Engineering*. New York: McGraw Hill Book Co.

McCutchan, J. H. and J. S. Pagano. 1968. "Enhancement of the Infectivity of Simian Virus to Deoxyribonucleic Acid with Diethyl Aminoethyl-Dextran," *J. Natl. Cancer Inst.*, 41:351.

McGregor, W. C. 1986. "Selection and Use of Ultrafiltration Membranes," in *Membrane Separations in Biotechnology*. W. C. McGregor, ed. New York: Marcel Dekker, Inc., pp. 1–36.

McKay, L. 1986. "Application of Genetic Engineering Techniques for Dairy Starter

Culture Improvement," in *Biotechnology in Food Processing*. S. K. Harlander and T. P. Labuza, eds. Park Ridge, NJ: Noyes Pub., pp. 145–155.

McKay, L. L. and K. A. Baldwin. 1984. "Conjugative 40-Megadalton Plasmid in *Streptococcus lactis* subsp. Diacetyl Lactis DRC3 Is Associated with Resistance to Nisin and Bacteriophage," *Appl. Environ. Microbiol.*, 47:68.

McKay, L. L., K. A. Baldwin and P. M. Walsh. 1980. "Conjugal Transfer of Genetic Information in Group N. *Streptococci*," *Appl. Environ. Microbiol.*, 40:84.

McKay, L. L., B. R. Cords and K. A. Baldwin. 1973. "Transduction of Lactose Metabolism in *Streptococcus lactis* C2," *J. Bacteriol.*, 115:810.

McKersie, B. D., T. Senaratna and S. R. Bowley. 1990. "Artificial Seeds," *Highlights* (Sept.):2.6.

Mellor, J., J. J. Dobson, N. A. Roberts, M. F. Tuite, J. S. Emtage, S. White, P. A. Lowe, T. Patel, A. J. Kingman and S. M. Kingman. 1983. "Efficient Synthesis of Enzymatically Active Calf Chymosin in Saccharomyces Cerevisiae," *Gene*, 24:1.

Mermelstein, N. H. 1989. "Continuous Fermenter Produces Natural Flavor Enhancers for Foods and Pet Foods," *Food Tech.*, 39(7):50.

Merten, O. W., G. E. Palfi and J. Steiner. 1986. "Online Determination of Biochemical/Physiological Parameters in the Fermentation of Animal Cells in a Continuous or Discontinuous Mode," *Adv. Biol. Processes*, 6:111.

Messing, J. 1983. "New M13 Vectors for Cloning," *Methods Enzymol.*, 101:120.

Messing, J. and J. Vieira. 1982. "A New Pair of M13 Vectors for Selecting Either DNA Strand of Double-Digest Restriction Fragments," *Gene*, 19:269.

Mierendorf, R. C. and D. Pfeffer. 1987. "Direct Sequencing of Denatured Plasmid DNA," *Methods Enzymol.*, 152:556.

Mimura, A., T. Kawana and R. Kodaira. 1969. "Biochemical Engineering Analysis of Carbon Fermentation I. Oxygen Transfer in the Oil Water System," *J. Fermentation Technol.*, 47:229.

Mittal, S. K. 1989. "Analysis of Bovine Herpesvirus 1 Thymidine Kinase," Ph.D. thesis, University of Cambridge, Cambridge, UK.

Miyagawa, K., H. Kimura, K. Nakahama, M. Kikuchi, M. Doi, S. Akiyama and Y. Nakao. 1986. "Cloning of the *Bacillus subtilis* IMP Dehydrogenase Gene and Its Application to Increased Production of Guanosine," *Biotechnol.*, 4:225.

Mogren, H. L., G. O. Hedenskog and L. E. Enebo. 1974. "Process for Extracting Protein from Microorganisms," U. S. patent 3,848,812.

Moir, D. T., J. Mao, J. J. Duncan, R. A. Smith, and T. Kohno. 1985. "Production of Calf Chymosin by the Yeast *Saccharomyces cerevisiae*," *Dev. Ind. Microbiol.*, 26:75.

Moo-Young, M., G. Van Dedem and A. Binder. 1979. "Design of Scraped Tubular Fermenters," *Biotechnol. Bioeng.*, 21:593.

Morris, C. E. 1982. "New Developments in Whey Processing," *Food Eng.*, 54:69.

Morris, C. E. 1986. "Breakthroughs in Biotechnology," *Food Eng.*, 58(2):50.

Morris, C. E. and A. E. Przybyla. 1989. "Biotechnology Blooms," *Food Eng.*, 61(7):51.

Morris, V. J. 1987. "New and Modified Polysaccharides," in *Food Biotechnology I*. R. D. King and P. S. J. Cheetham, eds. London: Elsevier Applied Science, pp. 193–248.

Mosbach, K. 1983. "New Biosensor Devices," in *Biotech 83: Proc. of the Int. Conf. on the Commercial Applications of Biotechnology*. Northwood, London: Online Pub., pp. 665–678.

Moser, A. 1982. "Bioreactors with Thin Layer Characteristics," *Biotech. Lett.*, 4:281.

Moshy, R. J. 1985. "Impact of Biotechnology on Food Product Development," *Food Technol.*, 39(10):113.

Moss, B. and C. Flexner. 1987. "Vaccinia Virus Expression Vectors," *Ann. Rev. Immunol.*, 5:305.

Moulin, G. and P. Galzy. 1984. "Whey, A Potential Substrate for Biotechnology," in *Biotechnology and Genetic Engineering Reviews I.* G. E. Russell, ed. Newcastle upon Tyne, UK: Intercept, pp. 347–374.

Mulder-Krieger, T., R. Verpoorte, A. Baerheim and J. J. C Scheffer. 1988. "Production of Essential Oils and Flavors in Plant Cell and Tissue Cultures–A Review," *Plant Cell, Tissue, and Organ Cultures*, 13:85.

Mulligan, R. C. and P. Berg. 1980. "Expression of a Bacterial Gene in Mammalian Cells," *Science*, 209:1422.

Mulligan, T. J. and R. D. Fox. 1976. "Treatment of Industrial Wastewaters," *Chem. Eng.*, 18(10):60.

Mullins, M. G. 1987. "Propagation and Genetic Improvement of Temperate Fruits: The Role of Tissue Culture," in *Plant Tissue and Cell Culture*. C. E. Green, D. A. Somers, W. P. Hackett and D. D. Biesboer, eds. New York: Alan R. Liss, Inc., pp. 395–406.

Munson, C. L. and C. J. King. 1984. "Factors Influencing Solvent Selection for Extraction of Aqueous Solution," *Industrial Engineering and Chemistry Process Design and Development*, 23(1):109–115.

Myer, H., O. Kappeli and A. Fiechter. 1985. "Growth Control in Microbial Cultures," *Annual Review of Microbiology*, 39:299–319.

Nakanishi, K., T. Kamikubo and R. Matsuno. 1985. "Continuous Synthesis of N-benzyloxycarbonyl-L-aspartyl-L-phenylalanine Methyl Ester with Immobilized Thromolysin in an Organic Solvent," *Biotechnol.*, 3:459.

Nakanishi, K., R. Matsuno, K. Torii, K. Yamamoto and T. Kamikubo. 1983. "Properties of Immobilized $\beta$-D-Galactosidase from *Bacillus circulans*," *Enzyme. Microb. Technol.*, 5:115.

Nanjo, M. and G. G. Guilbault. 1974. "Enzyme Electrode for L-Amino Acids and Glucose," *Anal. Chim. Acta.*, 73:367.

Nauman, E. B. 1985. "Laminar Flow Processes," in *Scale-Up of Chemical Processes*. A. Bisio and R. L. Kabel, eds. New York: John Wiley & Sons, pp. 406–430.

Navarro, J. M. and G. Durand. 1977. "Modification of Yeast Metabolism by Immobilization onto Porous Glass," *European J. of Applied Microbiol.*, 4:2434.

Naveh, D. 1985. "Scale-Up of Fermentation for Recombinant DNA Products," *Food Technol.*, 39(10):102.

Neidleman, S. 1986. "Enzymology and Food Processing," in *Biotechnology in Food Processing*. S. K. Harlander and T. P. Labuza, eds. Park Ridge, NJ: Noyes Pub., pp. 37–56.

Neumann, E., M. Schaefer-Ridder, Y. Wang and P. H. Hofschneider. 1982. "Gene Transfer into Mouse Lyoma Cells by Electroporation in High Electric Fields," *EMBO J.*, 1:841.

Nevins, D. J. 1987. "Why Tomato Biotechnology? A Potential to Accelerate the Applications," in *Tomato Biotechnology*. D. J. Nevins and R. A. Jones, eds. New York: Alan R. Liss, Inc., pp. 3–16.

Newell, N. and S. Gordon. 1986. "Profit Opportunities in Biotechnology for the Food

Processing Industry," in *Biotechnology in Food Processing*. S. K. Harlander and T. P. Labuza, eds. Park Ridge, NJ: Noyes Pub., pp. 297–311.

Newton, R. 1989. "The Automation of DNA Synthesis," *Am. Biotechnol. Lab*, 41(5):45.

Nilsson, H., A. A. Kerlund and K. Mosbach. 1973. "Biochim. Biophys." *Acta*, 320:529.

Nilsson, K. 1987. "Mammalian Cell Culture," in *Methods in Enzymol., Vol 135*. New York: Academic Press.

Nilsson, K., W. Scheirer, H. W. D. Katinger and K. Mosbach. 1987. "Entrapment of Animal Cells," in *Methods in Enzymology, Vol. 135*. New York: Academic Press, pp. 399–420.

Norrander, J., T. Kempe and J. Messing. 1983. "Construction of Improved M13 Vectors Using Oligodeoxynucleotide-Directed Mutagenesis," *Gene*, 26:101.

Nunberg, J. H., R. J. Kaufman, A. C. Y. Chang, S. N. Cohen and R. T. Schimke. 1980. "Structure and Genomic Organization of the Mouse Dihydrofolate Reductase Gene," *Cell*, 19:355.

Ohkubo, H., G. Vogeli, M. Mudryj, V. E. Avvedimento, M. Sullivan, I. Pastan and B. de Crombrugghe. 1980. "Isolation and Characterization of Overlapping Genomic Clones Covering the Chiken $\alpha 2$ (Type 1) Collagen Gene," *Nucleic Acids Res.*, 8:1823.

Okamoto, T., Y. Fujita and R. Irie. 1983. "Fusion of Protoplasts of *Streptococcus lactis*," *Agric. Biol. Chem.*, 47:2675.

Okayama, H. and P. Berg. 1982. "High-Efficiency Cloning of Full Length cDNA," *Mol. Cell. Biol.*, 1:161.

Okayama, H., M. Kawaichi, M. Brownstein, F. Lee, T. Yokota and K. Arai. 1988. "High-Efficiency Cloning of Full-Length cDNA; Construction and Screening of cDNA Expression Libraries for Mammalian Cells," *Methods in Enzymol.*, 154:1.

Oldshue, J. Y. 1983. "Transport Phenomena, Reactor Design and Scale Up," *Biotechnol. Adv.*, 1:17.

Oldshue, J. Y. 1985. "Transport Phenomena," in *Reactor Design and Scale-Up. Vol. 3. Biotechnology Advance*, pp. 219–237.

Olechno, J. D., J. M. Y. Tso, J. Thayer and A. Wainright. 1990. "Capillary Electrophoresis I: Separations," *Amer. Lab.* (Nov.):51.

Omata, T., A. Tanaka, T. Yamane and S. Fukui. 1979. "Immobilization of Microbial Cells and Enzymes with Hydrophobic Photo-Cross-Linkable Rennin Prepolymers," *Eur. J. Appl. Microbiol. Biotechnol.*, 6:207.

Opie, R. 1987. "Biosensors Set to Make an Impact," *Control & Instrumentation* (May):137.

Ouellette, R. P. and P. N. Cheremisinoff. 1985. *Applications of Biotechnology*. Lancaster, PA: Technomic Pub. Co. Inc.

Oultram, J. D. and M. Young. 1985. "Conjugal Transfer of Plasmid pAM$\beta$1 from *Streptococcus lactis* and *Bacillus subtilis* to *Clostridium acetobutylicum*," *FEMS Microbiol. Lett.*, 27:129.

Palva, I. 1982. "Molecular Cloning of $\alpha$-Amylase Gene from *Bacillus amyloliquefaciens* and Its Expressions in *B. subtilis*," *Gene*, 19:81.

Panchal, C. J., I. Russell, A. M. Sills and G. G. Stewart. 1984. "Genetic Manipulation of Brewing and Related Yeast Strains," *Food Technol.*, 38(2):99.

Pardun, H. 1972. "Preparation of Phosphatide Emulsifiers," U.S. patents no. 3,652,397 and 3,661,795.

Park, Y. K. and G. M. Pastor. 1988. "Recent Progress in the Immobilization of $\beta$-Galactosidase," in *Bioreactor Immobilized Enzymes and Cells—Fundamental and Applications*. M. Moo-Young, ed. New York: Elsevier Applied Sci., pp. 225–232.

Parkinson, G. 1983. "Reverse Osmosis: Trying for Wider Applications," *Chem. Eng.* (May):26–29.

Patel, P. R. 1985. "Enzyme Isolation and Purification," in *Biotechnology Applications and Research*. P. N. Cheremisinoff and R. P. Ouellette, eds. Lancaster, PA: Technomic Pub. Co., Inc., pp. 534–564.

Payne, G. F., M. L. Shuler and P. Brodelius. 1987. "Large-Scale Plant Cell Culture," in *Large-Scale Cell Culture Technology*. B. K. Lydersen, ed. New York: Hanser Pub., pp. 193–230.

Perry, R. H. and C. H. Chilton. 1984. "Chemical Engineers' Handbook." New York: McGraw-Hill Book Company.

Pickett, S. C., R. F. Johnston, M. F. Miller and D. L. Barker. 1990. "High Resolution Dynamic Imaging: A Method for Direct Monitoring of Gel Electrophoresis," *Am. Biotechnol. Lab.*, 8(3):34.

Picque, D. and G. Corrieu. 1988. "New Instruments for On-Line Viscosity Measurement of Fermentation Media," *Biotechnol. Bioeng.*, 31:19.

Pine, R. and P. C. Huang. 1987. "An Improved Method to Obtain a Large Number of Mutants in a Defined Region of DNA," *Methods Enzymol.*, 154:415.

Pitcher, W. H., Jr. 1980. "Immobilized Enzyme Engineering," in *Immobilized Enzymes for Food Processing*. W. H. Pitcher, Jr., ed. Boca Raton, FL: CRC Press, Inc., pp. 15–54.

Potrykus, I., J. Paszkowski, M. W. Saul, I. Negrutiu and R. D. Shillito. 1987. "Direct Gene Transfer to Plants: Fats and Future," in *Plant Tissue and Cell Culture*. C. E. Green, D. A. Somers, W. P. Hackett and D. D. Biesboer, eds. New York: Alan R. Liss, Inc.

Poulsen, P. B. 1984. "Current Applications of Immobilized Enzymes for Manufacturing Purposes," in *Biotechnology and Genetic Engineering Reviews I*. G. E. Russell, ed. Newcastle upon Tyne, UK: Intercept, pp. 121–140.

Powell, L. W. 1984. "Development in Immobilized Enzyme Technology," in *Biotechnology and Genetic Engineering Reviews II*. G. E. Russell, ed. Newcastle upon Tyne, UK: Intercept, pp. 409–438.

Prenosil, J. E. and H. Pedoisen. 1983. "Immobilized Plant Cell Reactors," *Enzyme Microb. Technol.*, 5(5):323.

Price, E. C. 1985. "The Microbiology of Anaerobic Digestion," in *Biotechnology Applications and Research*. P. N. Cheremisinoff and R. P. Ouellette, eds. Lancaster, PA: Technomic Pub. Co., Inc., pp. 52–59.

Pringle, J. R. and J. R. Mor. 1975. "Methods for Monitoring the Growth of Yeast Cultures and for Dealing with the Clumping Problem," in *Methods in Cell Biology, Vol. 11*. D. M. Prescott, ed. New York: Academic Press, pp. 131–168.

Quinn, J. J. 1983. "Adiabatic Drying," in *Fermentation and Biochemical Engineering Handbook*. H. C. Vogel, ed. Park Ridge, NJ: Noyes Pub., pp. 331–362.

Ramakrishna, S. V., V. P. Sreedharan and P. Prema. 1988. "Continuous Ethanol Production with Immobilized Yeast Cells in a Packed Bed Reactor," in *Bioreactor Immobilized Enzymes and Cells—Fundamentals and Applications*. M. Moo-Young, ed. New York: Elsevier Applied Sci., pp. 251–260.

Ratledge, C. 1984. "Microbial Oils and Fats—An Overview," in *Biotechnology for the*

*Oils and Fats Industry.* C. Ratledge, P. Dawson and J. Rattray, eds. Am. Oil Chemists Soc., pp. 119–127.

Rattray, J. B. M. 1984. "Biotechnology and the Fats and Oils Industry – An Overview," *J. Amer. Oil Chemist. Soc.*, 61(11):1701.

Rechnitz, G. A. 1981. "Bioselective Membrane Electrode Probes," *Science*, 214:287.

Rehg, T., C. Dorger and P. C. Chau. 1986. "Application of an Atomizer in Producing Small Alginate Gel Beads for Cell Immobilization," *Biotech. Lett.*, 8:111.

Rehm, H. J. and G. Reed. 1973. *Biotechnology, Vol. 4.* Frankfurt: VCH.

Reilly, P. J., 1980. "Potential and Use of Immobilized Carbohydrases," in *Immobilized Enzymes for Food Processing.* W. H. Pitcher, Jr., ed. Boca Raton, FL: CRC Press, Inc., pp. 113–152.

Rigby, P. W. J., M. Dieckmann, C. Rodes and P. Berg. 1977. "Labeling Deoxyribonucleic Acid to High Specific Activity *in vitro* by Nick Translation with DNA PolymeraseI," *J. Mol. Biol.*, 113:237.

Robert, L. S., R. D. Thompson and R. B. Flavell. 1989. "The Tissue-Specific Expression of a Wheat High-Molecular Weight Glutenin Gene in Transgenic Tobacco," *Plant Cell*, 1:569.

Roberts, R. J. 1988. "Restriction Enzymes and Their Isoschizomers," *Nucleic Acids Res. (suppl.)*, 16:271.

Rochefort, W. E., T. Rehg and P. C. Chau. 1986. "Trivalent Cation Stabilization of Alginate Gel for Cell Immobilization," *Biotech. Lett.*, 8:115.

Rols, J. L., J. S. Condoret, C. Fonade and G. Goma. 1990. "Mechanism of Enhanced Oxygen in Fermentation Using Emulsified Oxygen Vectors," *Biotechnol. and Bioeng.*, 35:427.

Rols, J. L. and G. Goma. 1989. "Enhancement of Oxygen Transfer Rates in Fermentation Using Oxygen-Vectors," *Biotech. Adv.*, 7:1.

Romos, M. S. and S. K. Harlander. 1990. "DNA Fingerprinting of *Lactococci* and *Streptococci* Used in Dairy Fermentations," in *Appl. Microbiol. Biotechnol.*, 34:368.

Rosevear, A. and C. A. Lambe. 1985. "Immobilized Plant Cells," *Adv. in Biochem. Engg./Biotech.*, 31:37.

Rudge, S. R. and M. R. Ladisch. 1986. "Process Considerations for Scale-Up of Liquid Chromatography and Electrophoresis," in *Separation, Recovery and Purification in Biotechnology.* J. A. Asenjo and J. Hong, eds. Washington, DC: Am. Chem. Soc., pp. 122–152.

Russell, I., R. Jones and G. Stewart. 1986. "The Genetic Modification of Brewer's Yeast and Other Industrial Yeast Strains," in *Biotechnology in Food Processing.* S. K. Harlander and T. P. Labuza, eds. Park Ridge, NJ: Noyes Pub., pp. 171–195.

Russel, I. and G. G. Stewart. 1980. "Transformation of Maltotriose Uptake Ability into a Haploid Strain of *Saccharomyces* sp.," *J. Inst. Brew.*, 86:55.

Russell, P. D. 1987. "Fermenter and Bioreactor Design," in *Food Biotechnology I.* R. D. King and P. S. J. Cheetham, eds. London: Elsevier Applied Science, pp. 1–48.

Sahai, O. P. and M. Knuth. 1985. "Commercializing Plant Tissue Culture Processes: Economics, Problems and Prospects," *Biotechnol. Progress*, 1(1):1.

Sambrook, J., E. F. Fritsch and T. Maniatis. 1989. "Molecular Cloning – A Laboratory Manual." New York: *Cold Spring Harbor Laboratory Press*.

Sandine, W. E. 1985. "The *Streptococci*: Milk Products," in *Bacterial Starter Cultures for Foods.* S. E. Gilliland, ed. Boca Raton, FL: CRC Press, pp. 5–24.

Sanger, F., S. Nicklen and A. R. Coulson. 1977. "DNA Sequencing with Chain Terminating Inhibitors," *Proc. Natl. Acad. Sci.*, 74:5463.

Scheibel, E. G. 1956. "Performance of an Internally Baffled Multistage Extraction Column," *AICHE. J.*, 2(1):74.

Schindler, J. and R. D. Schmid. 1982. "Fragrance or Aroma Chemicals—Microbial Synthesis and Enzymatic Transformation—A Review," *Process Biochem.*, 17(5):3.

Schleifer, K. H. 1990. "DNA Probes in Food Microbiology," *Food Biotechnol.*, 4(1):585.

Schmidt, H. L. and I. Saschewag. 1987. "Possibilities of Process Control with Biosensors," in *Biochemical Engineering*. H. Chmiel, W. P. Hammes and J. E. Bailey, eds. New York: Gustav Fischer Verlag, pp. 132–138.

Schmidt-Kastner, G. 1987. "Bioreactors—Production of Bioproducts and Biotransformation by Enzymes and Immobilized Enzymes," in *Biochemical Engineering*. H. Chmiel, W. P. Hammes, and J. E. Bailey, eds. New York: Gustav Fischer Verlag, pp. 111–131.

Schwartz, D. C. and C. R. Cantor. 1984. "Separation of Yeast Chromosome-Sized DNAs by Pulsed Field Gradient Gel Electrophoresis," *Cell*, 37:67.

Scowcroft, W. R. and P. J. Larkin. 1988. "Somaclonal Variation," in *Applications of Plant Cell and Tissue Culture*. G. Bock and J. Marsh, eds. Toronto: John Wiley and Sons, pp. 21–35.

Seki, T., E. H. Choi and D. Ryu. 1985. "Construction of Killer Wine Strain," *Appl. Environ. Microbiol.*, 49:1211.

Sfat, M. 1986. "Unitization of Fermented Foods: An Application of Fermentation Technology," in *Biotechnology in Food Processing*. S. K. Harlander and T. P. Labuza, eds. Park Ridge, NJ: Noyes Pub., pp. 223–236.

Shah, Y. T. and W. D. Deckwer. 1985. "Fluid-Fluid Reactors," in *Scale-Up of Chemical Processes*. A. Bisio and R. L. Kabel, eds. New York: John Wiley & Sons, pp. 201–252.

Sharp, W. R., D. A. Evans and P. V. Ammirato. 1984. "Plant Genetic Engineering: Designing Crops to Meet Food Industry Specifications," *Food Technol.*, 38(2):112.

Shay, L. K. and G. H. Wegner. 1985. "Improved Fermentation Process for Producing Torula Yeast," *Food Technol.*, 39(10):61.

Shewry, P. R., M. Kreis, M. M. Bunell and B. J. Miflin. 1987. "Improvement of the Processing Properties of British Crops by Genetic Engineering," in *Food Biotechnology I*. R. D. King and P. S. J. Cheetham, eds. London: Elsevier Applied Science, pp. 49–85.

Shifrin, N. S. 1984. "Oils from Micro Algae," in *Biotechnology for the Oils and Fats Industry*. C. Ratledge, P. Dawson and J. Rattray, eds. *Am. Oil Chemists Soc.*, pp. 145–162.

Shih, M. F., M. Arsenakis, P. Tiollais and B. Roizman. 1984. "Expression of Hepatitis B Virus S Gene by Herpes Simplex Virus Type 1 Vectors Carrying $\alpha$- and $\beta$-Regulated Gene Chimeras," *Proc. Natl. Acad. Sci.*, 81:5867.

Shinomiya, S., K. Yamane and T. Oshima. 1980. "Isolation of a *Bacillus subtilis* Transformant Producing Thermostable $\alpha$-Amylase by DNA from a Thermophilic Bacterium," *Biochem. Biophys. Res. Comm.*, 96:175.

Shoda, M. and Y. Ishikawa. 1981. "$CO_2$ Sensor for Fermentation Systems," *Biotech. Bioeng.*, 23:461.

Shoemaker, S. 1986. "The Use of Enzymes for Waste Management in the Food Industry," in *Biotechnology in Food Processing*. S. K. Harlander and T. P. Labuza, eds. Park Ridge, NJ: Noyes Pub., pp. 259–269.

Shomburg, D. 1990. "Methodological Aspects and Potential of Computer Aided Protein Engineering," *Food Biotechnol.*, 4(1):329–336.

Simon, E. S., A. Akiyama and G. Whitesides. 1989. "Synthesis without Cells," in *A Revolution in Biotechnology*. J. L. Marx, ed. Cambridge, UK: Cambridge Univ. Press, pp. 15–27.

Sinskey, A., S. James, D. Eassor, Jr. and C. Rha. 1986. "Biopolymers and Modified Polysaccharides," in *Biotechnology in Food Processing*. S. K. Harlander and T. P. Labuza, eds. Park Ridge, NJ: Noyes Pub., pp. 73–114.

Sinskey, A. and S. R. Tannenbaum. 1975. "Removal of Nucleic Acid," in *SCP—Single Cell Protein, Vol. II*. S. R. Tannenbaum and D. I. C. Wang, eds. Cambridge, MA: The MIT Press, pp. 158–178.

Sirkar, K. and R. Prasad. 1986. "Protein Ultrafiltration—Some Neglected Considerations," in *Membrane Separations in Biotechnology*. W. C. McGregor, ed. New York: Marcel Dekker Inc., pp. 37–60.

SivaRaman, H., B. S. Rao, A. V. Pundle and C. Siva Raman. 1982. "Continuous Ethanol Production by Yeast Cells Immobilized in Open Pore Gelation Matrix," *Biotech. Lett.*, 4:359.

Slininger, P. J., R. J. Petroski, R. J. Bothast, M. R. Ladish and M. R. Okos. 1989. "Measurement of Oxygen Solubility in Fermentation Media: A Calorimetric Method," *Biotechnol. and Bioeng.*, 33:578.

Slotin, L. A. 1984. "Biotechnology—A Perspective," *Agric. & Forestry Bull.* (June):24.

Smith, J. E. 1985. "Biotechnology Principles." Washington, DC: Amer. Soc. of Microbiol.

Snow, R. 1985. "Genetic Engineering of a Yeast Strain for Malolactic Fermentation of Wine," *Food Technol.*, 39(10):96.

Soares, M. E., A. G. Medina and P. J. Bailes. 1982. "Liquid-Liquid Equilibria for the System Water + Acetic Acid + Methyl Isoamyl Ketone," *Fluid Phase Equilibria*, 9:177.

Soderberg, A. C. 1983. "Fermentation Design," in *Fermentation and Biochemical Engineering Handbook*. H. C. Vogel, ed. Park Ridge, NJ: Noyes Pub., pp. 77–118.

Solnick, D. 1981. "Construction of an Adenovirus-SV40 Recombinant Producing SV40 T Antigen from an Adenovirus Late Promotor," *Cell*, 24:135.

Sonomoto, K., I. Jin, A. Tanaka and S. Fukui. 1980. "Application of Urethane Prepolymers for Immobilization of Biocatalysts: $\Delta'$-Dehydrogenation of Hydrocortisone by *Arthrobacter simplex* Cells Entrapped with Urethane Prepolymers," *Agric. Biol. Chem.*, 44(5):1119.

Southern, E. M. 1975. "Detection of Specific Sequences among DNA Fragments Separated by Gel Electrophoresis," *J. Mol. Biol.*, 98:503.

Stanbury, P. F. and A. Whitaker. 1984. "Principles of Fermentation Technology." New York: Pergamon Press.

Steiner, L. and S. Hartland. 1980. "Agitated Liquid/Liquid Extraction," *Chem. Eng. Progress*, 76(12):60.

Stewart, A. F., I. M. Willis and A. G. MacKinlay. 1984. "Nucleotide Sequences of Bovine $\alpha_{s1}$- and $x$-Casein cDNAs," *Nucl. Acid Res.*, 12:3895.

Stewart, G. G., C. J. Panchal and I. Russell. 1983. "Current Developments in The Genetic Manipulation of Brewing Yeast Strains—A Review," *J. Inst. Brew.*, 89(3):170.

Stewart, G. G., I. Russell and C. J. Panchal. 1982. "The Genetics of Alcohol Metabolism in Yeast," *Brewing Distilling Intl.*, 12(1):23.

Strathmann, H. 1987. "Membranes in Downstream Processing—State of the Art and Future Developments," in *Biochemical Engineering*. H. Chmiel, W. P. Hammes and J. E. Bailey, eds. New York: Gustav Fischer Verlag, pp. 288—300.

Strickland, M. and N. Strickland. 1990. "Free-Solution Capillary Electrophoresis Using Phosphate Buffer and Acidic pH," *Amer. Lab.* (Nov.): 60.

Subden, R. E. 1987. "Current Developments in Wine Yeasts," *CRC Critical Reviews in Biotechnol.*, 5(1):49.

Subramani, S., R. Mulligan and P. Berg. 1981. "Expression of the Mouse Dihydrofolate Reductase Complementary Deoxyribonucleic Acid in Simian 40 Vectors," *Mol. Cell. Biol.*, 1:854.

Sutcliffe, J. G. 1979. "Complete Nucleotide Sequence of the *Escherichia coli* Plasmid pBR322," *Cold Spring Harbor Symp. Quant. Biol.*, 43:77.

Suzuki, S. and I. Karube. 1981. "Method and Apparatus for Measuring Microorganism Activity," U. S. patent 4,288,544.

Suzuki, S., I. Satoh and I. Karube. 1982. "Recent Trends of Biosensors in Japan," *Applied Biochem. Biotech.*, 7:147.

Swaminathan, B. and S. A. Minnich. 1985. "Enzyme Immunoassay for the Detection of Salmonella," in *Biotechnology Applications and Research*. P. N. Cheremisinoff and R. P. Ouellette, eds. Lancaster, PA: Technomic Pub. Co., Inc., pp. 526—533.

Tabor, S. and C. C. Richardson. 1987. "DNA Sequence Analysis with a Modified Bacteriophage T7 DNA Polymerase," *Proc. Natl. Acad. Sci.*, 84:4767.

Tampion, J. and M. D. Tampion. 1987. "Immobilized Cells—Principles and Applications." New York: Cambridge Univ. Press.

Taylor, D. L. 1985. "Biotechnology Comes of Age in the Food Industry," *Food Eng.*, 57(9):28.

Teutonico, R. A. and D. Knorr. 1985. "Impact of Biotechnology on Nutritional Quality of Food Plants," *Food Technol.*, 39(10):127.

Thevenot, D. R., R. Sternberg, P. R. Coulet, J. Laurnet and D. C. Gautheron. 1979. "Enzyme Collagen Membrane for Electrochemical Determination of Glucose," *Anal. Chem.*, 51(1):96.

Thien, M. P., T. A. Hatton and D. I. C. Wang. 1986. "Liquid Emulsion Membranes and Their Applications in Biochemical Separations," in *Separation, Recovery and Purification in Biotechnology*. J. A. Asenjo and J. Hong, eds. Washington, DC: Am. Chem. Soc., pp. 67—77.

Thomas, P. S. 1980. "Hybridization of Denatured RNA and Small DNA Fragments Transferred to Nitrocellulose," *Proc. Natl. Acad. Sci.*, 77:5201.

Thompson, A. R. 1987. "Adsorption in Downstream Processing," in *Biochemical Engineering*. H. Chmiel, W. P. Hammes and J. E. Bailey, ed. New York: Gustav Fischer Verlag, pp. 252—260.

Thorpe, T. A. and S. Biondi. 1981. "Requirements for a Tissue Culture Facility," in *Plant Tissue Culture—Methods and Applications in Agriculture*. T. A. Thorpe, ed. New York: Academic Press, pp. 1—20.

Thummel, C., R. Tjian and T. Grodzicker. 1981. "Expression of SV40 T Antigen under Control of Adenovirus Promotors," *Cell*, 23:825.

Tipayang, P. and M. Kozaki. 1982. "Lactic Acid Production by a New *Lactobacillus* sp.,

*Lactobacillus vaccinostercus kozaki* and *okada* sp. Immobilized in Calcium Alginate," *J. Ferment. Technol.*, 60(6):595.

Tisserat, B. 1985. "Embryogenesis, Organogenesis and Plant Regeneration," in *Plant Cell Culture—A Practical Approach*. R. A. Dixion, ed. New York: Oxford Press, p. 79.

Tobian, J. A., L. Drinkard and M. Zasloff. 1985. "tRNA Nuclear Transport: Defining the Critical Regions of Human tRNA Met by Point Mutagenesis," *Cell*, 43:415.

Todd, D. B. 1983. "Solvent Extraction," in *Fermentation and Biochemical Engineering Handbook*. H. C. Vogel, ed. Park Ridge, NJ: Noyes Pub., pp. 175—201.

Tosa, T. et al. 1979. "Immobilization of Enzymes and Microbial Cells Using Carrageenan as Matrix," *Biotechnol. and Bioeng.*, 21:1697.

Towbin, H., T. Staehelin and J. Gordon. 1979. "Electrophoretic Transfer of Proteins from Polyacrylamide Gels to Nitrocellulose Sheets: Procedure and Some Applications," *Proc. Natl. Acad. Sci.*, 76:4350.

Tran-Minh, C. and G. Broun. 1975. "Construction and Study of Electrodes Using Cross-Linked Enzymes," *Anal. Chem.*, 47:1359.

Trivedi, N. B. 1986. "Baker's Yeast," *CRC Critical Reviews in Biotechnol.*, 4(1):75.

Trivedi, N. 1986. "Use of Microorganisms in the Production of Unique Ingredients," in *Biotechnology in Food Processing*. S. K. Harlander and T. P. Labuza, eds. Park Ridge, NJ: Noyes Pub.

Tsao, G. T. and L. F. Chen. 1977. U. S. patent 4,063,017.

Tsay, S. S. and K. Y. To. 1987. "Citric Acid Production Using Immobilized *conidia* of *Aspergillus nigar* TMB2022," *Biotechnol. and Bioeng.*, 29:297.

Tsuchida, T. and H. Momose. 1986. "Improvement of L-leucine-Producing Mutant of *Brevibacterium lactofermentum* 2256 by Genetically Desensitizing It to Alpha-Acetohydroxy Acid Synthetase," *Appl. Environ. Microbiol.*, 51:1024.

Turner, A. P. F., I. J. Higgins and A. Franklin. 1986. "Biosensors for the Food Industry," in *Food Biotechnology, Proc. of the Int. Symp., Quebec*. J. D. L. Noe, J. Goulet and J. Amiot, eds., pp. 51—68.

Turner, A. P. F. and A. Swain. 1988. "Commercial Perspectives for Diagnostics Using Biosensor Technologies," *Am. Biotechnol. Lab.*, 6(8):10.

Vaccaro, D. E. 1990. "Applications of Magnetic Separation: Cell Sorting," *Am. Biotechnol. Lab.*, 8(5):30.

Vaheri, A. and J. S. Pagano. 1965. "Infections Poliovirus RNA: A Sensitive Method of Assay," *Virology*, 27:434.

VanBrunt, J. 1985. "Scale-Up: The Next Hurdle," *Biotechnol.*, 3(5):419.

VanDam, A. F. 1978. "Oil in Water Emulsion and Process for the Preparation Thereof," U.S. patent no. 4,119,564.

Vasey, R. B. 1984. "Single Cell Protein," in *Biotechnology and Genetic Engineering Reviews II*. G. E. Russell, ed. Newcastle upon Tyne, UK: Intercept, pp. 285—312.

Venkat, K. 1986. "Large Scale Production of Monoclonal Antibodies by Immobilized by Bridoma/Mammalian Cell Systems," in *Proc. Int. Symp. on Immobilized Enzymes and Cells, Univ. of Waterloo*. Waterloo: GW Biotech.

Vieira, J. and J. Messing. 1982. "The pUC Plasmids, an M13mp7-Derived System for Insertion Mutagenesis and Sequencing with Synthetic Universal Primers," *Gene*, 19:259.

Vogeli, G., E. V. Avvedimento, M. Sullivan, J. V. Maizel, Jr., G. Lozano, S. L. Adams,

I. Pastan and B. de Crombrugghe. 1980. "Isolation and Characterization of Genomic DNA Coding for 22 Type 1 Collagen," *Nucleic Acids Res.*, 8:1823.

Von Elbe, J. and C. H. Amundson. 1977. U. S. patent 4,027,042.

Vorlop, K. D. and J. Klein. 1981. "Formation of Spherical Chitosan Biocatalysts by Ionotropic Gelation," *Biotech. Lett.*, 3:9.

Vuillemard, J. C. and J. Amiot. 1988. "Hydrolysis of Milk Proteins by Immobilized Cells," in *Bioreactor Immobilized Enzymes and Cells—Fundamental and Applications*. M. Moo-Young, ed. New York: Elsevier Applied Sci., pp. 213−224.

Wada, M., J. Kato and I. Chibata. 1981. "Continuous Production of Ethanol in High Concentration Using Immobilized Growing Yeast Cells," *Eur. J. Appl. Microbiol. and Biotechnol.*, 11:67.

Wagner, G. and R. D. Schmid. 1990. "Biosensors for Food Analysis," *Food Biotechnol.*, 4(1):215.

Wahl, G. M., M. Stern and G. R. Stark. 1979. "Efficient Transfer of Large DNA Fragments from Agarose Gel to Diazobenzyl-Oxymethyl-Paper and Rapid Hybridization by Using Dextran Sulfate," *Proc. Natl. Acad. Sci.*, 76:3683.

Wandrey, C. 1987. "The Growing Importance of Biochemical Engineering," in *Biochemical Engineering*. H. Chmiel, W. P. Hammes and J. E. Bailey, eds. New York: Gustav Fischer Verlag, pp. 43−68.

Wang, G., J. Shevitz, S. Weiss and N. Vosper. 1989. "Production of Gram Quantities of Monoclonal Antibody in a Hollow Fiber Bioreactor," *Amer. Biotechnol. Lab.*, 7(9):14.

Wang, H. Y. and D. J. Hettwer. 1982. "Cell Immobilization in $x$-Carrageenan with Tricalcium Phosphate," *Biotech. Bioeng.*, 24:1827.

Warden, D. and H. V. Thorne. 1968. "Infectivity of Polyoma Virus DNA for Mouse Embryo Cells in Presence of Diethylaminoethyl-Dextran," *J. Gen. Virol.*, 3:371.

Wasserman, B. P. 1984. "Thermostable Enzyme Production," *Food Technol.*, 38(2):78.

Wasserman, B. P. 1989. "Evolution of Enzyme Technology: Progress and Prospects," IFT Meeting.

Wasserman, B. P., T. J. Montville, E. L. Korwek and J. D. Hogan. 1988. "Food Biotechnology," *Food Technol.*, 42(1):133.

Wayman, M., A. D. Jenkins and A. G. Kormendy. 1984. "Bacterial Production of Fats and Oils," in *Biotechnology for the Oils and Fats Industry*. C. Ratledge, P. Dawson and J. Rattray, eds. Am. Oil Chemists Soc., pp. 129−143.

Webb, C., H. Fakuda and B. Atkinson. 1986. "The Production of Cellulase in a Spouted Bed Fermenter Using Cell Immobilized in Biomass Support Particles," *Biotech. Bioeng.*, 28:41.

Weetal, H. H. 1975. "Immobilized Enzymes and Their Application in the Food and Beverage Industry," *Process Biochem.*, 10:3.

Werner, R. G., W. Merk and F. Walz. 1987. "Fermentation with Immobilized Cell Cultures," in *Biochemical Engineering*. H. Chmiel, W. P. Hammes and J. E. Bailey, eds. New York: Gustav Fischer Verlag, pp. 327−332.

Wernerspach, D. 1986. "Benchtop Fermentation Systems: New Equipment for an Ancient Process," *Am. Biotechnol. Lab.*, 4(11):22.

West, S. 1988. "The Enzyme Maze," *Food Technol.*, 42(4):98.

Wetzel, R. 1986. "Protein Engineering—Potential Applications in Food Processing," in *Biotechnology in Food Processing*. S. K. Harlander and T. P. Labuza, eds. Park Ridge, NJ: Noyes Pub., pp. 57−71.

Whitaker, R. J. and D. A. Evans. 1987. "Plant Biotechnology and the Production of Flavor Compounds," *Food Technol.*, 41(9):86.

White, T. J., J. H. Meade, S. P. Shoemaker, K. E. Koths and M. A. Innis. 1984. "Enzyme Cloning for the Food Fermentation Industry," *Food Technol.*, 38(2):90.

White, W. C. and G. G. Guilbault. 1978. "Lysine Specific Enzyme Electrode for Determination of Lysine in Grains and Foodstuffs," *Anal. Chem.*, 50:1481.

Wigler, M., S. Silverstein, L. S. Lee, A. Pellicer, Y. C. Cheng and R. Axel. 1977. "Transfer of Purified Herpesvirus Thymidine Kinase Gene to Cultured Mouse Cells," *Cell*, 11:223.

Wilson, R. F. 1984. "Effect of Genetic Selection Upon Lipid Metabolism in Soybeans," in *Biotechnology for the Oils and Fats Industry*. C. Ratledge, P. Dawson and J. Rattray, eds. Am. Oil Chemists Soc., pp. 77−87.

Wilson, R. J. H. 1977. "Introducing Gases into Fermentation Liquids," U.S. patent 4,041,180.

Winquist, F., B. Danielsson, I. Lundstrom and K. Mosbach. 1982. "Use of Hydrogen-Sensitive Pd-MOS Materials in Biochemical Analysis," *Appl. Biochem. Biotechnol.*, 7:135.

Yanich-Perron, C., J. Vieira and J. Messing. 1985. "Improved M13 Phage Cloning Vectors and Host Strains: Nucleotide Sequences of the M13mp18 and pUC19 Vectors," *Gene*, 33:103.

Young, R. A. and R. W. Davis. 1983. "Efficient Isolation of Genes by Using Antibody Probes," *Proc. Natl. Acad. Sci.*, 80:1194.

Zajic, J. E. and W. Seffens. 1984. "Approaches to Hydrophilic-Lipophilic Balance Evaluation of Microbes," in *Biotechnology for the Oils and Fats Industry*. C. Ratledge, P. Dawson and J. Rattray, eds. Am. Oil Chemists Soc., pp. 241−253.

Zappe, H., D. T. Jones and D. R. Woods. 1986. "Cloning and Expression of *Clostridium acetobutylicum* Endoglucanase, Cellobiase and Amino Acid Biosynthesis Genes in *E. coli*," *J. Gen. Microbiol.*, 132:1367.

Zimmermann, R. H. and O. C. Broome. 1980. "Blueberry Micropropagation," *U.S. Dept. Agric., Sci. Educ. Admin.*, ARR-NE-11, pp. 44−47.

Zimmermann, U. 1982. "Electric Field-Mediated Fusion and Related Electrical Phenomena," *Biochem. Biophys. Acta*, 694:227.

Zimmerman, U. and P. Scheurich. 1981. "High-Frequency Fusion of Plant Protoplasts by Electric Fields," *Planta*, 151:26.

Zoller, M., Jr. and M. Smith. 1983. "Oligonucleotide-Directed Mutagenesis of DNA Fragments Cloned into M13 Vectors," *Methods Enzymol.*, 100:468−500.

Zoller, M. N. and M. Smith. 1987. "Oligonucleotide-Directed Mutagenesis: A Simple Method Using Two Oligonucleotide Primers and a Single-Stranded DNA Template," *Methods Enzymol.*, 154:329.

Zwietering, Th. N. 1958 "Suspension of Solid Particles in Liquid by Agitators," *Chem. Eng. Sci.*, 8:244.

# INDEX

α-Amylase, 320
α-Complementation, 21, 31
β-Galactosidase, 21, 24, 31, 43, 137, 144–145, 316
β-Glucan, 113, 136
β-Lucanase, 84
β-Mercaptoethanol, 80
β-Particles, 83
κ-Casein, 129
λgt10, 25
λgt11, 24–26
0.5 TBE mix, 75
2,3-Butanediol, 316–317
2.5 TBE mix, 75
2TY, 41, 44
40% Acrylamide, 74
5-Bromo-4-chloro-3-indolyl-β-galactoside, 21, 24, 31, 43, 72
Acetic acid, 4, 106, 156–158, 181
Activation energy, 143
Acylation, 139
Adenine, 7
Adenosine, 7
Adenovirus, 86
Adsorption, 215–216
Aeration, 287–289
Agar, 149, 160–161
Agarose, 149–150, 160
Agarose gel preparation, 36
Agitation, 287–289
Agitation and mixing, 261, 263, 264
Agitator speed, 275
*Agrobacterium tumefaciens*, 84
Air sterilization, 268
Alcohol, 167, 181
Aldehydes, 137
Alfalfa, 88

Alginate, 4, 111–112, 152–154, 160–161
Alginate gelatin, 155–156
Alkaline phosphatase (*see* Calf intestinal alkaline phosphatase)
Alkaline SDS, 45
Alkaloids, 102
Amino acids, 11, 319
Aminoglycoside phosphotransferase, 86
Ammonium persulphate solution, 75
Ampicillin, 21
Amylase, 135, 312, 319
    α-, 135, 146
    β-, 135, 146
    fungal, 135
Amyloglucosidases, 135
Annealing, 122
Anti-antibodies, 83
Anti-codon, 11
Antigen, 176
Antimicrobial, 110–111
Antioxidant, 4
Antiserum, 83
Aqueous two-phase systems, 240
Artificial seed, 95–96
Ascorbic acid, 4
Asepsis in fermenter design, 269–270
Aseptic
    inoculation, 269–270
    operation, 269
    sampling point, 270
    seals, 269
Aspartame, 144, 317
Automation of DNA synthesis, 86
Autonomously replicating sequences, 84

Autoradiography of sequencing gel, 78
Auxin, 90
Axial flow impeller, 263

Bacterial lysis buffer, 44−45
Bacteriocin, 18, 110−111
Bacteriophage λ, 21, 24−26
Bacteriophage λ exonuclease, 20
Bacteriophage polymerase, 20
Bacteriophage T4 DNA ligase, 19, 30, 61, 63
Bacteriophage T4 DNA polymerase, 19
Bacteriophage T4 polynucleotide kinase, 19, 63
Bacteriophage transformation, 43−44
Baked goods, 3, 5, 134
Baker's yeast, 112, 307, 313−314
Bal-31 (*see* Nuclease Bal-31)
Beef, 3, 134
Beer, 1, 3, 134, 245, 306−307, 315−316
Beta-carotene, 4, 106
Beverage processing, 306−310
Beverages, 134
Bioluminescence, 178
Biomass components, 277
Biomass measurement, 276−277
    dry or wet weight, 276
    nephelometry, 276−277
    turbidimetry, 276−277
Biopolymers, 111−113
Biosensor,
    acetic acid, 181, 184, 185
    alcohol, 181, 185
    amperometric, 173−175
    amygdalin, 181
    arginine, 184
    aspartate, 184
    based on ammonia, 180, 183−184
    based on $H_2O_2$, 179−180
    based on $O_2$, 179
    based on pH, 180
    BOD, 178, 184, 185
    calorimetric, 174−175
    characteristics, 181−182, 184

    cholesterol, 181, 184, 188
    $CO_2$, 185
    conductimetric, 176−177
    D-amino acid, 181, 188
    DNA probe, 183−185
    enzyme probe, 178−182
    ethanol, 184, 185
    formic acid, 184
    glucose, 177−181, 184, 186
    glutamine, 183, 184, 188
    introduction, 171−172
    L-amino acid, 181
    L-asparagine, 182
    L-glutamic acid, 182, 184, 186
    L-glutamine, 182
    L-tyrosine, 182
    lactic acid, 182
    material, 171
    meat freshness, 179, 187
    methane, 186
    microbe, 183−185
    microbial count, 177−178
    nicotinic acid, 184, 186
    nitrate, 182, 184
    nitrite, 182
    nystatin, 184, 187
    optical, 175−176
    organelle, 183−185
    piezoelectric, 176−177
    potentiometric, 172−173
    redox, 187
    succinic acid, 182
    sugars, 184
    sulphate, 182
    techniques, 172−177
    urea, 182, 187
    uric acid, 182
    vitamin $B_1$, 184, 187
Biosurfactants, 115−116
Biosynthesis, 87
Biotechnology−definition, 1
Biozan, 112
Blunt end, 16
BOD, 178, 184
Bovine embryo, 129
Bovine growth hormone, 84
Brewer's yeast, 307−308

Brewing efficiency, 308
Bromelain, 135, 321
Buckingham pi theorem, 286
Buffered-gradient polyacrylamide
    gel, 76–77
Buffers and solutions for SDS-PAGE,
    80
Butanol, 321
Butter, 3
Butter flavor, 105

Caesium chloride density gradient
    centrifugation, 47
Calcium phosphate precipitation, 85
Calf intestinal alkaline phosphatase,
    19, 40
Callus, 88, 101–102
Callus route, 91
Canola oil, 314
Capillary transfer of DNA, 54
Cassava starch, 153
Catalases, 137
cDNA cloning, 60–64
Cell adsorption, 157–159
Cell covalent bonding, 156–157
Cell culture, 94
Cell disruption, 192–194
    by bead mill, 193
    by lysis, 192
    by pressure, 192
    by ultrasonic, 192
Cell encapsulation, 156
Cell meiosis, 88
Cell separation, 191
Cellobiose, 136
Cellulase, 135–136, 319
Cellulose, 112, 135, 154–155
Centrifugation, 191, 198–201, 243
Centrifuge, 198–201, 243
    chamber, 198
    characteristics, 201
    decanter, 198
    nozzle bowl, 198
    selection, 200
    self-cleaning, 198
    tubular, 198
Cereal processing, 312–314

Charged modified membranes, 170
Chase mix, 75
Cheese, 3, 5, 251, 301–303
Cheese culture, 301–303
Chelators, 138
Chem-FET, 180–181
Chemical modification, 5
Chitin, 143–144
Chitosan, 112, 145, 154
Chloramphenicol, 18
Chromatography, 215–219, 243
    adsorption, 216–217
    affinity, 216–217
    centrifugal partition, 218
    CPC, 218
    efficiency of, 216
    gel filtration, 216–217
    HPCPC, 219
    HPLC, 218
    ion exchange, 216, 218–219
    reverse phase, 216, 219
Chrysogenin, 106
Chymosin, 129, 134, 144
Citric acid, 4, 106, 152, 155
Citronellol, 106
Clonal propagation, 87
Cloning, 113
Cloning in cosmids, 26, 28–30
Cloning in plants, 84–85
Cloning in yeast, 83
ColEI, 18, 26
Coagulation, 191
Coagulation, milk, 139
Cocoa butter, 113, 314
Coconut oil, 115
Codon, 11
Cofactors, 137
Coffee, 3, 4
Collagen, 151–152
Colony count, 177
Color, 107
Column diameter, 291
Computer-aided protein engineering,
    132
Concatemers, 30
Concentration of DNA or RNA,
    34–35

Coomassie brilliant blue, 82
Corn, 88
Corn oil, 115
*Cos* sites, 30
Cosmids, 26–30
Cross linking, 319
Crown gall tumor, 85
Crystallization, 200
    design, 200
    equipment, 200
    selection, 200
CT solution, 59
Culture stability, 291
Curdlan, 4, 106, 111–112
Cytidine, 7
Cytokinin, 90
Cytosine, 7

D-Arabitol, 106
Dairy processing, 295–305
Damkohler number, 147
De-emulsification, 115–116
DEAE cellulose paper, 49
DEAE-dextran, 85
Decanter, 191
Deletion mutagenesis, 119
Deletion mutagenesis by Bal-31,
    119–121
Denaturation solution, 53
Denhardt's solution, 56
Deoxyribonuclease I, 20, 51
Deoxyribonucleic acid, 1, 67
Deoxyribose, 7
Detoxification, 4
Dextran, 106, 112
Dextran sulphate, 56
Diacetyl, 105–106
Dideoxynucleotide chain termination
    technique, 65–66
Dihydrofolate reductase, 26, 86
Direct shoot route, 91
Dissolved oxygen, 265–268,
    271–272
Dissolved oxygen tension, 249
Distillation, 223, 241
Distribution coefficient, 224–225,
    227

DNA (*see* Deoxyribonucleic acid)
DNA cloning vectors, 18–31
DNA ligation, 41, 72
DNA probe, 183–185
DNA purification from agarose gel,
    48–49
DNA sequencing, 64–78
DNAase I (*see* Deoxyribonuclease I)
dNTP, 50, 51, 52
dNTP: ddNTP mixes, 74
Double helix, 7
Down-stream processing, 189–244
Down-stream processing techniques
    selection, 243–244
Dryer,
    drum, 242
    flash, 241
    fluid bed, 241
    rotary, 241–242
    selection, 242
    spray, 241
    tray, 241
    vacuum, 242
    vacuum freeze, 242
Drying, 241–242
$d_{s1}$-casein, 129
Dynamic similarity, 285–286

*E. coli* DNA polymerase I, 19, 51,
    58, 60–61, 63
*E. coli dut*$^-$, 125
*E. cole ung*$^-$ 126
Electric impulse, 101
Electrodialysis, 202, 209–210, 243,
    305
Electro-osmotic flow, 213
Electrophoresis, 211–214
    capillary (CE), 212–214
    gel (GE), 212
    HPCE, 214
Electrophoretic mobility, 213
    transfer of DNA, 54
    velocity, 213
Electroporation, 85
Eluting solutions, 101
Embryogenesis, 95–96
Emulsan, 112

Emulsification, 115–116
Emulsifier, 106, 115, 116
End-labeling reaction, 50
Endocellulase, 135
Endogalacturonase, 136
Endohemicellulase, 136
Endoprotease, 134
Enzyme, 4, 133–170, 319–321
    activity, 138–139
    chemical modification, 139
    cloning, 139
    database, 139
    electrode, 174–175, 178–182
    immobilization, 140–147
    immunoassay, 137
    inhibitors, 138, 140
    microbial, 137
    modifications, 139
    process development, 140
    protection, 138
    redox, 137
    selection, 139–140
    task, 133–134
    thermistor, 174–175
Enzyme-effected transistors, 175, 180
Enzyme FET (ENFET), 173, 175
Equilibrium coefficient, 225
Erucic acid, 4, 314
Essential oils, 102
Ester, 136, 137
Ethanol, 151, 153, 157, 304,
    321–322
Ethanol production, 316
Ethidium bromide, 37
Ethyl alcohol, 178
Evaporation, 242–243
Exchange reaction, 19
EXO III (*see* Exonuclease III)
Exo-cellobiohydrolase, 136
Exocellulase, 136
Exogalacturonase, 136
Exohemicellulase, 136
Exonuclease III, 20
Exonuclease III deletion, 67
Exoprotease, 134
Explant material, 93–94
Extraction,

    liquid solid, 222–237
    solid liquid, 219–222

Fat and oil, 5, 113–115, 136
Fatty acid esters, 106
Feed,
    aid, 194–195
    axial, 197
    cake, 194–196
    design, 195
    equipment, 197
    medium, 194, 196
    process variables, 196
    resistance, 195
    rotary, 196
    slurry, 194
    testing, 195
Fermentation, 245–282
    aeration, 258, 261, 263
    air sterilization, 268
    apparent viscosity, 249–251
    asepsis, 269–270
    biomass measurement, 276–277
    cations, 247
    design, 279–282
    equipment, 252
    factors affecting, 245–251
    fungi or molds, 251–252
    instrumentation and control,
        270–278
    introduction, 245
    *lactobacilli*, 251
    media sterilization, 268–269
    microbial growth, 275–276
    moisture, 247
    Monod eq., 247
    nutrients, 247
    oxygen, 248–249
    oxygen transfer, 263–268
    pH, 247
    process control, 277–278
    rheology, 249–251
    temperature, 246
Fermentation products, 281–282
    extracellular, 282
    intracellular, 282
    secondary metabolites, 282

Fermenter, 245−246, 252−281
    aeration, 258, 261, 263
    agitation and mixing, 261, 263
    air lift, 165, 167, 252−253
    air sterilization, 268
    all glass, 254
    asepsis, 269−270
    batch, 163, 254−256
    biomass measurement, 276−277
    bubble column, 294
    continuous stirred, 255−257
    continuous stirred tank, 163,
        255−257, 280, 293−294
    CSTR, 163, 255−256
    deep shaft, 257−258
    design, 279−282
    fluidized bed, 163, 167, 257, 293
    gas lift, 254
    hollow fiber, 160, 168, 259
    immobilized packed bed, 256,
        280
    instrumentation and control,
        270−278
    media sterilization, 268−269
    membrane, 160, 168, 259−260
    microbial growth, 275−276
    oxygen transfer, 263−268
    process control, 277−278
    roto, 257
    scraped tubular, 258
    selection, 261−262
    solid state, 257−258
    stirred tank, 167, 254−257
    tower, 253
    trickle bed, 257
Fermenter design, 279−282
FET, 172−173, 175
Ficin, 135
Field effect transistor, 172−173, 175
Filamentous bacteriophages, 30−32
Filling recessed 3′ ends, 50
Filter,
    crossflow, 197
    microstrainer, 198
    plate-frame, 197
    selection, 198
Filtrate, 194
Filtration, 194−199, 243

Flavor, 101−102, 105−107
Flavor modifier, 134
Flocculation, 191
Flotation, 191
Flow rates, 275
Foam control, 271
Foam fractionation, 240, 243
Foaming, 271
Food biotechnology, 1
Formamide dyes mix, 74
Fructose, 313
Fumaric acid, 4
Functional properties, 3

Galacturonic acid, 136
Gametic fusion, 88
Gametoclonal variation, 88
Gamma-decalactone, 106
Gel electrophoresis for DNA, 35−38
Gel drying, 82
Gellangum, 111
Gene, 7
Gene transfer into mammalian cells,
    85−86
Genetic code, 11
Genetic engineering, 4, 7
Geometric similarity, 285−287
Geranoil, 106
Gluconic acid, 145
Glucose, 136, 313
Glucose isomerase, 4, 313
Glucose oxidases, 137
Glucosinolate, 314
Glutamic acid, 4, 106−107, 178
Gluten, 313
Glycerol, 106, 314
Glycosylation, 11, 79
GT solution, 59
Guanine, 7
Guanosine, 7
Guanylic acid, 107

Handerson-Hasselbalch equation, 273
Heat transfer, 280, 291
Hemicelluloses, 136
Herpes virus, 86
Heteroduplex, 122
HETS, 237

HFCS, 313
High fructose corn syrup, 313
Hollow fiber reactor, 160
Host, 322
Human hepatitis B antigen, 84
Hybrid DNA, 6
Hybridization, 55–58
Hybridization mix, 56
Hydrogen bonding, 7
Hydrolysis, 135–137
Hydrophile-lipophile balance, 115
Hydroxylamine, 127

Ice cream, 317
Immobilization,
    adsorption, 141–142
    advantages, 140
    animal cell, 148–159, 315–317
    applications, 315–317
    cell adsorption, 157–159
    cell covalent bonding, 156–157
    cell encapsulation, 156
    cell entrapment, 148–156
    changes in cells, 168–169
    covalent attachment, 142–143
    cross linking, 143–145
    diffusional approach, 147
    electrostatic forces, 148
    encapsulation, 145–146
    entrapment, 145, 159–162
    enzymes, 140–147
    examining cells, 169
    kinetics, 147–148
    microbial, 148–159
    on agar, 149
    on agarose, 149–150
    on alginate-gelatin, 155–156
    on alginates, 152–154
    on carrageenan, 150–151, 156
    on celluloses, 154–155
    on chitosan, 154
    on inorganic supports, 157–158
    on ion-exchange supports,
        158–159
    on organic supports, 158
    on polyacrylamides, 154, 161
    on polyurethane, 161
    on proteins, 151–152

plant cell, 159–162
protoplasts, 162
reactors, 163–168
selection, 147
techniques, 141–147
viability of cells, 169
Immobilized cell applications,
    315–317
Immobilized cell reactors, 165–168
Immobilized cells, 102
Immobilized enzyme reactors,
    163–165
Immunochemical FET (IMFET), 173
Immunoprecipitation, 83
Impeller,
    axial flow, 263
    radial flow, 263
In vitro translation, 58
Independently replicating vector, 84
Inoculation, 278
Inorganic salts, 90
Inorganic supports, 157–158
Inosinic acid, 4, 107
Instrumentation and control,
    270–278
Integrating vector, 84
Intensifying screen, 56
Interesterification, 314–315
Interfacial area, 264–267
Interferon, 84
Invertase, 144, 314
Ion selective FET (ISFET), 173, 175
Ion exchange, 239–240, 243
Ion-exchange membranes, 170
Ion-exchange supports, 158–159
Ion-pairing liquid chromatography,
    214
IPTG (see Isopropyl-thio
    β-D-galactoside)
Isolation of high molecular weight
    DNA, 31
Isolation of high molecular weight
    DNA from mammalian cells,
    33–34
Isolation of high molecular weight
    DNA from plant cells, 33
Isolation of plasmid, 44–48

Isomerases, 136
Isopropyl-thio-$\beta$-D-galactoside, 21, 31, 43, 73
Isoschizsomer, 16

Jelly, 317
Juice, 317

$\varkappa$-Carrageenan, 145, 150−151, 156, 160
Kanamycin, 85
Katal, 138
Kefir, 3
Kinematic similarity, 285
Kinetics, 280−281
Klenow fragment, 19, 50
Klenow mix, 75

L-Amino acid, 322
L-Aspartic acid, 317
L-Lysine, 180
Lactase, 137, 144−145
Lactic acid, 4, 106, 153, 302−304
Lactic acid bacteria, 111
*Lactobacilli*, 299−301
Lactones, 107
Lactose, 137, 144, 299, 302−304
Laminarin, 136
Large fragment of *E. coli* DNA polymerase (*see* Klenow fragment)
Large-scale plasmid preparation, 45−48
Leader peptide (*see* Signal peptide)
LEM, 202−204
Leucine, 106
Lichenan, 136
Ligand, 217
Ligation (*see* DNA ligation)
Ligation buffer, 41
Linoleic acid, 114, 314−315
Lipase, 136−137, 304, 314, 319
Liposomes, 85
Lipoxidae, 137
Lipoxygenase, 314
Liquid-liquid equilibrium, 225, 235
Liquid-liquid extraction, 222−238

agitation, 228
binary, 215
controlled cycle, 232, 234
countercurrent, 230−234
crosscurrent, 229, 232
design criteria, 232−238
distribution coefficient, 224−225, 227
double solvent, 232−233
equilibrium coefficient, 225
HETS, 237
liquid-liquid equilibrium, 225, 235
phase diagram, 223, 225, 236
phase ratio, 237
separation factor, 225
solvent selection, 226−228
solvent-feed ratio, 225
stage efficiency, 234, 237
stripping tower, 235
ternary, 215
tie line, 236
triangle method, 226
Liquid-liquid membrane, 202−204
Loading mix, 18, 37
Low-calorie foods, 5
Lowry method, 277
LVDT, 273
Lysine, 4, 106, 319
Lysogeny, 24
Lysozyme, 45, 320

M13, 30, 31, 71
M13mp18, 32
M13mp19, 32
Magnetic separation, 214−215
Maillard browning, 137
Malic acid, 4
Malolactic, 153
Malt, 3
Maltodextrins, 312
Mannitol, 106
Material selection, 279−280
Mayonnaise, 137
Meat fermentation, 305−306
   freshness, 179, 187
   processing, 305−306

Media formulas, 91–93
preparation, 91–92, 291
  sterilization, 268–269
Meiotic, 95
Membrane,
  anion-permeable, 209
  binding, 169–170
  cation-permeable, 209
  charged modified, 170
  electrodialysis, 209
  fouling, 202
  installation, 203
  ion exchange, 170
  liquid-liquid, 202–204
  module, 202
  nitrocellulose, 169–170
  polysulfone, 170
  reactor, 160
  RO, 207
  separation, 200–211
  UF, 204–205
Membrane binding, 169-170
Membrane module,
  capillary, 202–203
  hollow fiber, 202–203, 205, 207
  plate and frame, 202–203
  spiral wound, 202–203
  tubular, 202–203
Membrane reactor, 160
Meristem cloning, 87, 88
Meristematic cells, 89
Messenger RNA, 7, 59
Methanol, 106
Methionine, 4
Methotrexate, 85
Methoxylamine, 127
Methylbutanol, 106
Micellar electrokinetic capillary
  chromatography, 214
Michaelis-Manten constant, 142,
  147–148
Microbe probe, 183–185
Microbial,
  biological methods, 276
  count sensor, 177
  electronic methods, 275–276
  growth, 275–276

microscopic methods, 275
Microdialysis, 211
Microfiltration, 202–203, 243, 305
Micropropagation, 87, 88
Micro-injection, 86
Milk, 3, 111, 143, 295–305
Milk proteins, 302
Mini-bomb cell disruptor, 193–194
Mitotic, 95
Mixing, 287–289
Monascin, 106
Monoclonal antibody, 83
Monod equation, 247–248, 281
Monosodium glutamate, 107
mRNA (*see* Messenger RNA)
MS medium, 92, 94, 96
Mung bean nuclease, 20
Mutagen, 127
Mutagenesis, 130–132, 319, 298,
  300–302
Mutagenesis, site-directed, 139
Mutagenesis using degenerate
  oligonucleotides, 128
Mutagenic oligonucleotide, 122
Mutton, 3

*N*-Acetyl tripeptide, 106
*N, N, N',*
  *N'*-tetramethyl-ethylenediamine
  (*see* TEMED)
Narginase, 143
Neutralization solution, 53
Nick translation, 50–52
Nick translation buffer, 51
Nisin, 4, 106, 111
Nitrocellulose filter, 53
Nitrocellulose membranes, 169–170
NMR, 129
Northern blotting, 58
Nozzle separators, 191
Nuclease BAl-31, 20
Nucleoside, 7
Nutrient, 278
Nutrition, 90
Nylon membrane, 53

Oil and fat processing, 314–315

Oil modification, 115
Oleic acid, 114, 314−315
Oligo(dT) [*see* Poly(dT)]
Oligonucleotide, 319
Oligonucleotide-mediated
    mutagenesis, 121−126
Oligosaccharides, 135
Organelle probe, 183−185
Organic acids, 90
Organic nitrogen, 90
Organic supports, 158
Organogenesis, 88, 95−96
Osmotic flow, 207
Ovalbumin, 5
Overall growth rate, 246
Oxidation, 116−117
Oxidoreductases, 137−138, 176
Oxygen,
    diffusion coefficient, 267
    galvanic type, 265
    polarographic type, 265,
        267−268
    probe, 265−268, 271−272
    scavengers, 137
    solubility, 267−268
    transfer, 263−268, 290−291
    transfer rate, 264−267, 281
    uptake rate, 246, 281
    vector, 264

Palmitic acid, 114
Palmitoleic acid, 114
Papain, 135, 321
Papilloma virus, 86
Particle gun, 86
pBR322, 21, 22
Pd-MOSFET, 180
Pectin, 136, 137
Pectinase, 137
Pectinestarase, 136
Peddle-type stirrer, 263
PEG solution, 73
Pepsin, 135
Peroxidase, 138
Pervaporation, 202, 203, 208−209
pH, 272−273
pH control, 272−273

Phase diagram, 223, 225, 236
Phase ratio, 237
Phenylmethylsulfonyl fluoride (*see*
    PMSF)
Phosphatase buffer, 40
Phospho-fractokinase, 314
Phosphodiester bond, 38
Phospholipase, 136−137
Phospholipid, 115
Phosphorylation, 11, 79
Photo-multiplier, 175
Photodiode, 175
Phytohormone, 90
Pickle, 4, 251
Pies, 317
Piezoelectric, 175−178
Pigment, 102, 106, 318
Pipes and fittings, 279
pJB8, 26−27
Plant cell,
    cell, 87
    cell fusion, 89
    cell synthesis, 101
    culturing, 101−103
    immobilization, 159−162
Plant regeneration, 94−95
Plant tissue culture, 87
Plasmids, 18−21
Plasmid transformation, 43
pMBI, 18
PMSF, 79
Poly(A) tail, 59, 64
Poly(dT), 58−60
Polyacrylamide, 154, 161
Polyethylene glycol, 56
Polynucleotide, 7
Polypeptide, 134
Polysaccharide, 111−113
Polysulfone membrane, 170
Polyurethane, 161
Pork, 3
Potassium glutamate buffer, 16
Precipitation, 237, 239
Prefiltration, 199
Prehybridization mix, 55
Pressure, 273
Primer, 68

Proα2 collagen, 26
Process control, 277−278
Product concentration, 189
Product recovery, 189
Proline, 106
Promoter, 84
Pronase, 34
Propeller-type stirrer, 263
Protease, 134−135, 319
Protein, 151−152
Protein A, 83
Protein engineering, 5, 128−129, 321
Protein synthesis, 130
Proteinase-K, 33
Protoplast,
    culture, 98−99
    culture medium, 99
    fusion, 89, 97−102
    immobilization, 162
    isolation, 98−99
Protruding end, 16
pUC18, 21, 23
pUC19, 21, 23
Pullulan, 111−112
Pullulanase, 135
Pulse-field gel electrophoresis, 35
Purification, 190
Purification of labeled DNA, 52
Purine, 7
Pyrazine, 107
Pyrimidine, 7
Pyruvate kinase, 314

Quartz crystal microbalance (QCM),
    177

Radial flow impeller, 263
Radiation sterilization, 269
Radiolabeled probe, 53
Radiolabeling of mammalian cell
    proteins, 79
Raffinate, 227
Random mutagenesis, 126−128
Random mutation, 5
Rapeseed oil, 4
Rare mating, 308−309
Reactor,

air lift, 165−167, 252−253
batch, 163, 254−256
bubble column, 294
continuous flow, 163−166
continuous stirred tank, 163,
    255−257, 293−294, 315
CSTR, 163, 255−256
fluidized bed, 163, 167, 257, 293
for immobilized cells, 165−168
for immobilized enzymes and
    cells, 163−168
hollow fiber, 160, 168, 259
membrane, 160, 168, 259−260
packed bed column, 163, 315
plate and frame, 163
plug flow, 163−164
rotating thin film, 167
spouted bed, 163−165
stirred tank, 167, 254−257
STR, 167
tapered, 167
Reactor scale-up, 292−294
Recombinant DNA (rDNA), 2, 4−6,
    107, 112−113, 298−299, 301,
    319
Redox probe, 275
Removal of terminal phosphate from
    linear DNA, 38, 40
Rennin, 129, 134, 144
Replacement synthesis, 61, 64
Replicative form, 30, 31
Replicon, 18
Reporter gene, 119
Residence time distribution, 289−290
Resistance temperature detector, 274
Resolving gel mix, 80
Resonant frequency, 177
Restriction endonuclease, 11
Restriction enzyme, 11, 13−15
Restriction enzymes cleaving DNA,
    16−18
Retrovirus, 86
Reuterin, 111
Reverse osmosis, 202−207, 243,
    305, 307
Reverse transcriptase, 19, 58, 60, 62,
    64

Riboflavin, 4, 107
Ribonucleic acid, 7
Ribose, 7
Ribosomal RNA, 7, 58
Ribosome, 11
RNA (*see* Ribonucleic acid)
RNA analysis, 58
RNA extraction, 59
RNA polymerase, 7
RNAase A, 34
RNAase H, 58, 61, 63−64
rRNA (*see* Ribosomal RNA)
RTD, 274
Rules of thumb, 291−292
Running of sequencing gel, 77

S1 nucleae mapping, 58
*Saccharomyces cerevisiae*, 83
Salad dressing, 317
Salmon sperm DNA, 56
Sample buffer, 79
Sauerberey equation, 176
Sauerkraut, 3
Sausage, 3, 251
Scale-up, 283−294
    aeration, 287−289
    agitation, 287−289
    introduction, 283−284
    mixing, 287−289
    oxygen transfer, 290−291
    reactor, 292−294
    relationships, 286
    required data, 284−285
    rules of thumb, 291−292
Scleroglucan, 111−112
SDS-PAGE, 78−83
SDS-polyacrylamide gel
    electrophoresis (*see* SDS-PAGE)
Secondary metabolites, 88
Sedimentation, 191
Seebeck coefficient, 273
Separation factor, 203−204, 225
Sepharose, 83
Sequenase, 69
Sequencing gel (*see* Buffered-
    gradient polyacrylamide gel)
Sequencing (*see* DNA sequencing)

Sequencing reaction, 74−76
Sesquiterpenes, 106
Shikonin, 318
SI nuclease, 20, 60
Signal peptide, 130
Simian virus 40, 86
Single cell oil, 113−115
Single cell protein, 4, 108−110
Single-stranded template preparation,
    73
Sintered-type sparger, 264
Site-directed mutagenesis, 320−321
    (*see* Oligonucleotide-mediated
    mutagenesis)
Site-directed mutagenesis without
    phenotypic selection, 125
Small-scale plasmid DNA
    preparation, 44−45
SOB, 42
Sodium bisulphite, 127
Solid liquid extraction, 219−222
    compositions, 219−220
    design, 220−222
    extractor size, 220
    extractor type, 220
    solvent, 219
    temperature, 219
Solid state temperature sensor, 274
Solvent,
    agitation, 228
    density, 227
    recoverability, 227
    selection, 226−228
    selectivity, 226
    solubility, 227
Solvent-feed ratio, 225
Somaclonal variation, 88, 96, 102
Somatic embryogenesis, 88, 95−96
Sonication of DNA, 70
Southern blotting, 53−55
Soy sauce, 3, 134
Soybean, 3
Soybean oil, 315
Specific growth rate, 246, 248−249,
    281
Specific ion probes, 275
Spheroplast, 84

Spheroplast fusion, 306–308
SSC, 54, 55
Stable gene expression, 85
Stacking gel mix, 81
Starch, 135
Starter culture, 295–302, 305–306
Stearic acid, 114
Sterilization, 89–90, 93–94, 291
Stern-Volmer equation, 272
Stirrer,
    peddle-type, 263
    propeller-type, 263
    turbine, 263
Stop codon, 11
*Streptococci*, 297–299
Sucrose inversion, 316
Sulphation, 11, 79
Sunflower, 314
Supercritical extraction, 239
Surface acoustic wave (SAW),
    176–177
Surfactant, 106, 111, 116
Sweetener, 318
Swelling, 135
Synthetic oligonucleotide, 121

T-DNA, 84
T4 DNA ligase (*see* Bacteriophage
    T4 DNA ligase)
TAE buffer, 36
Taq DNA polymerase, 69
TBE buffer, 37
TEMED, 76, 80, 81
Temperature, 273–274
Template, 68
TEN buffer, 49
Terminal deoxynucleotidyl
    transferase, 19, 62–63
Terpenes, 107
Tetracycline, 21
Tetramethyl pyrazine, 106
TFB, 42
Thaumatin, 5, 129, 318
*Thaumatococcus daniellii Benth*, 129
Thermistor, 274
Thermocouples, 273–274
Thermolysin, 144

Thermometric enzyme-linked
    immunosorbent assay
    (TE-LISA), 175
Thymidine, 7
Thymidine kinase, 86
Thymine, 7
Ti plasmid, 84
Tissue grinder, 194
TM buffer, 74
Torula yeast, 108
TPA, 177
Transeliminases, 136
Transfer RNA, 7, 59
Transformation, 307–308, 319
Transformation of bacterial cells,
    41–44
Transient gene expression, 85
Translation, 11
Triton solution, 45
tRNA (*see* Transfer RNA)
Tryptophan, 4
Turbidity, 177
Turbine stirrer, 263
Ultra-centrifugation, 198–201
Ultrafiltration, 202–206, 243,
    305–307
Universal primer, 68
Upstream regulatory elements, 119
Uracil, 7
Uracil-*N*-glycosylase, 126
Urea, 180
Urea electrode, 178, 182, 187
Uridine, 7
UTPase, 125
UV illuminator, 37

Vaccinia virus, 86
Vacuum transfer of DNA, 54
Vanilla, 102
Vegetable and fruit processing,
    310–312
Vinegar, 3, 134
Virus vectors, 86
Viscosity, 274–275
Vitamins, 90, 102, 107

Waste,
    agricultural, 109

Waste *(continued)*
    brewery, 109
    dairy, 109
    meat, 109
Wax-ester, 116–117
Western blotting, 83
Wet milling, 313
Wheat, 88
Whey, 144, 167
Whey processing, 302–303
Wine, 3, 134, 137, 307–310
Wine yeast, 309–310

X-gal *(see* 5-bromo-4-chloro-3-

indolyl-$\beta$-galactoside)
X-ray crystallography, 129
Xanthan, 4, 106, 111–112
Xanthine-guanine phosphoribosyl
    transferase, 86
Xenozymes, 5
XM-6, 112
Xylitol, 106

Yeast oil, 114
Yogurt, 3, 295–296

Zygotic embryos, 95